Polymer Electrolytes and their Composites for Energy Storage/Conversion Devices

Polymer Electrolytes and their Composites for Energy Storage/Conversion Devices presents a state-of-the-art overview of the research and development in the use of polymers as electrolyte materials for various applications. It covers types of polymer electrolytes, ion dynamics, and the role of dielectric parameters and a review of applications. Divided into two parts, the first part of the book focuses on the types of polymer electrolytes, ion dynamics, and the role of dielectric parameters, while the second part provides a critical review of applications based on polymer electrolytes and their composites. This book:

- Presents the fundamentals of polymer composites for energy storage/conversion devices
- Explores the ion dynamics and dielectric properties role in polymer electrolytes
- Provides detailed preparation methods and important characterization techniques to evaluate the electrolyte potential
- Reviews analysis of current updates in polymer electrolytes
- Includes various applications in supercapacitor, battery, fuel cell, and electrochromic windows

The book is aimed at researchers and graduate students in physics, materials science, chemistry, materials engineering, energy storage, engineering physics, and industry.

Emerging Materials and Technologies
Series Editor: Boris I. Kharissov

Green Synthesized Iron-Based Nanomaterials
Application and Potential Risk
Piyal Mondal and Mihir Kumar Purkait

Polymer Nanocomposites in Supercapacitors
Soney C George, Sam John and Sreelakshmi Rajeevan

Polymer Electrolytes and their Composites for Energy Storage/Conversion Devices
Edited by Achchhe Lal Sharma, Anil Arya, and Anurag Gaur

Hybrid Polymeric Nanocomposites from Agricultural Waste
Sefiu Adekunle Bello

Photoelectrochemical Generation of Fuels
Edited by Anirban Das, Gyandeshwar Kumar Rao and Kasinath Ojha

Emergent Micro- and Nanomaterials for Optical, Infrared, and Terahertz Applications
Edited by Song Sun, Wei Tan, and Su-Huai Wei

Gas Sensors: Manufacturing, Materials, and Technologies
Edited by Ankur Gupta, Mahesh Kumar, Rajeev Kumar Singh and Shantanu Bhattacharya

Environmental Biotechnology
Fundamentals to Modern Techniques
Sibi G

Emerging Two Dimensional Materials and Applications
Edited by Arun Kumar Singh, Ram Sevak Singh and Anar Singh

For more information about this series, please visit: www.routledge.com/Emerging-Materials-and-Technologies/book-series/CRCEMT

Polymer Electrolytes and their Composites for Energy Storage/Conversion Devices

Edited by Achchhe Lal Sharma,
Anil Arya, and Anurag Gaur

CRC Press
Taylor & Francis Group
Boca Raton New York London

CRC Press is an imprint of the
Taylor & Francis Group, an **informa** business

First edition published 2023
by CRC Press
6000 Broken Sound Parkway NW, Suite 300, Boca Raton, FL 33487–2742

and by CRC Press
4 Park Square, Milton Park, Abingdon, Oxon, OX14 4RN

CRC Press is an imprint of Taylor & Francis Group, LLC

ISBN: 978-1-032-07759-8 (hbk)
ISBN: 978-1-032-07760-4 (pbk)
ISBN: 978-1-003-20866-2 (ebk)

DOI: 10.1201/9781003208662

Typeset in Times
by Apex CoVantage, LLC

Contents

Editor Biographies

Achchhe Lal Sharma is Assistant Professor in the Department of Physics, Central University of Punjab, Bathinda. Dr. Sharma has authored and co-authored over 100 publications in peer-reviewed journals including research publications, review articles, conference proceedings, and nine book chapters for international publishers. He has 11 years of research experience and his research interests include the development of electrode and electrolyte material for lithium-ion batteries and supercapacitors. His h-index is 22 and i10-index 35 with over 1500 citations, and he has served as a reviewer for several high-impact journals and given more than 24 talks at various conferences/colleges and universities. He pursued his PhD at the Department of Physics, Indian Institute of Technology, Kharagpur. He has been granted the Best Research award twice by Central University of Punjab, Bathinda. Dr. Sharma's research interests include the development of nanostructured materials/composites/nanocomposites for application in renewable energy storage/conversion devices (batteries and supercapacitors).

Anil Arya is serving as Dr. D. S. Kothari Post-Doctoral Fellow in the Department of Physics, Kurukshetra University, Kurukshetra. Dr. Arya has co-authored 50 research publications, review articles, conference proceedings, and nine book chapters. His h-index is 18 and i10-index 28 with over 1220 citations, and he has served as a reviewer for several electrochemistry-related journals. He pursued his PhD at the Department of Physics, Central University of Punjab, Bathinda. Prior to joining the PhD program, he received his bachelor's in science degree from University College, Kurukshetra University (Kurukshetra), and master's in physics degree from the Central University of Punjab, Bathinda. Dr. Arya was also featured in the World's Top 2% Scientists: Stanford University USA List 2021, published by Elsevier BV. Dr. Arya's research interests include the synthesis of electrode/electrolyte materials for energy storage/conversion devices.

Anurag Gaur is Assistant Professor in the Department of Physics, National Institute of Technology, Kurukshetra. Dr. Gaur has published more than 130 research articles in peer-reviewed, reputed journals and has led or been involved in over eight national and international projects funded by various government agencies (e.g. SERB-DST, CSIR, etc.). He has 15 years of research experience in nanomaterials synthesis and developed supercapacitors, lithium-ion batteries, hydroelectric cells, and spintronic devices. He has guided 7 PhD and 50 M.Tech students in their thesis work. His h-index is 27 and i10-index 61 with 2150 citations, and he has filed three patents. He has served as a reviewer for several high-impact journals and delivered over 65 talks at various national/international conferences. He did his PhD at Indian Institute of Technology, Roorkee in 2007. He has been granted the Best Faculty award by National Institute of Technology, Kurukshetra in 2018. Dr. Gaur's research interests include energy storage devices and green energy production through water splitting.

Contributors

Anurag Gaur
Department of Physics
National Institute of Technology
Kurukshetra, India

Anil Arya
Department of Physics
Kurukshetra University
Kurukshetra, India

Soubhagya Ranjan Bisoi
Department of Chemistry
Siksha 'O' Anusandhan, Deemed to Be University
Bhubaneswar, Odisha, India

Sujeet K. Chaurasia
Centre for Nanoscience and Technology
Prof. Rajendra Singh (Rajju Bhaiya) Institute of Physical Sciences for Study & Research, V.B.S. Purvanchal University
Jaunpur, India

Avirup Das
Department of Physics, VIT Bhopal University,
Bhopal, Madhya Pradesh, India

Simran Kour
Department of Physics
Central University of Punjab
Bathinda, Punjab, India

Vijay Kumar
Department of Physics
Institute of Integrated and Honors Studies, Kurukshetra University
Kurukshetra, India

Vikram Kunwar
Department of Physics
Graphic Era, Deemed to Be University
Dehradun, Uttrakhand, India

Ravi Vikash Pateriya
Department of Physics
Central University of Punjab
Bathinda, India

Atma Rai
Department of Applied Sciences
Women Institute of Technology
Sudhowala, Dehradun, India

Pooja Rawat
Department of Physics, Institute of Science, Banaras Hindu University
Varanasi, UP, India

Naresh Kumar Sahoo
Department of Chemistry
Siksha 'O' Anusandhan, Deemed to Be University
Bhubaneswar, Odisha, India

Prasanta Kumar Sahoo
Department of Mechanical Engineering
Siksha 'O' Anusandhan, Deemed to Be University
Bhubaneswar, Odisha, India

A. L. Saroj
Department of Physics, Institute of Science
Banaras Hindu University
Varanasi, UP, India

A. L. Sharma
Department of Physics
Central University of Punjab
Bathinda, Punjab, India

Annu Sharma
Department of Physics
Kurukshetra University
Kurukshetra, India

Manish Pratap Singh
Department of Physics
Faculty of Engineering and Technology, V.B.S. Purvanchal University
Jaunpur, India

Manoj K. Singh
Department of Applied Science & Humanities
Rajkiya Engineering College Banda
Uttar Pradesh, India

Ankur Soam
Department of Mechanical Engineering
Siksha 'O' Anusandhan, Deemed to Be University
Bhubaneswar, Odisha, India

Shweta Tanwar
Department of Physics
Central University of Punjab
Bathinda, Punjab, India

Preface

The first part of the book begins by discussing the brief introduction of polymer composites, classification, and fundamentals. Hybrid polymer composites and key characteristic features are discussed in detail. Ion dynamics and dielectric relaxation are explored in polymer composites by discussing the physical models proposed to get insights into ion transport. Synthesis methods and characterization techniques used to examine the suitability of polymer composites are reviewed in detail. Then the second part provides a glimpse of polymer-composite-based supercapacitors, lithium-ion batteries, electrochromic potential windows, dye-densitized solar cells, and Fuel cells. This part will also enable the reader to know about the challenges that remain and need to be resolved for future devices. In brief, this book will be an essential guide for researchers, scientists, and advanced students in polymer science, composites, nanocomposites, and materials science. It is also a precious book for engineers, R&D professionals, and scientists working toward the development of polymer-electrolytes-cum-separators for energy storage/conversion devices.

A. L. Sharma, PhD
Anil Arya, PhD
Anurag Gaur, PhD

Introduction

The depletion of traditional sources of energy due to the global energy demand has focused the attention of researchers on the development of sustainable and renewable energy sources. So, the need of the hour is to switch from conventional/traditional to renewable/clean-green sources of energy completely for a broad range of applications. Some efficient renewable sources of energy are solar, wind, hydro, geothermal, etc. The energy generated from these sources needs to be stored in energy storage devices (batteries, supercapacitors, fuel cells, dye-densitized solar cells (DSSCs), etc.). The electrolyte is an important component of any energy device, and in existing devices, the liquid electrolyte is used. An Electrolyte acts as a carpet for ions. Since the first report in 1973, polymer electrolytes/composites emerged as attractive candidates for application in energy storage/conversion devices. The polymer composites and their types affect the ion dynamics and hence the overall performance of the device. Polymer composites have potential to allow the development of efficient materials that can fulfill the demand of society to design flexible devices. Overall, polymer composites will enable the creation of a sustainable lifestyle, which is the aspiration of the planet and society.

The present book presents a state-of-the-art overview of the R&D in the development of polymers as electrolyte materials for supercapacitors, lithium-ion batteries, electrochromic potential windows, dye-densitized solar cells, fuel cells. For a better guide to the eye of readers, the book will be divided into two parts. The first part explores the fundamentals of polymer composites, classification, and ion dynamics within the polymer matrix. This part also provides a glimpse of different synthesis methods used for polymer composite formation, and crucial characterization techniques required to evaluate the electrolyte performance and check its potential for a particular application. The second part explores the application part and will provide a comprehensive analysis of current development in polymer electrolytes for supercapacitors, lithium-ion batteries, electrochromic potential windows, and fuel cells. In brief, this book will be a treasured locus for the students, researchers working in the energy area, as well as for manufacturers working toward the development of polymer-electrolytes-cum-separators.

A. L. Sharma, PhD
Anil Arya, PhD
Anurag Gaur, PhD

Part I

Fundamental

1 Polymer and Their Composites
An Overview

Anil Arya, Annu Sharma, A. L. Sharma, and Vijay Kumar

CONTENTS

DOI: 10.1201/9781003208662-2

1.1 INTRODUCTION

Energy is very important in our daily lives because it is a basic human need. Most of the energy contribution is from fossil fuels, coal, etc., and it causes serious environmental concerns globally. To reduce the dependency on traditional sources of energy, various renewable sources of energy such as hydro energy, wind energy, solar energy, etc. have emerged as alternatives. The energy obtained from these sources needs to be stored in a device so that it can be used as per demand. The important energy storage/conversion devices that have dominated the energy sector are batteries, supercapacitors, and fuel cells. The electrolyte is an integral component of any device, and most commercial devices are using liquid electrolyte. Some of the key drawbacks of the liquid electrolyte-based device are bulky size, flammability, separator need, and poor safety. To fulfil the need of highly efficient and cost-effective high-energy density storage/conversion devices, polymer composites have been tried as electrolytes-cum-separators. The first report of ionic conductivity in polymer electrolytes/composites was by Armand (Armand 1979). Peter V. Wright and Fenton examined poly(ethylene oxide) (PEO) with alkali iodide salts and reported the conductivity followed by use of PEO/alkaline salt compound in batteries as an ionic conductor by Armand in 1978. This polymer matrix favored the ion migration and exhibited good mechanical flexibility and good interface contact in batteries (Armand 1994; Wright 1975; Armand et al. 2011a; Fenton 1973). Some key features of polymer composites are high flexibility, light weight, cost-effectiveness, and better contact with electrodes, and they can be used to design devices with varied architecture/shape geometry. The development of flexible and soft material as a separator enables the development of devices with varied shape geometry. The simultaneous presence of the crystalline phase and amorphous phase indicates their unique characteristics (Arya et al. 2017). Generally, the amorphous phase favors faster ion transport and results in higher ionic conductivity (Sequeira and Santos 2010; Cameron 1988). This chapter presents an overview of the polymer composites, their classification based on additive species. Then the selection criteria for polymer, salt, nanofi ller, nanoclay, and solvent are discussed. Finally various fundamental ion transport model proposed by researchers to get insights into ion dynamics are summarized.

1.2 POLYMER COMPOSITES

Polymer composites (PCs) were introduced for the first time in the 1970s, and soon became the strong candidates to be used as electrolytes in energy storage/conversion devices. In general, they are comprised of a macromolecule matrix with low lattice energy salt dissolved in low viscosity and high dielectric constant organic solvents. The interaction between the electron-rich group in the polymer backbones and the salt favors salt dissociation and facilitates the cation (Li, Na, K) migration. Such type of migration occurs via coordinating sites provided by polymer chains and is termed as a hopping (intrachain/interchain) mechanism. This mechanism generates the ionic conductivity within the matrix (Figure 1.1). The ionic radii of an anion is much larger than a cation for better dissociation, and the anion remains immobilized in the matrix due to its large size. Smaller cation radii of salt and the high dielectric constant (ϵ) of the host polymer is beneficial for better composite formation with optimum properties. The ionic conduction is attributed to segmental motion of polymer chains with strong Lewis-type acid-base interaction between the cation and donor atom (Muldoon et al. 2015). As polymer composites play a dual role both as a separator and a composite in an energy storage device, they must possess some characteristic properties (Marcinek et al. 2015). Figure 1.2 shows the transport, stability, and other properties of polymer composites. Polymer composites are also termed polymer electrolytes (PEs).

Polymer composites overcome many drawbacks of liquid electrolytes in terms of shape flexibility, size, and weight. Ion transport is faster in an amorphous phase in comparison to a crystalline phase, and higher conductivity by two or three orders of magnitude is achieved. Due to this, polymer composites are used in various applications. Figure 1.2 shows the properties, advantages, and applications of polymer

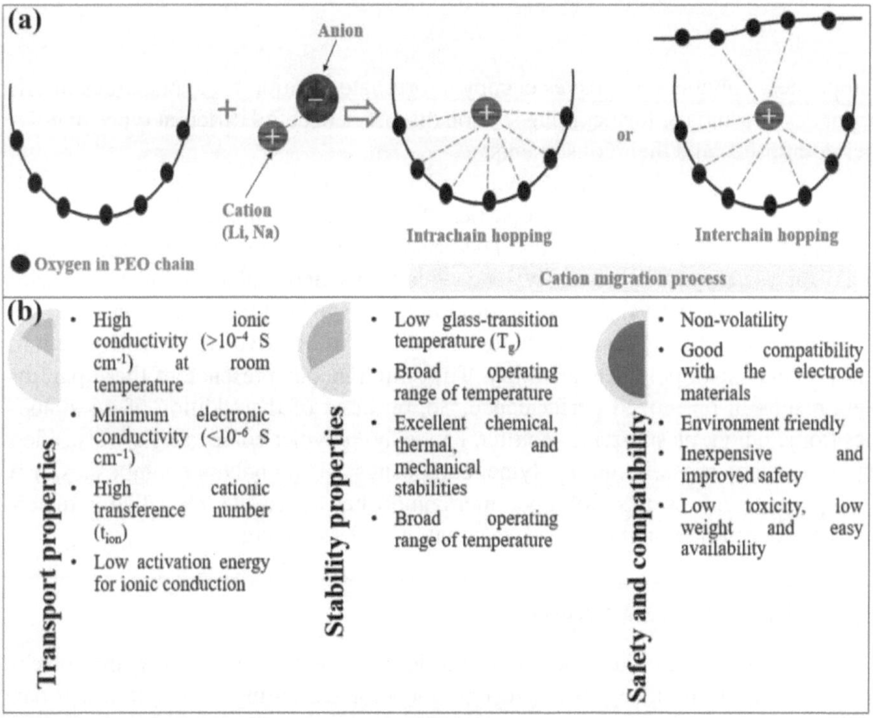

FIGURE 1.1 (a) Schematic of ion dynamics in a polymer composite; (b) important characteristics of polymer composites.

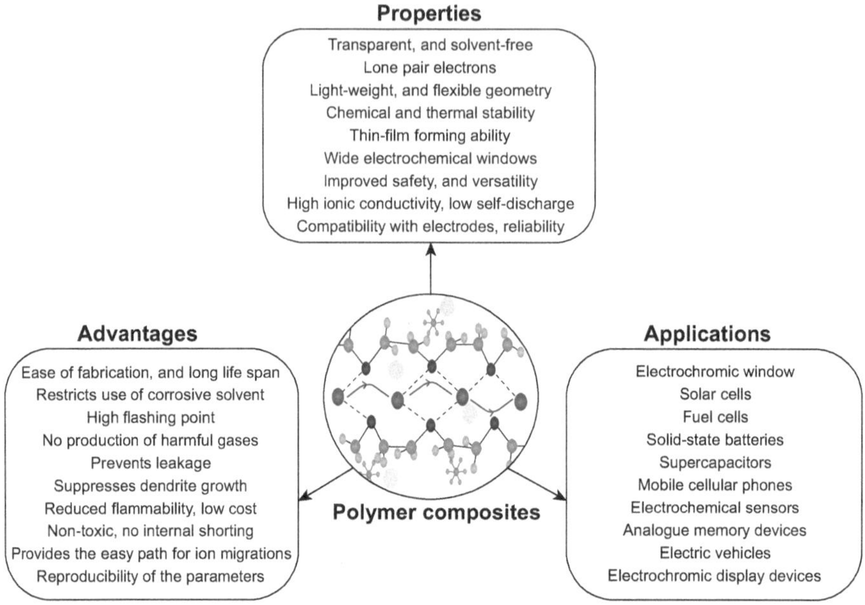

FIGURE 1.2 Properties, advantages, and applications of polymer composites/composites.

composites. Polymer composites occupy a fortunate position for applications in different devices and the forthcoming section discusses in detail different types of polymer composites and their constituents.

Various polymer hosts have been explored to achieve the desired electrical, mechanical, and ion transport properties. This section discusses the classification of polymer composites/electrolytes. The pristine polymer is of an insulating nature, and displays very low conductivity. Its conductivity can be enhanced by doping with nanoparticles, nanoclays, etc. Three important tasks to be done by nanoparticles are (i) improve salt dissociation, (ii) better dispersion of additives for effective action, and (iii) enhanced polymer flexibility. The simultaneous presence of these parameters results in optimized performance. So, on basis of the addition of guest species (ionic liquid, plasticizer, nanofiller, nanoclay) polymer composites are classified into three categories, (i) liquid polymer composites, (ii) gel polymer composites, (iii) solid polymer composites. Solid polymer composites are further classified in different types depending on the type of host matrix and architecture.

1.2.1 Liquid Polymer Composite

Liquid polymer composites consist of a lithium salt dissolved in an organic solvent (e.g. $LiPF_6$/EC) and a separator prevents the short-circuiting of electrodes. It can be classified into non-aqueous liquid (NALE) or aqueous liquid composite (ALE) depending on the nature of the solvent. The former one has high ionic conductivity and is obtained by dissolving alkali salts in solvents (EC, PC, etc.). Along with this, they demonstrate a large electrochemical stability window, which is a crucial parameter for energy devices (Cheon et al. 2003; Liang et al. 2011; Choi et al. 2007). While, ALE has more ionic conductivity than NALE and is non-flammable, inexpensive, and can be used safely at high temperatures. Nowadays, ionic liquids (ILs) have emerged as new guest species. ILs are organic-salts-like materials with a low melting point and exist in liquid form at room temperature because of weak ion coordination/poor packing of atoms. Some key advantages are (i) they can be used in a wide temperature range (300–400 °C), and (ii) they have wide electrochemical stability. These make them a strong candidate for energy devices (e.g. batteries, fuel cells, supercapacitors, etc.). By changing the ratio of cation to the anion, and interchanging different cations or anions, properties can be tuned as per requirement (Earle et al. 2006; Gebbie et al. 2013; Armand et al. 2011b).

1.2.2 Gel Polymer Composite

Gel polymer composites or electrolytes (GPEs) emerged as an alternative to previous composites, due to the their poor mechanical properties. GPE is an intermediate stage between the liquid composite and solid polymer composite. So, GPEs are composed of a polymer matrix, liquid solvent as a plasticizer (ethylene carbonate, EC; propylene carbonate, PC; diethyl carbonate, DEC; dimethyl carbonate, DMC) and lithium salt. Based on sample preparation, gels are categorized in two ways. When the liquid composite is confined in a polymer matrix without any bond formation between polymer and solvent, then it is termed as a *physical gel*. When a cross-linker leads to the formation of the chemical bond between the functional group of polymers and cross-linker agent, then it is called a *chemical gel* (Hellio and Djabourov 2006). This type of polymer composite exhibits both cohesive properties of solids and diffusive properties of liquids. High ionic conductivity, a wide electrochemical stability window (ESW), and good compatibility with electrodes strengthens their candidature compared to liquid electrolytes (Sequeira and Santos 2010; Kokorin 2011). The addition of plasticizers also lowers the glass transition temperature and favors faster ion dynamics. Although, they demonstrate better ionic conductivity, but still lack the desirable mechanical stability (Xiao et al. 2012). In gel polymer composites, PVdF-co-HFP-based polymer composites (electrospinning and non-woven) are fascinating candidates due to the advantage associated with them, better flexibility and a broad electrochemical stability window. But, there remains one issue—that is, leakage of composite even after the addition of ionic liquid (Jeong et al. 2012; Pitawala et al. 2014). So, oligomeric ionic liquid-type gel polymer composites based on oligomeric ionic liquids, PVdF-co-HFP, and the composite solution was prepared (Kuo et al. 2016). The prepared polymer composite was flexible, and possessed porous morphology with an ionic conductivity of 0.12×10^{-3} S cm^{-1} (at RT) as compared to PVdF-HFP gel polymer composite. The electrochemical voltage stability window was about 4.5 V.

1.2.3 Solid Polymer Composite

Solid polymer composites/electrolytes are the new generation electrolytes and has inherent potential to dominate the energy sector by replacing the traditional

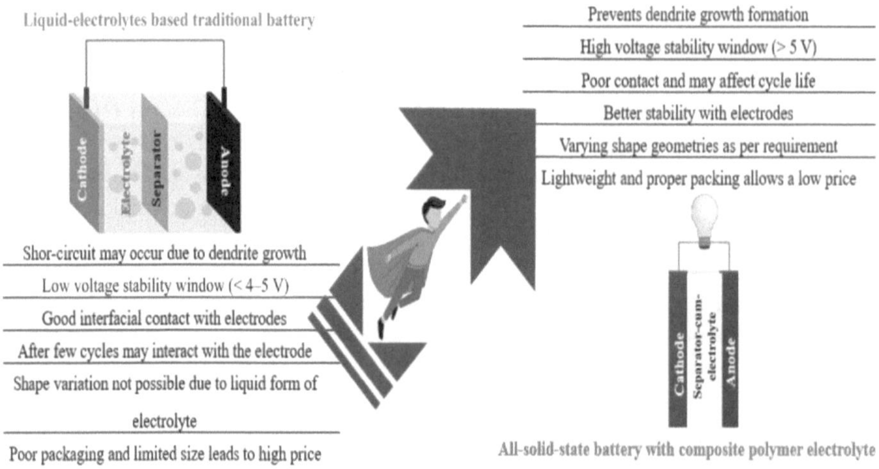

FIGURE 1.3 Advantages of polymer composite over the liquid composite-based battery.

electrolytes. Some of the important advantages that strengthen their candidature are high durability, safety, high energy density, great flexibility for cell design, negligible reactivity towards the electrodes, and reduced packaging cost. A broad electrochemical stability window and enhanced thermal stability make them suitable candidates for high-temperature energy storage devices (Zhang et al. 2007). Figure 1.3 displays the advantages of composite/solid polymer composite.

Generally, solid polymer electrolytes (SPEs) consist of the salt dissolved in a polymer matrix. However, the optimized electrolyte must have good ionic conductivity and mechanical stability. The ionic conductivity of SPEs is not comparable to the existing electrolytes. SPEs having optimized conductivity and stability properties need to be developed for the potential for applications in solid-state ionic devices (Jeddi et al. 2010). So, enhancement in conductivity is linked to the salt dissociation and several cation-conducting sites. Various strategies have been adopted to enhance salt dissociation, conducting sites, and polymer flexibility. The forthcoming section discusses the important strategies to enhance electrical properties.

1.3 STRATEGY TO TUNE THE PROPERTIES OF SOLID POLYMER COMPOSITE

Rapid ion dynamics in polymer composites is an essential requirement for application in batteries, supercapacitors, fuel cells, etc., and is linked to the glass transition temperature of the polymer, free ions in a polymer matrix, low crystallinity, and faster segmental motion. Further insights in the composite matrix are explored by examining the various modification strategies to tune the ion dynamics which influence the electrical properties. This section provides a different class of polymer composites with different architecture to achieve the optimum properties. Figure 1.4 shows the important types of polymer-electrolytes-cum-composites. On the

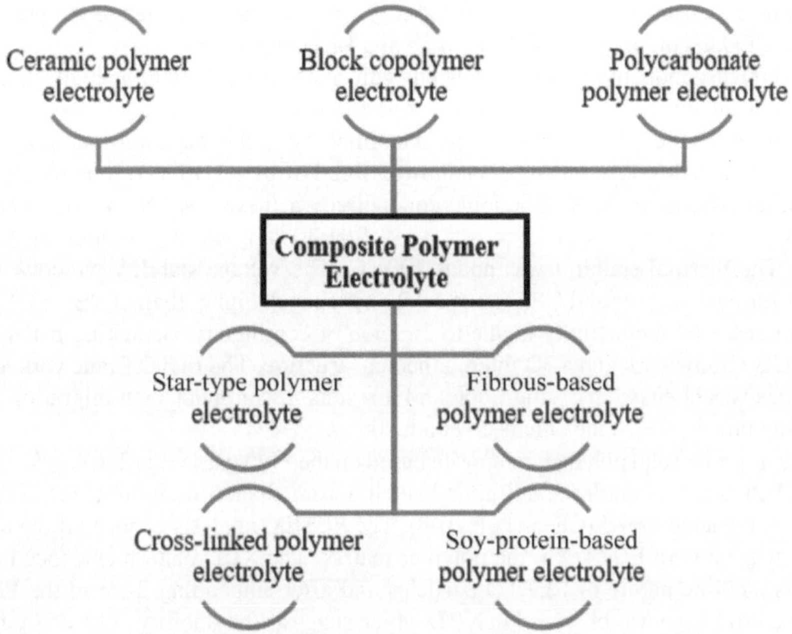

FIGURE 1.4 Classification of polymer composites.

basis of host matrix/dopant, polymer composites are ceramic polymer electrolytes, soy-protein-based electrolytes, polycarbonate polymer electrolytes. On the basis of architecture, polymer composites are block copolymers, star type, fibrous based, cross-linked types.

1.3.1 Ceramic (LATP/LLTO/LLZTO/Li_3PS_4) Polymer Composite

Ceramic polymer composite provides improved conductivity as compared to solid polymer composite. The Lewis acid-based interaction of the surface group of nanofiller with a polymer and salt enhances the ion dynamics. The addition of nanofiller enhances the segmental motion of the polymer chain owing to disruption of the polymer chain which suggests enhanced amorphous content for ion migration. The surface groups (-OH) associated with the nanofiller favors strong Lewis acid-base interaction and create the conducting pathways for ion migration (Zhang et al. 2018; Liang et al. 2018). Prof. Goodenough's group developed a new sandwich structure (polymer/ceramic/polymer composite) and it demonstrated enhanced properties (Zhou et al. 2016). To resolve the issue of the low ionic conductivity and chemical instability of Ti_4^+ ions in LLTO against lithium metal, Li et al. [Li et al. 2018] developed a polymer composite with sandwiched structure (PVdF/LLTO-PEO/PVdF). The 10 wt. % LLTO-based system (interlayer-II) shows the highest ionic conductivity of 2.1×10^{-4} S cm^{-1} (at RT). The ionic conductivity possessed by the sandwiched structure is ~3.01×10^{-3} S cm^{-1} with voltage stability window close to

5 V. The enhancement of the voltage stability window was attributed to the prevention of PEO decomposition by PVdF layer and LLTO nanowire.

Although conductivity gets enhanced with nanofiller addition, at high content nanofiller agglomeration masks the ion migration. This agglomeration needs to be suppressed for the effective task of the nanofiller. So, a flexible composite polymer composite was developed using 3D nanostructured hydrogel (LLTO) frameworks as nanofiller (Bae et al. 2018). The ionic conductivity achieved was 8.8×10^{-5} S cm^{-1} (at 25 °C) and increased to 1.5×10^{-4} S cm^{-1} (at 30 °C) with the addition of 3.0 g PVA. The thermal stability was about 400 °C. The voltage stability window was larger for the LLTO-based CPE (i.e. 4.5 V) and is much higher than of PEO (3.9 V). The increase in conductivity is due to creation of continuous conducting paths for the LLTO framework and a 3D interconnected structure. The LLTO framework also provides vacant sites for Li migration and it results in continuous ion migration via hopping proves, hence the enhanced conductivity.

A composite solid polymer composite based on the PEO garnet $Li_{6.4}La_3Zr_{1.4}Ta_{0.6}O_{12}$ (LLZTO) as the nanofiller and lithium bis(trifluoromethanesulfonyl)imide (LiTFSI) as the salt was reported (Chen et al. 2018). The FESEM analysis confirmed the uniform dispersion of LLZTO in the polymer matrix. The XRD pattern evidenced the poly-crystalline nature of LLZTO particles and after embedding them in the PEO matrix, no change was observed in XRD which suggests the stability of LLZTO. The highest ionic conductivity was 1.17×10^{-4} S cm^{-1} (at 30 °C) for 10 wt. % LLZTO particles. The increase in the conductivity was attributed to the increased free volume and faster chain movement owing to the uniform dispersion of LLZTO particles. The voltage stability window was about 5.0 V (vs Li/Li^+) and is within the safe limit of battery operation. Recently Liu et al. (Liu et al. 2019) reported the preparation of the composite polymer composite based on PEO as polymer matrix and $Li_{1.4}Al_{0.4}Ti_{1.6}(PO_4)_3$ (LATP) as nanoparticles with an average size ~140 nm. The highest ionic conductivity was observed for the PEO-LATP01 CPE which is about 1.15×10^{-5} S cm^{-1} (at 30 °C), and reaches 7.03×10^{-4} S cm^{-1} (at 80 °C). The voltage stability window was close to 4.8 V.

Several strategies have been adopted to enhance the characteristics parameters of the Li-ion battery, but still there is scope for improvement. The network structure of the LLTO may affect the ion dynamics parameters. So, a three-dimensional (3D) LLTO network has been investigated (Wang et al. 2018). The unique feature of this approach is (i) interconnected LLTO effectively enhances the electrical and mechanical properties as well as suppresses the dendrite growth; (ii) a combination of hot press and quenching process results in the formation of dense and self-standing CPE. The composite polymer electrolyte (CPE) comprises PEO as host matrix, LiTFSI as salt, and a three-dimensional (3D) LLTO nanofiber network (synthesized the first time). The highest conductivity value was about 1.8×10^{-4} S cm^{-1} at RT, cation transport number (t_+) 0.33, and voltage stability window of about 4.5 V vs. Li/Li^+.

1.3.2 Soy-Protein-Based Polymer Composite

These are newly advanced polymer composites that show enhanced properties as compared to the existing polymer composites. The soy protein (SP) is denatured before use. The denaturation process results in unfolding the chain, and the negative acid group present in the SP results in Li-ion adsorption. Here, the polymer chains are surrounded by the PEO. The electron-rich site of polymer absorbs Li-ions as well as interacts with the ammonium group having a positive charge. It disrupts the crystalline arrangement of PEO and an amorphous phase is formed (Ji et al. 2012). In the CPE matrix, two possible interactions occur: (i) cation interaction with backbone oxygen in protein, and (ii) formation of anion clusters owing to the presence of electrostatic interactions between anion and positive charge side groups (e.g. Lysine (Lys) and arginine (Arg)) of protein. It may be concluded that the modification of nanofiller with SP creates new ion conduction channels.

The addition of nanofiller effectively promotes ion dissociation and creates ion-conducting pathways to the cation. However, still, there are restrictions to the enhancement of the conductivity up to the desirable limit. Surface functionalization of the nanofiller is an alternative approach that will facilitate the faster ion migration. The soy protein is an attractive candidate owing to the ability to transfer the cations via the functional group present in it (Zhu et al. 2016). Also, modification of nanofiller with SP reduces the nanofiller agglomeration tendency which is crucial for effective role of nanofiller.

Fu et al. (Fu et al. 2016) investigated the protein-based ion conductor (PIC) by adding $LiClO_4$ salt in the denatured soy protein. The increase of ionic conductivity with temperature from ~ 10^{-9} to 10^{-5} S cm^{-1} is observed. The Li^+ transference number is 0.94 and confirms the ionic nature of PIC. Another study by the same group reported the manipulation of protein configuration by nanofiller and it shows effective enhancement in the ionic conductivity and mechanical properties. Fu et al. (Fu et al. 2018a) prepared the polymer composite by the modification of SP with two types of hybrid-nanofiller: (i) TiO_2-(SP-close) hybrid, and (ii) TiO_2-(SP-open) hybrid. The temperature-dependent ionic conductivity and for the TiO_2/(SP-open) conductivity is comparable to liquid composite at 90 °C (i.e. 7 × 10^{-4} S cm^{-1}). The voltage stability window is very high, 5.2 V for TiO_2/(SP-close)–CPE and 5.4 V for TiO_2/(SP-open)–CPE.

Both protein configurations, and protein-TiO_2 interaction affects the ion migration due to altered environment. Another report by Fu et al. (Fu et al. 2018b) highlights the preparation of the core-shell protein@TiO_2 hybrid NWs, and their effect on the solid polymer composite (pure ultrahigh-molecular-weight PEO; UHMWPEO-$LiClO_4$) is examined in detail. The thickness of the protein coating is approximately 8 nm as evidenced by HRTEM. The XRD diffractograms of SP and TiO_2 nanowire concluded that both the SP and TiO_2 NW are integrated well; also there is a decrease in peak intensity of crystalline TiO_2 NW. It infers that SP coating enhances the amorphous phase that is a very critical requirement for fast ion transport. FTIR spectra confirm the in-situ growth of the protein layer on the TiO_2 surface (SP@TiO_2). The highest ionic conductivity was exhibited by the CPE with core-shell SP@TiO_2 NWs and is about 1.1 × 10^{-4} S cm^{-1} (at 10 wt. % SP@TiO_2 loading). This conductivity value is larger than the SP-TiO_2 nanoparticles-based composite. This enhancement is attributed to the high surface area of the nanowire and the high aspect ratio which creates continuous conducting paths. The conductivity increases with temperature and reached to 2 × 10^{-3} S cm^{-1} (at 80 °C). The highest voltage stability window is 5.3 V for the 10 wt. % SP@TiO_2-based composite polymer composite. The cation transference number is 0.62 for the 10 wt. % SP@TiO_2-based composite polymer composite is higher than pure PEO ($t_{Li}^+ = 0.41$).

1.3.3 Polycarbonate Polymer Composite

Nowadays, polycarbonates (PC)-based polymer composites are new emergent materials owing to their amorphous nature, high polymer chain flexibility, and high dielectric constant. The combination of these properties results in enhanced thermal stability, voltage stability window, and cation transference number. To achieve the high conductivity and flexibility, PCs are made up of aliphatic backbones and a few—poly(vinylene carbonate) (PVC), poly(propylene carbonate) (PPC), poly(trimethylene carbonate) (PTMC), and poly(ethylene carbonate) (PEC)—are preferred (Mindemark et al. 2018; Zhang et al. 2018). Also, polycarbonates (PCs) demonstrate good salt solubility due to the presence of a highly polar carbonate group (-O-(C=O)-O-) (Xu 2004). Polyurethane (PU)-based composites are also gaining attention and comprise of two segments, (i) soft segment (SS), (ii) hard segment (HS). SS makes polyurethane stretchable and flexible (beneficial for faster ion migration), while HS provides improved mechanical properties (Xiao et al. 2017; Xu et al. 2017; Karimi et al. 2017).

Polycarbonate-based polyurethanes (PCPU) with different contents of polycarbonate diol (PCDL)—1,6-hexamethylene diisocyanate (HDI), diethylene glycol (DEG)—were prepared via addition polymerization reaction by Bao et al. (Bao et al. 2018). Then the effect of the soft and hard segments of PU was examined on the polycarbonate-based polyurethanes (PCPU), and LITFSI was used as salt. DSC examination displayed the presence of two T_g, one associated with a soft segment ($T_{g,s} = -40 - 50$ °C), and another with a hard segment ($T_{h,s} = 42 - 48$ °C). A decrease in $T_{g,s}$ with the increase of hard segment content was observed. The improved tensile strength on the addition of the hard was attributed to the following reasons, (i) interaction via hydrogen bonding between neighboring units of 1,6-hexamethylene diisocyanate (HDI) units, and (ii) HS acting as reinforcing filler (Lee et al. 2013; Wang et al. 2010). The highest ionic conductivity was 2.2×10^{-6} S cm^{-1} (at 25 °C), 1.58×10^{-5} S cm^{-1} (at 60 °C), and 1.12×10^{-4} S cm^{-1} (at 80 °C) for 20 wt. % salt content. The most of contribution to the conductivity was from the soft segments owing to enhanced segmental motion. The voltage stability window was 4.5 V at 80 °C and within the limit for practical applications.

Most of the polymer composites are prepared via the solution cast method, but at a large scale, this synthesis method is not efficient due to the use of volatile organic solvents which may harm the environment. So, the polymerization reaction is more efficient and allows the synthesis of environmentally friendly waterborne polyurethane (WPU) (Karimi et al. 2017). A WPE as polymer matrix was prepared by polymerization of polyethylene glycol (PEG), hexamethylene diisocyanate (HDI), diethylene glycol (DEG), dimethylol propionic acid (DMPA), LiTFSI as salt, and water as a solvent by Cong et al. (Cong et al. 2018). The voltage stability window was about 4.8 V (at 60 °C) and is sufficient for the solid-state Li-ion battery applications. Recently, a poly(propylene carbonate)/$Li_{6.75}La_3Zr_{1.75}Ta_{0.25}O_{12}$-based composite was reported by Zhang et al. (Zhang et al. 2017). The highest conductivity was 5.2×10^{-4} S cm^{-1}

(at 20 °C) and is due to the formation of conductive paths owing to the PPC matrix and LLZTO bulk interface. The high voltage stability window (~ 4.6 V) and high cation transference number (~ 0.75) suggest a perfect correlation between ion dynamics properties.

1.3.4 Block Copolymer Composite

Tuning the architecture of the polymer composite, physiochemical properties can be changed, and one such composite type is block copolymer (BCP) composite having covalently bound polymers. BCP enables us to synthesize polymer electrolytes with noble ion conduction pathways (from any electron-rich group in the polymer), and results in improved properties as compared to individual polymers (Young et al. 2014; Giacomelli et al. 2010; Young et al. 2012).

BCP composites have the potential to provide better performance even at room temperature as compared to solid polymer electrolytes which operate efficiently only at an elevated temperature range (~70 °C). Operating near room temperature range leads to safety, cost reduction, and better Li compatibility. Based on this approach, a comb polymer matrix grafted with soft and disordered polyether moieties (Jeffamine®) and a popular salt, lithium bis (fluorosulfonyl)imide (LiFSI), was synthesized via the solution cast technique by Aldalur et al. (Aldalur et al. 2018). The ionic conductivity is 5.6×10^{-4} S cm^{-1} (at 70 °C), 2×10^{-4} S cm^{-1} (at 40 °C) with cation transference number of 0.16. The lower voltage stability window (4 V) for the Jeffamine-based polymer composite as compared to PEO is attributed to the presence of PPO segment which has a high possibility of oxidation.

Another strategy to alter the architecture is by preparation of semi-interpenetrating network polymer composites. An in-situ plasticized solid-state polymer composite with a double-network (DN-SPE) was prepared via facile polymerization by Duan et al. (Duan et al. 2018). The DN-SPE indicates decrease of crystallinity, and highest ionic conductivity was 5.3×10^{-5} S cm^{-1} (PEGDE-PEGDA-1000), owing to the mutual plasticization of the double network (PEGDE:PEGDA). A high value of cation transport number (0.44) and broad voltage stability window (4.7 V) validate the suitability of the polymer composite for battery applications.

1.3.5 Star-Type Polymer Composite

The star polymer's architecture provides enhanced physical and topological properties as compared to the linear polymer. The outer spheres of arms in a star polymer enhance the electrical properties (ion mobility, ionic conductivity), while the connection of covalent bonds between the core and arms enhances the stability of the external environment (Ren et al. 2013). The various branching points disorders the polymer crystallization and enhances the free volume for segmental motion and hence faster ion dynamics. Along with this, this architecture improves the salt dissociation and mechanical properties can be improved by introducing a rigid framework within star polymer configuration (Zhang et al. 2016; Xu et al. 2018).

The star-shaped copolymers having POSS segments are an attractive candidate due to the unique multiple-chain-ended structure. A star structure with octavinyl octasilsesquioxane (OV-POSS) and poly(ethylene glycol) methyl ether methacrylate (PEGMEM) by one-step free radical polymerization was synthesized by Zhang et al. (Zhang et al. 2016). The star-shaped PE shows improved free volume and shows a higher conductivity of 1.13×10^{-4} S cm^{-1} than the linear copolymer composite (LCP5.1) which shows about 5.63×10^{-5} S cm^{-1} (at 25 °C). Also, the cation transference number was higher for SCP5.1 ($t_+ = 0.35$) than LCP5.1 ($t_+ = 0.19$). The voltage stability window for SCP5.1 is about 5.31 and 5.04 V (vs. Li/Li^+) at 25 and 80 °C. Another star polymer composite using poly(ethylene glycol) dimethacrylate (PEGDMA) as a monomer and ethylene glycol dimethacrylate (EGDMA) as a crosslinker was prepared by Xiao et al. (Xiao et al. 2019). The synthesized composite is flexible (tensile stress = 1.67 MPa, strain = 300%) and exhibits ionic conductivity about 1.48×10^{-5} S cm^{-1} (at 20 °C) for $PEGDMA_{550}$ (Li-SPE550-Li). This high value of conductivity is attributed to the favorable topological architecture which boosts the ion migration. The voltage stability window is about 5.4 V and the cation transference number is 0.3.

1.3.6 Fibrous-Based Polymer Composite

Fibrous polymer membrane-based polymer composite shows improved performance as compared to traditional polymer composites. In fiber, the energy barrier of particle-particle junctions is low as compared to nanofiller that will enhance ion transport. Also, the better interfacial contact between polymer and fiber results in improved electrochemical stability. The high surface area of nanofiber effectively enhances the ion conduction by creating the continuous ion conduction pathways in the polymer matrix (Zhang et al. 2011; Zhu et al. 2018; Li et al. 2019). A PAN electrospun fibrous membrane (PVdF-PAN-ESFMs) was prepared using the electrospinning technique reported by Gopalan et al. (Gopalan et al. 2008). The 1M $LiClO_4$-PC solution was used as the composite. FESEM analysis evidenced the formation of PVdF-ESFM fibers interconnected with a large number of voids and the uniform diameter was 600 nm. The highest ionic conductivity was 7.8×10^{-3} S cm^{-1} (at 25 °C) and is higher than the PVdF-based composites which are attributed to the elimination of crystalline domains after PAN addition (Song et al. 2004). The voltage stability window of the PVdF-PAN (25)-ESFM was 5.1 V and is much higher than the PVdF (i.e. 4.38 V) and PAN (i.e. 4.25 V) membranes.

As it is well known that the blending approach is an important approach to enhance the electrical and mechanical properties. But, one issue remaining is the poor compatibility of blend polymer with the PEO. So, a new strategy was adopted to synthesize the solid polymer composite. It comprises two chemically dissimilar polymer segments, (i) aromatic polymer segment and (ii) host polymer matrix. The former provides sufficient mechanical/thermal stability, while the latter facilities flexibility and low crystallinity. Lu et al. (Lu et al. 2013) developed a self-standing solid polymer composite membrane where polysulfone (PSF) plays a role as a former one and PEO as the latter one. The salt used was lithium bis(trifluoromethanesulfonyl)imide (LiTFSI) and with low content of SN is used for synthesis via one-step condensation copolymerization. The highest ionic conductivity was 1.6 × 10^{-4} S cm^{-1} (at RT) and increased to 1.14 × 10^{-3} S cm^{-1} (at 80 °C) for PSF-PEO_{35} + LiTFSI + SN system. The voltage stability window of the prepared system was 4.2 V vs. Li/Li^+ and is in a desirable range.

Another strategy to enhance the properties of polymer composite is the copolymerization technique where both composite uptake as well as the amorphous content get improved. Shi et al. (Shi et al. 2018) prepared a gel polymer composite by blending the PEO and PMMA with copolymer P(VDF-HFP) (PE-PM-PVH) by solution cast technique using a liquid composite of $LiPF_6$ – EC + DMC (1:1 v/v) as the plasticizer. FESEM analysis evidenced the creation and growth of the pores after PEO-PMMA blending with P(VDF-HFP) (Porosity = 58%) as compared to pristine P(VDF-HFP) (Porosity = 30%). The ionic conductivity (σ) of the PE-PM-PVH membrane is 0.81 mS cm^{-1} and is higher than pristine P(VDF-HFP) which has 0.25 mS cm^{-1}. The cation transport Number ($t_{Li}{}^+$) of PE-PM-PVH polymer membrane was higher (0.72) than the pristine P(VDF-HFP) membrane (0.29). The voltage stability window of the blend was higher (~5.0 V) than the pristine P(VDF-HFP) membrane (~4.5 V).

1.3.7 Cross-Linked Polymer Composite

Cross-linking of polymer composites is an attractive strategy to prepare new polymer composites with different architectures. The cross-linking approach enhances dimensional stability and dynamic storage modulus (Lu et al. 2017; Kim et al. 2010). The chemical cross-linking of polymer needs to be done in such a way that balanced ionic conductivity and mechanical strength may be achieved. The cross-linker prevents polymer crystallization and facilitates faster segmental mobility (Lin et al. 2018).

Shin et al. (Shin et al. 2016) prepared the cross-linked composite polymer composite (CLCPE) based on methacrylate-functionalized SiO_2 (MA-SiO_2) nanoparticles, PAN membrane, and gel composite precursor containing tri(ethylene glycol) diacrylate (TEGDA). The ionic conductivity of the CLCPE with non-porous MA-SiO_2 particles and mesoporous MA-SiO_2 particles are 1.1*10^{-3} S cm^{-1} and 1.8 × 10^{-3} S cm^{-1}, respectively. The highest conductivity is due to porosity in MA-SiO_2 nanoparticles. The effect of the cross-linker on the properties of the solid polymer composite was examined by Youcef et al. (Youcef et al. 2016). They reported the preparation of the cross-linked polymer composite (CLPE) by UV-induced cross-linking of poly(ethylene glycol) diacrylate (PEGDA) and divinylbenzene (DVB) within a poly(ethylene oxide) (PEO) matrix. There was no effect of the DVB on the conductivity and for 10% DVB, ionic conductivity was high (1.4 × 10^{-4} S cm^{-1}) for practical applications. The voltage stability window of the composite was close to 5.0 V. The cation transference number was 0.23. Zhang et al. (Zhang et al. 2019) demonstrated the preparation of flexible cross-linked SPE with PEO, TEGDMA, and TEGDME (PTT) SPE with LiTFSI salt. In this work, the electrode/composite has been prepared by in-situ UV-derived dual-reaction to minimize the interfacial resistance, and low molecular weight cognate monomers have been introduced to enhance the conductivity and reduce the crystallinity. The prepared PTT-SPE membrane is transparent and flexible, and the benefits of this will be reflected in electrical properties. The PTT-SPE membrane exhibits ionic conductivity of about 0.27 mS cm^{-1} and is 30 times higher than PEO-SPE. The cation transport number is also higher and is about 0.56 and is favorable in the elimination of polarization. The voltage window of the PTT-SPE is about 5.38 V.

1.4 SELECTION CRITERIA AND PROPERTIES OF POLYMER HOST/SALT/NANOFILLER/SOLVENT

1.4.1 Properties for Polymer Host

Since the first report of ion conduction in 1973 for PEO/Li^+ salt complexes, PEO (poly(ethylene oxide)) is one of the strong candidates for the polymer composites, but a low ionic conductivity value (10^{-8} S cm^{-1}) hinders its use. Various strategies have been used to enhance the conductivity by the addition of ionic liquid, nanofiller, plasticizer, etc. (Bruce 1995). PAN (polyacrylonitrile) emerged as an alternative to PEO and has two unique features, (i) small thermal resistance, (ii) flame retardant behavior. Two types of gel composites—by taking PAN as host, EC/DMC as plasticizers with $LiPF_6$ or $LiCF_3SO_3$ as salt—were reported by Appetecchi et al. (Appetecchi et al. 1999). The blend of PEO and PAN provides enhanced properties as compared to individual polymers, also improves the interfacial properties at the electrode-composite interface (Choi et al. 2000). Another polymer capable of reducing cost and enhancing interfacial stability is poly(methyl methacrylate) (PMMA), but the poor mechanical flexibility of film restricts its use as composites in energy storage devices. Some of the reports suggested that copolymerization of PMMA with another polymer improves the mechanical and electrical properties of the polymer composite. Porous PDMS-CNT nanocomposites with PMMA improved the flexibility and control phase separation between PDMS and PMMA, as observed by Lee et al. (Lee et al. 2012). Nowadays, poly(vinylidene difluoride) (PVdF) has grabbed attention due to its semi-crystalline nature, high dielectric constant (ϵ = 8.4), and the presence of strong electron-withdrawing functional groups (-C-F). The presence of functional groups supports more dissolution of lithium salts and subsequently supports the high concentration of charge carriers. While, high dielectric constant helps for more significant dissolution of lithium salts and consequently supports the high concentration of charge carriers, and the functional group induces a net dipole moment (Esterly 2002). Ionic conductivity can be tuned up to that of liquid composite by adding nanofiller, plasticizer, etc. (Choe et al. 1995). Another attractive candidate that has been less discussed is polyvinyl pyrrolidone (PVP). The amorphous nature and presence of the carbonyl group (C-O) in the side chains of PVP make it suitable for polymer composites. One key advantage is that it is highly soluble in polar solvents such as alcohol. So, various polymer hosts, as well as blends, have been examined for suitability in polymer composites. Two important parameters for any polymer host are dielectric constant and glass transition temperature. Table 1.1 shows some commonly used polymer hosts with their glass transition and melting temperature.

TABLE 1.1
Properties of Mostly Used Polymer Host in Polymer Composites.

Abbreviation	Polymer Name	T_g (°C)	T_m (°C)	Formula
PEO	Poly(ethylene) oxide	−67	65	$(-CH_2CH_2O-)_n$
PMMA	Poly(methyl methacrylate)	105	160	$(C_5O_2H_8)_n$
PVC	Poly(vinyl chloride)	81	160	$(C_2H_3Cl)_n$
PVA	Poly(vinyl alcohol)	85	230	(C_2H_4O)
PAN	Poly(acrylonitrile)	125	317	$(C_3H_3N)_n$
PEMA	Poly(ethyl methacrylate)	66	160	$[CH_2C(CH_3)\ (CO_2C_2H_5)]_n$
PS	Polystyrene	100	240	$(C_8H_8)n$
PPO	Poly(propylene oxide)	−60		$-(CH(-CH_3)CH_2O)_n$
PVdF-HFP	Poly(vinylidene fluoridehexafluoropropylene)	−65	135	$-(CH_2CF_2)_n\ [CF_2CF(CF_3)]_m-$
PDMS	Poly(dimethylsiloxane)	−127	−40	$-[SiO(-CH_3)_2]n$
PVdF	Poly(vinylidene fluoride)	−40	171	$-(CH_2CF_2)n-$

1.4.2 Solvents for Polymer Composite

Both conductive species and solvent interact with ion-dipole interaction in polymer matrix (Eiamlamai 2015). In the case of salt, the solute role is played by ionic species and the polar solvent by dipoles. The interaction force tells the energy of solvation required in the case of ionic solutes and is inversely proportional to the squared distance between solute and solvent molecule. So, a smaller ion results in more interaction force with greater dissolution in the solvent. So, polar solvents such as water, nitrile, amide, etc. are used for better solvation due to the high dielectric constant as suggested in a model by Born. The main properties of a solvent (Ponrouch et al. 2015) which are required for more dissolution of ions and high mobility are summarized in Table 1.2.

Table 1.2 shows a list of commonly used solvents and their properties. Acetonitrile is a good solvent due to its moderate dielectric constant and low viscosity. DMF is a mostly used solvent and is suitable for almost all polymers in electrochemical devices, but its high boiling temperature sometimes restricts its use. Two solvents can be mixed, one having a high dielectric constant and the other having low viscosity for faster ion conduction.

The key requirement for a solvent is a high dielectric constant and high flashing point (FP); the former helps in the dissociation of more salt and later improves the safety of the energy storage device. Linear carbonates DEC and DMC have low dielectric constant and low viscosity (Tamura et al. 2010). High viscosity may be due to the large mutual interaction between solvent particles. If both EC and PC are added together then they can dissolve more lithium salt as compared to a single solvent (Li and Balbuena 1999). Figure 1.5 depicts the key characteristics of polymer host, salt and solvent.

1.4.3 Properties of Salt

Salt is one of two major components of any composite. Amongst the properties directly affecting the salt is the solubility in the solvent for the formation of more

TABLE 1.2
Properties of solvents for polymer electrolytes.

S. No.	Abbreviation	Solvents	Mol. Wt. (g/mol)	Boiling point (°C)	Dielectric Constant (at 25 °C)	Viscosity/cP	Dipole Moment (Debye)	Density (g/cm3)
1	ACN (C_2H_3N)	Acetonitrile	41.05	81.6	35.95	0.341	3.92	0.377
2	THF (C_4H_8O)	Tetrahydrofuran	72.11	66	7.39	0.46	1.63	0.880
3	DMF (C_3H_7NO)	Dimethylformamide	73.10	158	36.71	0.796	3.82	0.944
4	DMSO (C_2H_6OS)	Dimethyl sulfoxide	78.13	189	46.45	1.991	3.96	1.095
5	DMC ($C_3H_6O_3$)	Dimethyl carbonate	90.08	90.01	3.1	0.59	0.91	1.07
6	EC ($C_3H_4O_3$)	Ethylene carbonate	88.06	248	89.6	1.85	4.9	1.322
7	PC ($C_4H_6O_3$)	Propylene carbonate	102.1	241	64.4	2.53	4.9	1.19
8	DME ($C_4H_{10}O_2$)	Dimethoxyethane	90.12	84.0	7.20	0.455	1.30	0.859
9	DEC ($C_5H_{10}O_3$)	Diethyl carbonate	118.13	126	2.82	0.75	0.9	0.975
10	NMP ($C_5H_{9N}O$)	N-Methyl-2-pyrrolidone	99.13	202–204	32.55	1.65	4.1	1.03

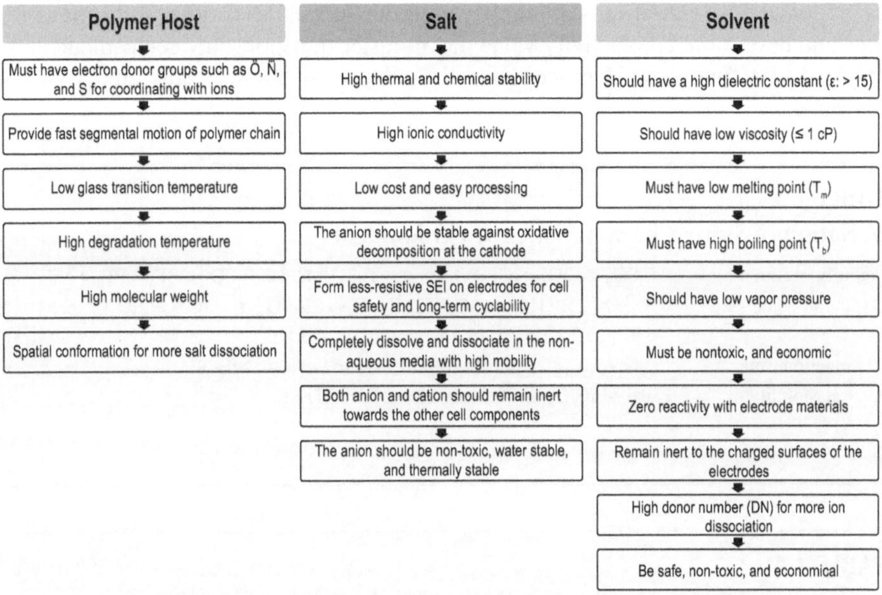

FIGURE 1.5 Key characteristics of polymer host, salt, and solvent.

free charge carriers, wide potential stability window, non-toxic nature, chemical and thermal stability (Aravindan et al. 2011). Salt is chosen such that it does not form ion pairs or ion aggregation and also easily get dissociated in the solvent with the polymer. So, the salt with small cation radii are preferred for faster dissociation of salt due to electric charge delocalization, low basicity, and screened negative charge fulfilling our requirement of cationic transference number (t_{cation}) equal to unity (Grünebaum et al. 2014). Also at a high concentration of salt, the unavailability of suitable coordinating sites in the polymer host hinders the ion motion and affects the polymer segmental motion. At a low concentration of salt, conductivity is almost independent of salt content, but with an increase in this more free charge carriers participate after dissociation in the conduction process. Gray et al. (Gray 1997) proposed a concept for solvation of cations in polymer composites and suggested that the formation of a coordination bond due to partial sharing of the lone pair of electrons and energy change in PEs is dominated by cation solvation, while solvation with anion is less due to weak interaction of the polymer with the anion. $LiClO_4$ was the most common lithium salt used in lithium primary batteries. However, the high oxidizability of anions (ClO_4^-) caused some safety issues in secondary batteries; thus $LiPF_6$ replaced $LiClO_4$ in the newer designs and became the major material for composites, which exhibits better overall performance including higher ionic conductivity, more solubility, and broad chemical stability (Sloop et al. 2001). Variation of some properties of lithium salts is shown here in Table 1.3.

LiTFSI has excellent thermal stability with a decomposition temperature of 360 °C. The high dispersion of the anionic charge of lithium salt makes it more easily ionized. Li-based salts are mostly used due to smaller cationic radii than Na ions and have more conductivity value due to faster transport between cathode and anode. Some fundamental properties and structures of lithium salts are given in Table 1.4.

TABLE 1.3
Variation of Some Characteristics of Lithium Salts

Stability of Anion	$CF_3SO_3 < ClO_4 < (CF_3SO_2)_2N < (CF_3SO_2)_3C < C_4F_9SO_3 < BF_4 < PF_6 < AsF_6$
Oxidation Potential	$[CF_3SO_3]^-$(*5.9 V*) *A*< ClO_4^-(*6.0 V*) < $[N(CF_3SO_2)_2]^-$ = $[C(CF_3SO_2)_3]^-$(*6.1 V*) < BF_4^-(*6.2 V*) < PF_6^-(*6.3 V*) < AsF_6^- < (*6.5 V*).
Dissociation Constant	LiTf< $LiBF_4$< $LiClO_4$< $LiPF_6$< $LiAsF_6$< LiTFSI< LiBOB
Av. Ion Mobility	$LiBF_4$>$LiClO_4$>$LiPF_6$>$LiAsF_6$> LiTf >LiTFSI

Source: Aravindan et al. 2011, Copyright Wiley.

TABLE 1.4
Properties of Commonly Used Lithium Salts for Studies on Polymer Composites

Lithium Salt (Abbreviations)	Main Characteristics
$LiClO_4$	• Broad electrochemical stability window • Low solubility in commonly used carbonate-type solvents
$LiBF_4$	• Broad electrochemical stability window • Low solubility in commonly used carbonate-type solvents
$LiPF_6$	• High ionic conductivity, favors SEI formation, passivates Al substrate at the cathode side • Decomposes in the presence of moisture and reacts with composites at elevated temperatures resulting in the formation of HF
LiFSI	• Higher ionic conductivity compared to LiTFSI, high electrochemical stability • Unable to form passivation layers on Al current collectors (in the presence of LiCl), but purified LiCl free salt passivates Al collectors
LiBETI	• High solubility and high ionic conductivity, high electrochemical stability • Unable to form passivation layers on Al current collectors
LiBOB	• High electrochemical stability and long-term stability • Form highly resistive SEI-films (low conductivity in comparison to $LiPF_6$ and LiTFSI)
LiDFOB	• High electrochemical stability and cycling behavior, able to form passivation layer on Al current collectors • Lower solubility in carbonate-type solvents compared to LiTFSI and $LiPF_6$, but higher than LiBOB
LiTFSI	• High solubility and high ionic conductivity, high electrochemical stability • Unable to form passivation layers on Al current collectors (Al-degradation and corrosion)

Source: Reprinted with permission from Grünebaum et al. [2014, Copyright Elsevier].

1.4.4 Inorganic Fillers/Clay

Polymer composites can be prepared by the addition or incorporation of nanofiller and intercalation of clay in the polymer matrix. The addition of nanofiller/clay not only enhances the conductivity but also improves the mechanical properties such as the physical strength of PEs.

1.4.4.1 Inorganic Filler

The particle size of the filler is a very important factor that affects ion conduction. Ferroelectric fillers such as $BaTiO_3$ reduce the interfacial resistance between electrode composite surfaces due to the permanent dipole of ferroelectric materials (Shanmukaraj et al. 2008). The filler acts as a solid plasticizer that enhances the transport properties and reduces the crystallinity due to an increase in the dielectric constant. Fillers with Lewis acid surface group interact with both polymer and ion and reduce the ion coupling (Moskwiak et al. 2006). Various filters are presently used: $BaTiO_3$, Al_2O_3, TiO_2, SiO_2, and CeO_2, resulting in enhanced ionic conductivity and better thermal and structural properties. Plasticizers such as EC/PC help in the dissociation of salt and enhance the conductivity, but their use is limited due to their volatile and flammable nature. Different classes of filler are shown in Figure 1.6 (Nunes-Pereira et al. 2015).

One constraint with nanofiller is the chances of agglomeration due to their high surface energy (200–5000 dyn cm^{-1}), high van der Waals forces, or high electrostatic forces, resulting in the poor dispersion in the polymer matrix (surface energy = 10 and 50 dyn cm^{-1}), and hence poor electrical properties (Lewin et al. 2005; Luo et al. 2019). To overcome this large difference in energy, nanofillers are coated with low-energy surface materials, siloxane coupling reagents or polymers, whereby a thin layer can reduce the surface energy of inorganic fillers. So, nanofiller dispersion in the polymer matrix can be improved with this approach. Further, ion transport can be controlled easily by examining the electrostatic interactions, hydrogen bonding, or dipole-dipole interactions (Figure 1.7a).

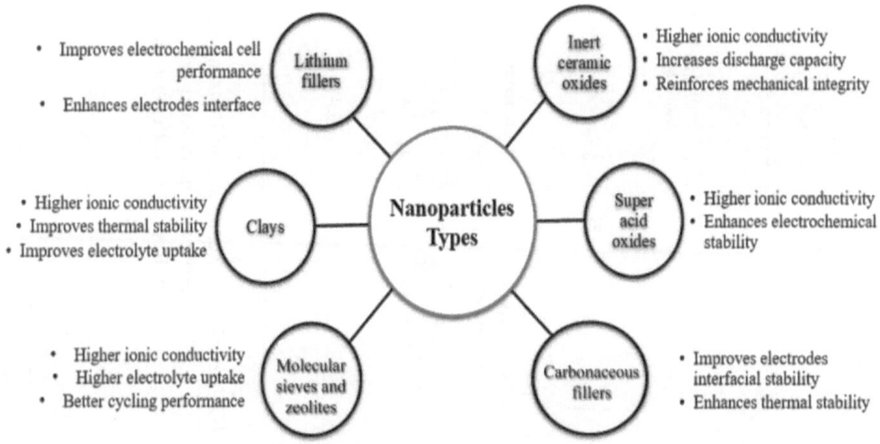

FIGURE 1.6 Main effects of each class of filler within a polymer matrix.

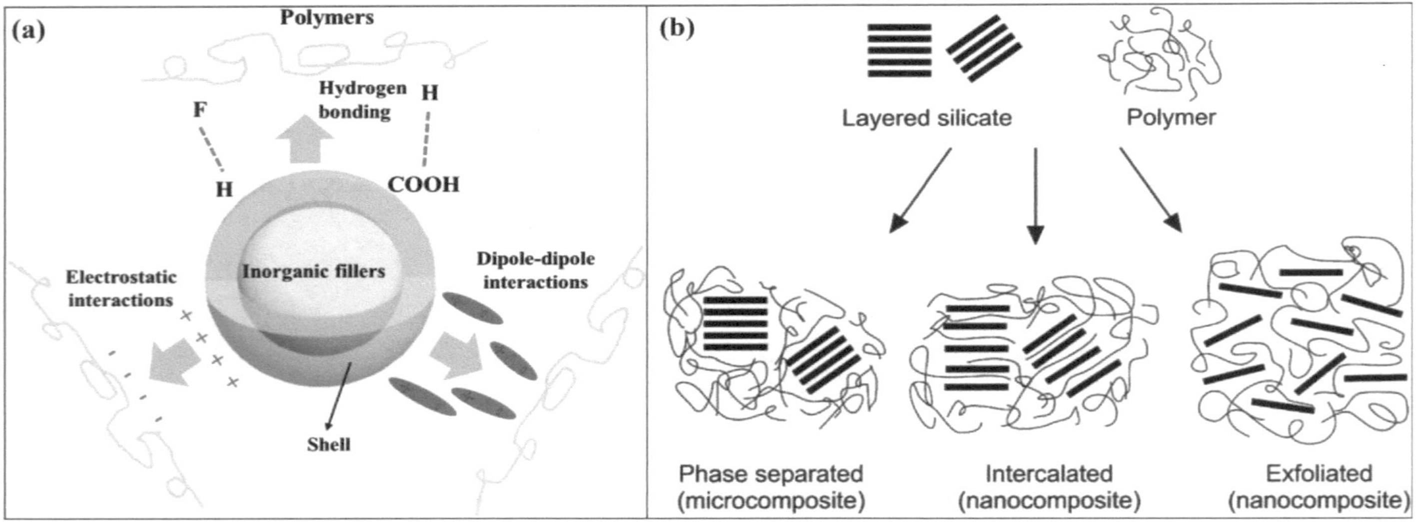

FIGURE 1.7 (a) Schematic of the range of interactions between the core-shell structured filler and polymers matrix; (b) types of composite derived from interaction between clays and polymers. Source: (a) Reprinted with permission from Luo et al. [2019, Copyright RSC]. (b) Reprinted with permission from Alexandre and Dubois [2000, Copyright Elsevier].

1.4.4.2 Clay

Instead of filler, solid smectite group clay minerals are preferred for the reinforcement of polymer matrices. The key advantages are the unique structure, high strength, and high aspect ratio of each platelet. In polymer composites, cation coordinated polymer chains get intercalated inside the negatively charged clay galleries. Two benefits are (i) the anion due to bulky size remains outside clay galleries, hence single ion conduction, and (ii) polymer intercalation enhances the stability (thermal, chemical) of the polymer. Key characteristics of the clay are as given:

1. High aspect ratio
2. High cation exchange capacity (CEC)
3. Large specific surface area
4. Appropriate interlayer charge
5. Length scale

Nanoclay is used for preparing highly conductive polymer nanocomposite (PNC). Three important categories depending on polymer chain intercalation in clay galleries are shown in Figure 1.7b (Alexandre and Dubois 2000). A phase-separated composite is formed when no intercalation occurs, and electrical properties remain the same as traditional composite. Intercalated polymer nanocomposites are termed when cation coordinated polymer chains get accommodated inside the clay galleries (Figure 1.7b). The presence of a negative charge layer in clay galleries increases the ion conduction as well as prevents polymer degradation. When clay layers are uniformly dispersed in the polymer matrix, then nanocomposite is referred to as exfoliated polymer nanocomposites (Figure 1.7c). Polymer intercalation may be confirmed with the XRD by examining the interlayer spacing. An increase in interlayer spacing (a shift toward lower angle) for nanoclay-based PNC as compared to pristine nanoclay is an indication of intercalation.

1.5 CONCEPT OF ION TRANSPORT IN POLYMER COMPOSITES

The phenomena of ion transport in polymer composites is complex and depends on various factors such as cation size, the dielectric constant of the host polymer, ion-pairing, amorphous phase character, and conduction by both mobile ions (cations and anions). Several models such as the free volume theory, configurational entropy model, VTF behavior, WLF behavior, EMT approach, etc., have been proposed to interpret the ion conduction in polymer composites. Two main mechanisms for the cation conduction in the solid compound are the Grotthuss mechanism (proton hopping or structural diffusion) and the vehicle mechanism (Chen et al. 2016; Luo et al. 2013). The plastic crystal doped ion diffusion involves coupling to certain, rotational, motions while in polymer composites structural dynamic occurs due to the whole polymer chain (MacFarlane and Forsyth 2001).

1.5.1 Activation Energy

The ion transport in the polymer composite is influenced by the temperature change and the boost of ion migration with an increase in temperature is attributed to the polymer flexibility and thermal activation of charge carriers. Thermal activation promotes faster ion conduction via oxygen sites (in polymer chain) due to lower potential barrier or activation energy. The temperature-dependent enhancement of the ionic conductivity is elaborated with two mechanisms: Arrhenius behavior and Vogel-Tamman-Fulcher (VTF) behavior. The $\sigma(T)$ curve in the case of polymer composites depicts the following patterns of behavior (MacCallum 1988):

1. Arrhenius behavior is used for low temperatures whereas VTF behavior is at higher temperatures
2. Arrhenius behavior occurs throughout, but with two different activation energies, high E_a closer to T_g and a smaller E_a at higher temperatures
3. The VTF behavior for temperature which is slightly greater than T_g
4. Behavior is very unlike either Arrhenius or VTF at all temperatures.

1.5.1.1 Arrhenius Behavior

With the increase of temperature, polymer flexibility increases along with amorphous content. Overall ion migration gets enhanced due to the faster segmental motion of polymer chains which favors faster cation migration. Such type of behavior occurs only in the polymer composites where ion migration is supposed to occur via the amorphous phase content (Agrawal and Pandey 2008). The Arrhenius behavior is expressed as $\sigma = \sigma_0 \exp(-E_a/kT)$, where, σ_0 is the pre-exponential factor, k is the Boltzmann constant, and E_a is the activation energy. The activation energy is acquired from the slope of linear-least square fitting of the log σ vs. 1/T plot, and the lower value of the activation energy indicates the fast ion migration. The Arrhenius behavior indicates the ion transport occurs via hopping mechanism decoupled from the polymer-chain breathing and occurs below the glass transition temperature.

1.5.1.2 Vogel-Tamman-Fulcher (VTF) Behavior

In this approach, the segmental motion of the polymer chain is the key player in enhancing the ion migration inside the composite matrix. On the application of temperature, ions cross the energy barrier and are thermally activated, which swim in the free volume provided by the polymer matrix. Initially, the VTF equation was proposed to explore the viscous properties of supercooled liquids, then extended to polymer composites, and is valid above the glass transition temperature (Norman et al. 2012). The ion diffusion is possible only when the free volume is provided by the polymer chains for ion migration in the system (Ratner et al. 2000). The expression for VTF behavior is $\sigma = AT^{-1/2} \exp\left[-B/\left(T - T_0\right)\right]$, where, σ is the ionic conductivity, A is the pre-exponential factor related to the conductivity and number of charge carriers, B is the pseudo-activation energy for the conductivity, and T_0 is the temperature close to the T_g of material. The σ vs. 1/T plot is a non-linear plot and indicates long-range migration via ion hopping. The VTF conductivity vs. temperature plot is non-linear and suggests the ion migration via the coupled hopping mechanism, relaxation/breathing with the segmental motion of the polymer chain.

1.5.2 Proposed Ion Transport Model

Various models have been proposed to get insights into ion transport in polymer composites. Some of the crucial and impactful models are discussed in the following sections. Armand et al. (Armand et al. 1979) for the first time explained the ion transport mechanism in PEO-alkali metal salt complexes (PEO-KSCN, PEO-NaSCN). In the classical PEO-salt complexes, transport of the cation was suggested to occur by hopping through the helices and the anions were assumed to be almost immobile and placed outside the helices. Molecular dynamics simulations suggest that the Li^+ ions are complexed to PEO through approximately five ether oxygens of a PEO chain and that the mobility of the cations is decreased considerably by this complexation (Muller-Plathe and Gunsteren 1995). Consequently, the mobility of the Li cations is related to the motions of the complexing segments of the PEO chain. Thus, the cation transport is described as the motion of the Li^+ species between complexation sites assisted by the segmental motion of the PEO matrix. Later, Papke et al. (Papke et al. 1982) suggested that the transport of the cation, complexed within the helical regions, occurs through the disordered amorphous phase of the polymer composite.

1.5.2.1 Free Volume Model

It is one of the simplest transport mechanisms used to understand the polymer ion transport and segmental mobility given by Cohen and Turnbull in 1959 (Cohen and Turnbull 1959). It is based on the assumption that motion in a polymer composite depends on the redistribution of free volume associated with the end of a polymer chain in the matrix, not on the thermally activated process. The increase in temperature expands the material and the creation of voids leads to free volume enhancement which provides more free space to polymer segmental motion. This increase in free volume (V_f) increases the ability of the polymer chain to rotate freely which directly affects the ion transport and obeys the following relation expressed by equation 1.1:

$$V_f = V_g\left[0.025 + \alpha\left(T - T_g\right)\right] \tag{1.1}$$

Where V_g is the critical volume per mole at T_g, α is the thermal expansivity, and T represents the temperature during the experiment. The biggest drawback faced by this model is that it ignores the kinetic effects associated with long-chain molecules and does not involve microscopic behavior like dielectric relaxation, the effect of the molecular weight in transport properties, etc.

1.5.2.2 Configurational Entropy Model

Gibbs and Marzio (Gibbs and Marzio 1958) proposed this model to overcome the drawbacks of the free volume model. The entropy model is based on the cooperative rearrangement of the polymer chains while the free volume model includes transport via the formation of voids. The average probability of chain arrangement is given by the equation 1.2:

$$W = A\,exp\left(-\frac{\Delta\mu\, S_c^*}{k_B T S_c}\right) \tag{1.2}$$

Where A is pre-exponential constant, k_B is Boltzmann's constant, T is the absolute temperature (K), $S_c{}^*$ is minimum configurational entropy required for the rearrangement, S_c is configurational entropy at temperature T, and $\Delta\mu$ is apparent activation energy opposing the rearrangement of polymer segmental unit (kJ/mol).

1.5.2.3 Vogel-Tamman-Fulcher (VTF) Model

The Vogel-Tamman-Fulcher model was developed by Vogel, Tamman, and Fulcher in the early part of the 20th century, for diffusion in the glassy system or viscosity of supercooled liquids. Here the ionic transport mechanism is attributed to the segmental motion of polymer chains via formation and destruction of a coordinating sphere of the solvated ion (Bruce and Gray 1995, Gong et al. 2012). In this created free volume ions diffuse under the influence of the electric field. The VTF equation derived using the entropy model can be expressed by equation 1.3:

$$\sigma = \frac{A}{\sqrt{T}} exp\left[-\frac{B}{T - T_o}\right] \tag{1.3}$$

Where A is a constant proportional to the number of carrier ions, B is the pseudo-activation energy of the ion, T_o is the equilibrium glass transition temperature or pseudo-transition temperature ($<T_g$).

As it is well known, ion transportation occurring above the glass transition temperature (T_o) is due to relaxation and transport processes as suggested by the VTF model (Bruce and Vincent 1993; Luo et al. 2011). This model is based on the assumption that the transport in polymer composites occurs in the amorphous phase content of the host polymer. As above the glass transition temperature, disorder takes place in polymer chains which changes the entropy and promotes ion transport. Then Williams (Williams et al. 1955) developed extension of the VTF equation (i.e. WLF equation), to get insights of ion dynamics in polymers and other glass-forming materials. Effect of viscosity and inherent relaxation processes was considered. This modified version of VTF model can be explained by equation 1.4:

$$loga_T = \left[\frac{\{-C(T - T_s)\}}{(C_2 + T - T_s)}\right] \tag{1.4}$$

Where a_T is shift factor and is the ratio of the mechanical relaxation process at ordinary temperature T to some reference temperature T_s.

The temperature dependence of conductivity which obeys the VTF equation confirms the free-volume ion transport mechanism in polymer composites and concludes that the micro-Brownian motion of the polymer chain segment affects the ionic transport which supports the transport in the amorphous phase. Further, investigations were done using WLF theory to clarify the ionic conductivity mechanism by using the relationship between ionic conductivity and fractional free volume $f_r(T)$

(Ji et al. 2012). It concludes that ionic conductivity increases with an increase in fractional free volume and can be described by the following equations 1.5 and 1.6:

$$f_r(T) = f_r(T_g) + \alpha(T - T_g) \tag{1.5}$$

$$\log_{10}\sigma(T) = log_{10}\sigma(T_g) + C_1\left[1 - \frac{f_r(T_g)}{f_r(T)}\right] \tag{1.6}$$

1.5.2.4 Angell's Decoupling Theory

The ion transport mechanism in polymer composites was later enlightened by C. A. Angell (Angell 1986) based on the assumption that coupling between transport and relaxation plays an active role in ion transport. Angell generalized the ion transport mechanism using a new term decoupling index (R_τ) denoted by equation 1.7:

$$Decoupling\ Index\ (R_\tau) = \frac{\text{Structural Relaxation Time}(\tau_s)}{\text{Electrical Relaxation Time}(\tau_\sigma)} \tag{1.7}$$

Here, τ_s refers to the segmental relaxation or viscosity, and (τ_σ) is inversely proportional to the conductivity. A larger value of R_τ reflects highly decoupled charge-conducting modes by the viscous motion of the matrix. If $R_\tau = 1$ then the ionic motion and the structural relaxation occur at the same time, and for polymer-salt complexes $R_\tau < 1$, which indicates rapid structural relaxation rather than the conductivity relaxation. The relaxation process, which allows the rearrangement of the polymer structure, may or may not help in ion dynamics and may be attributed to the inter-ionic interaction, which results in ion immobilization (Bruce 1995). Dam et al. (Dam et al. 2015) reported a solid polymer composite and correlated the segmental and conductivity relaxation time using decoupling index. Value of R_τ in the range 5 to 48 suggests that ion conduction is coupled to the segmental relaxation.

1.5.2.5 Amorphous Phase Model

Various reports on the study of polymer composites suggest that the addition of nanofiller decreases the crystallinity and amorphous phase content increases, which leads to faster ion transport. This is the assumption for the amorphous phase model (Wieczorek et al. 1989). Ion transport is sustained by polymer segmental motion and is faster within the amorphous region than within the crystalline region of the polymer host. The presence of large numbers of nucleation centers in the polymer composites promotes the crystallization process and, as a consequence, a bigger disorder, typical of a liquid state, is frozen in the newly formed solid polymeric matrix on solvent evaporation.

1.5.2.6 Effective Medium Theory

The effective medium theory was originally developed by Nan and Smith in 1991 (Nan and Smith 1991) for polymer composites based on the simple effective medium equation (Landauer 1952). It can be applied for explaining the electrical properties in polymer composites, composite solid composites, etc. (Wieczorek et al. 1994). The improvement in the conductivity is obtained with the addition of nanofiller and may be due to the change in microstructure at the composite-filler interface. According to the theory, there are three components present in a composite polymer composite with different electrical properties: (i) the highly conductive interface layers coating on the surface of the grain, (ii) the dispersed insulating fillers in the polymer host matrix, and (iii) the dispersed filler throughout the polymer composite host matrix and the highly conducting interface layer make a matrix phase which consists of overlapping/touching units that provide high-conducting pathways in the polymer matrix.

According to the Maxwell Garnett rule (Maxwell 1881), the equivalent conductivity (σ_c) of the CPEs unit can be expressed by equation 1.8:

$$\sigma_c = \frac{2\sigma_1 + \sigma_2 + 2Y(\sigma_2 - \sigma_1)}{2\sigma_1 + \sigma_2 - Y(\sigma_2 - \sigma_1)} \tag{1.8}$$

Here σ_1 and σ_2 are, respectively, the conductivity of the interface layer and that of dispersed insulating grain. Y is the volume fraction of dispersed grain in a composite unit and for spherical geometry can be calculated by equation 1.9:

$$Y = \frac{1}{\left(1 + \frac{t}{R}\right)^3} \tag{1.9}$$

Where t is the thickness of the interface layer and R is the radius of dispersed grain. The t/R parameter is based on the assumption that an increase in amorphous phase content in CPEs in comparison to the pristine system is due to the amorphous presence in the surface (Shukla and Agrawal 1999). This model was further improved by Wieczorek et al. (Wieczorek et al. 1989) and reported the presence of three phases in a CPE explained as a highly conductive layer covering the surface of the grain (phase 1), dispersed filler grain (phase 2), and the bulk polymeric composite (phase 3). The formation of a high conductive pathway composed of the surface layers (i.e. phase 1), is proposed to be the possible region of enhanced conductivity of the composite polymer composite.

1.5.3 Proposed Transport Mechanism in Dispersed Polymer Composite

In solid polymer composites, alkali metal salts are dissolved in the polymer matrix and salt gets dissociated in the cation and anions owing to the electrostatic interactions. The cation migrates via the coordinating sites provided by the polymer chain and anion; due to its large size is immobilized with the polymer backbone. The combined effect of the hopping and segmental motion results in ion transport. The four possible processes of cation transport are shown in Figure 1.8a, and are (i) intrachain hopping, (ii) interchain hopping, (iii) intrachain hopping via ion cluster, (iv) interchain hopping via ion cluster (Xue et al. 2015).

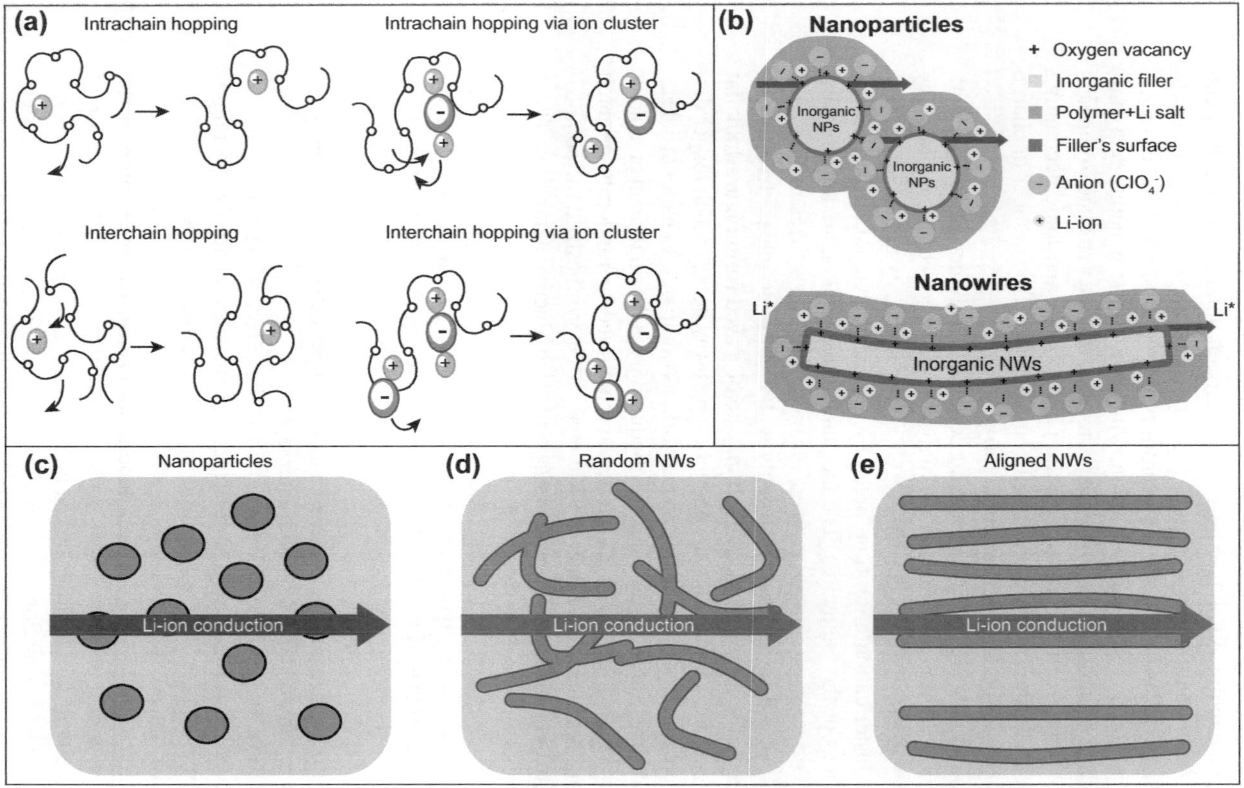

FIGURE 1.8 (a) Mechanism of ion transport in PEO; (b) schematic illustration for Li-ion transport in the composite polymer electrolytes with nanoparticle and nanowire fillers; (c) the comparison of possible Li-ion conduction pathways; (a–c) Li-ion conduction pathways in composite polymer electrolytes with nanoparticles (c), random nanowires (d), and aligned nanowires (e). Compared with isolated nanoparticles, random nanowires could supply a more continuous fast conduction pathway for Li ions.

Source: (a) Reproduced with permission from Xue et al. [2015, Copyright Royal Society of Chemistry]. (b) Reproduced with permission from Liu et al. [2016, Copyright ACS]. (c–e) Reproduced with permission from Liu et al. [2017, Copyright Nature].

The ionic conductivity of the polymer composites depends on the equilibrium of the Lewis acid-base interactions. The nature of the interaction is dipole-dipole interaction and the polymer phase acts as an ion transport medium. The Lewis acid group of the nanofiller may compete with the Lewis acid lithium cation for the formation of a complex with the polymer host. The hydrogen bonding present between the host polymer and nanofiller promotes fast ion transport by perturbing the chain and increasing the free space volume. The ceramic surface group may act as cross-linking centers for the host polymer and anion, which lowers the structure reorganization tendency and promotes faster ion transport. Also, the filler surface group acts as Lewis acid-base interaction centers with the polymer/salt, and ionic decoupling promotes more salt dissociation which directly enhances the free number of charge carriers and hence conductivity (Croce et al. 2001).

Liu et al. (Liu et al. 2016) prepared a solid composite polymer electrolyte with $Y2O_3$-doped ZrO_2 (YSZ) nanowires and the highest ionic conductivity exhibited by YSZ nanowires was 1.07×10^{-5} S cm^{-1} at 30 °C. The conductivity enhancement was attributed to the faster cation migration forced by positively charged oxygen vacancies on the surfaces of the nanowires. The mechanism for ion conduction is shown in Figure 1.8b. The positive-charged oxygen vacancies on the surfaces of the fillers act as Lewis acid sites that can interact strongly with anions and release free cations which participate in the ionic conductivity. It may be observed from Figure 1.8b that nanowires offer a more continuous ion-conducting pathway across much longer distance and generate an effective percolation network, as compared to nanoparticles.

Another important strategy to enhance the conductivity is the alignment of nanowires along the normal direction of electrodes as compared to randomly oriented nanowires. Using this approach, Liu et al. (Liu et al. 2017) synthesized the composite polymer electrolyte with well-aligned inorganic Li^+-conductive nanowires. These aligned nanowires result in an increase of conductivity up to 6.05×10^{-5} S cm^{-1} at 30 °C. Figure 1.8c compares the ion migration in the composite polymer electrolytes doped with the nanoparticle, random nanowire, and aligned nanowire. The enhancement in conductivity was attributed to the absence of crossing junctions (for aligned nanowires) that formed in aggregated nanoparticles or random nanowires with wire-to-wire junctions. It was concluded that the surface region of LLTO nanowires with their high specific surface area and large aspect ratio provides a fast pathway for Li-ion diffusion to long distances without interruption.

Arya et al. (Arya and Sharma 2018) examined the effect of nanofillers on ion transport, and concluded that an electron-rich group provides a path for ion migration in the polymer matrix. When nanofiller is added to the polymer salt system, then there are two means by which it can affect the ionic transport or conductivity: (i) direct interaction of cations with nanofiller surface, (ii) interaction of the nanofiller surface with a polymer chain. At low nanofiller content, dissociation of free ions occurs owing to nanofiller interaction with an anion via hydrogen bonding (Figure 1.9a). The interaction between anion and nanofiller dominates at lower nanofiller content and anions can be held on the nanofiller surface, and cation easily migrates via the conduction path provided by the polymer chain. Initially, the affinity for anions is more toward the nanofiller surface than the cation. At high nanofiller content decrease in conductivity is due to anion trapping, and is due to

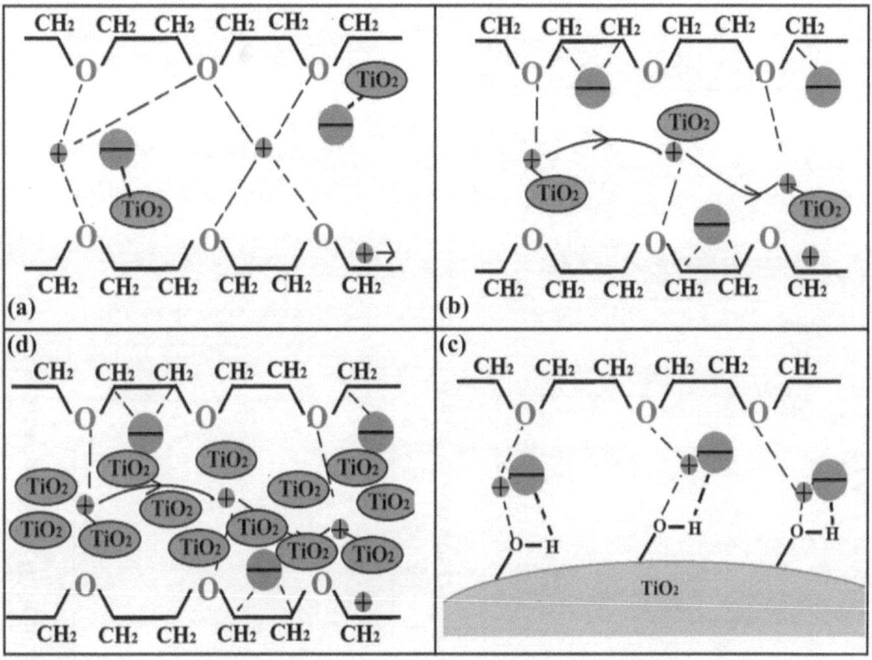

FIGURE 1.9 Ion transport in nanofiller-dispersed polymer composites.

the increased ability of ion-pairing, as nanofiller also interacts weakly with the ether group, hence salt is not dissociated properly (Figure 1.9c). Figure 1.9d displays the role of the surface group of a nanofiller. The surface group of a nanofiller provides additional conducting sites for cation migration as well as supports salt dissociation. The acidic nature of nanofiller at optimum concentration obstructs ion-pairing due to the interaction of TiO_2 with PF_6^- ($TiO_2 + PF_6^- \leftrightarrow TiO_2{:}PF_6^-$) and a local electric field is formed which assists in salt dissociation.

In an intercalated type of polymer composites, polymer structures occupy the gallery of clay due to an interaction between the clay surface and polymer electron donor group due to the creation of a surface charge (Figure 1.10). The complexes of PEO molecules confined within the PMMA domain enter between the MMT platelets and increase the clay gallery spacing. The anion size is larger than the gallery spacing, and having Coulombic repulsion between the anion and negative surface charge of clay restricts their entry into it, which results in the elimination of concentration polarization (Figure 1.10a). The various interaction between polymer/salt/filler like EO-Li^+, C=Ö-Li^+, MMT-Li^+, and C=Ö-Li^+-EO makes nanocomposites and anions remain uncoordinated somewhere in the polymer backbone (Figure 1.10b) (Choudhary and Sengwa 2015).

Another approach to package the nanotubes within insulating clay layers to form effective 3D nanofillers was reported by Tang et al. (Tang et al. 2012). The authors reported that the growth of CNTs within clay interlayers results in exfoliation of clay

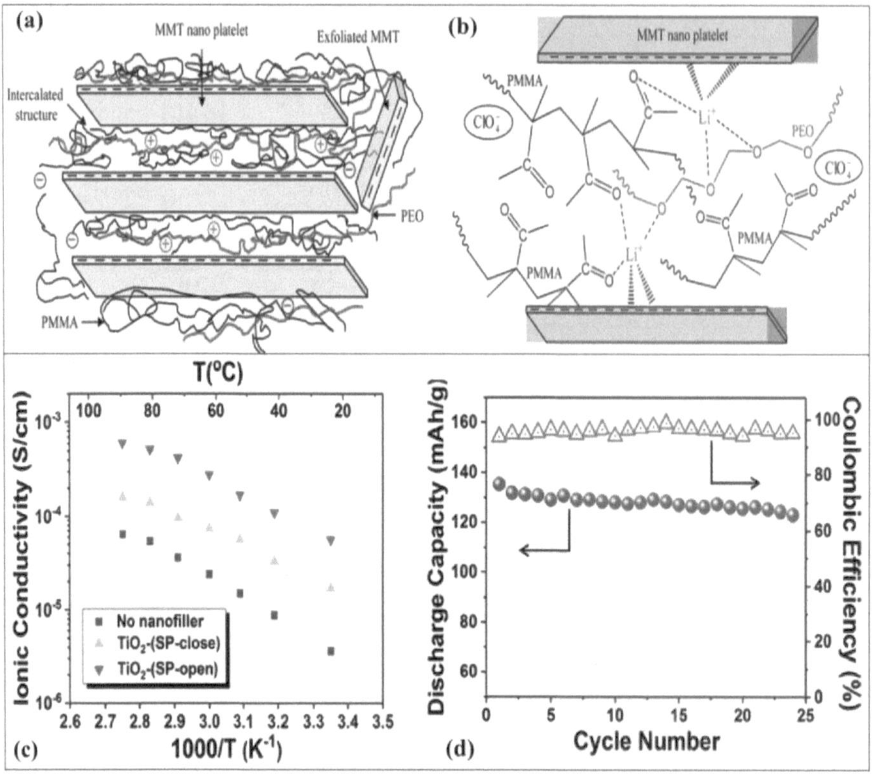

FIGURE 1.10 Schematic representation of (a) intercalated and exfoliated MMT structures, and (b) ion-dipolar and ion-ion interactions in cation-coordinated polymers blend intercalated MMT structures in the (PMMA-PEO)-$LiClO_4$-x wt % MMT nanocomposite composite film. (c) Arrhenius plots of ionic conductivity of CPEs with the TiO_2/SP hybrid nanofillers as compared with the pure PEO–$LiClO_4$ electrolyte; (d) performance and Coulombic efficiency of $LiCoO_2$/CPE–TiO_2/(SP-open) hybrid/Li cells at 0.1 C tested at 65 °C.

Source: (a–b) Reproduced with permission from Choudhary and Sengwa [2015, Copyright Wiley]. (c–d) Reproduced with permission from Fu et al. [2018, Copyright ACS].

platelets, and a hybrid nanofiller with a high aspect ratio was obtained. The lithium cation interacts with both the high negative charge of the electron cloud in the outer surface of CNTs as well as the negative oxygen atoms on the clay. This dual interaction mechanism leads to the separation of the contact-ion pairs and is beneficial for high ionic conductivity.

The filler/clay-dependent changes in electrical properties of the polymer composite were explored with a mechanism (Sharma and Thakur 2011). In a polymer matrix, the Li^+ cation is coordinated with electron-rich nitrogen ($-C\equiv\ddot{N}$) sites of the polymer host matrix due to polymer-ion-clay interaction and the bulky anion (PF_6^-) is leftover and hung somewhere in the host polymer backbone (PAN) as an uncoordinated free anion. Also, the presence of a negative charge at the clay surface galleries interacts with cations by coulombic interaction, which favors polymer chains access into clay galleries and is thermodynamically feasible ($\Delta G = -ve$). The net conductivity enhancement, at low clay loading, is due to the direct interaction of clay platelets with the Li^+ coordinated polymer, and at high clay content, redistribution leads to exfoliated to intercalated nanocomposite structure.

Another important biopolymer electrolyte that has gained attention is soy protein (SP). It also acts as a nanofiller to enhance the salt dissociation and cation conduction attributed to the presence of functional groups in the protein backbone. So, the combination of these two approaches will be more efficient in promoting ion dynamics and modified nanoparticles will be more effective (Fu et al. 2018). Wang et al. (Wang et al. 2016) manipulated the protein structure using ceramic TiO_2 nanoparticles. The optimized protein on the nanoparticle shows various interactions at the protein nanoparticle interface. Since, the protein backbone has various functional groups that provide strong interaction to substrate surface such as charge-charge interactions, hydrogen bonding, van der Waals force, π–π interactions. In SP-based polymer electrolytes, denaturation environment and procedures play a crucial role in deciding the interactions between SP and nanofiller. This surface modification of nanofiller is an effective strategy to create ion conduction pathways. Fu et al. (Fu et al. 2018) synthesized a novel protein-ceramic hybrid nanofiller by coating ion-conductive soy proteins onto TiO_2 nanoparticles via a controlled denaturation process. The composite polymer electrolyte with 5 wt % TiO_2/(SP-open) hybrid nanofiller has the highest ionic conductivity of 6×10^{-5} S cm^{-1} at room temperature and is about 1 magnitude higher than that of the pure PEO-$LiClO_4$ electrolyte and 5 times higher than that of the polymer electrolyte with TiO_2/(SP-close) hybrid or untreated TiO_2 nanofiller. Conductivity for hybrid TiO_2/(SP-open)-based polymer electrolytes becomes comparable to liquid electrolyte at 90 °C (i.e. 7×10^{-4} S cm^{-1}) (Figure 1.10c). Figure 1.10d shows the discharge capacity and coulombic efficiency versus cycle number at 0.1 C. Based on experimental and simulation results, an ion conduction model was proposed (Figure 1.11). The SP coating on the TiO_2 surface results in better dispersion of nanofiller. The presence of a functional group in SP interacts with PEO which enhances the flexibility and mechanical strength. Also, the denatured SP can interact with lithium salts through strong electrostatic interactions between backbone oxygen of protein and cation and favors better salt dissociation. So, the nanofiller modification enhances the ion migration via better dispersion of nanofiller and the creation of additional ion conduction channels.

1.5.4 Ion Transport Parameters

The electrical properties, especially the ionic conductivity in polymer composites have three key performance parameters: (i) diffusion constant, D; (ii) mobility, μ; and (iii) charge carrier concentration, n. Both μ (designates the degree of ease with

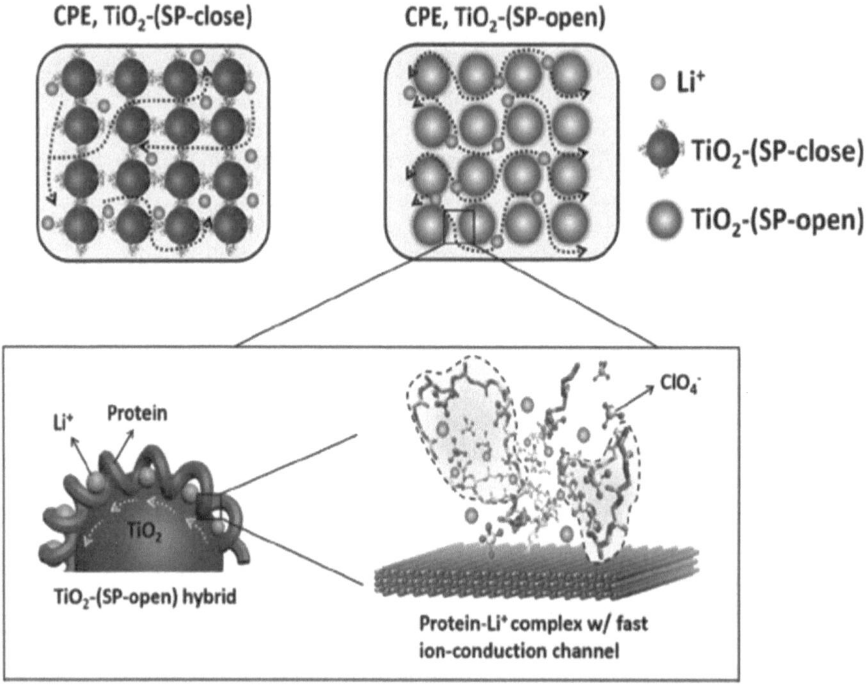

FIGURE 1.11 Schematics of the fast ion-conduction channel formed in the TiO_2/(SP-open) hybrid and ion transport pathways in CPEs with the TiO_2/(SP-open) hybrid compared with the TiO_2/(SP-close) hybrid.

Source: Reproduced with permission from Fu et al. [2018, Copyright ACS].

which ions pass through media on the application of external electrical field), and the *D* (the ease with which ions pass through media under a concentration gradient) are linked to the number of free charges available in the polymer matrix. Impedance and FTIR study provides direct calculation of number density of mobile ions and their mobility for performance and development of a good solid polymer composite. To estimate the value of transport parameters (n, μ, and D), three approaches are available (Table 1.4): (i) the Bandara and Mellander (B-M) approach (Bandara et al. 2011; Jonsson et al. 2006); (ii) Impedance Spectroscopy (IS) approach (Arof et al. 2014); and (iii) the FTIR method (Ericson et al. 2000; Petrowsky and Frech 2008). Table 1.5 summarizes the information related to all three approaches and provides significant evidence of ion dynamics in polymer composites.

ACKNOWLEDGMENT

Anil Arya is thankful to the UGC for providing the DS Kothari Post Doc Fellowship via award letter number BSR/PH/20–21/0156.

TABLE 1.5
Approaches to Obtaining the Number Density (*n*), Mobility (*μ*), and Diffusion Coefficient (D).

Method	B-M Approach	IS Approach	FTIR Method
Units			
D (cm^2s^{-1})	$D = \frac{d^2}{\tau_2 \delta^2}$	$D = \frac{(k_2 \varepsilon_r \varepsilon_o Ad)^2}{\tau_2}$	$D = \frac{\mu k_B T}{e}$
μ ($cm^2V^{-1}s^{-1}$)	$\mu = \frac{ed^2}{kT\tau_2\delta^2}$	$\mu = \frac{e(k_2 \varepsilon_r \varepsilon_o Ad)^2}{k_B T \tau_2}$	$\mu = \frac{\sigma}{ne}$
N (cm^{-3})	$n = \frac{\sigma kT\tau_2\delta^2}{e^2 d^2}$	$n = \frac{\sigma k_B T \tau_2}{(ek_2 \varepsilon_r \varepsilon_o Ad)^2}$	$n = \frac{M \times N_A}{V_{Total}} \times \text{free ion area}(\%)$
Parameters	τ_2 is time constant corresponding to the maximum dissipative loss curve, $\delta = d / \lambda$, λ is the thickness of the electrical double layer, d is half-thickness of the polymer composite	K_2 and K_1 are obtained from the trial and error on the Nyquist plot. The value of τ_2 was taken at the frequency corresponding to a minimum in Z'' (i.e. at Z'' 0, k_B is the Boltzmann constant (1.38×10^{-23} J K^{-1}) and T is the absolute temperature)	M is the number of moles of salt used, N_A is Avogadro's number (6.02×10^{23} mol^{-1}), V_{Total} is the total volume of the polymer composite, and σ is dc conductivity, e is the electric charge (1.602×10^{-19} C), k_B is the Boltzmann constant (1.38×10^{-23} J K^{-1}), T is the absolute temperature

REFERENCES

Agrawal R., & Pandey G. (2008). Solid polymer electrolytes: Materials designing and all-solid-state battery applications: An overview. *Journal of Physics D: Applied Physics*, *41*(22), 223001.

Aldalur I., Martinez-Ibañez M., Piszcz M., Rodriguez-Martinez L. M., Zhang H., & Armand M. (2018). Lowering the operational temperature of all-solid-state lithium polymer cell with highly conductive and interfacially robust solid polymer electrolytes. *Journal of Power Sources*, *383*, 144–149.

Alexandre M., & Dubois P. (2000). Polymer-layered silicate nanocomposites: Preparation, properties and uses of a new class of materials. *Materials Science and Engineering: R: Reports*, *28*(1–2), 1–63.

Angell C. (1986). Recent developments in fast ion transport in glassy and amorphous materials. *Solid State Ionics*, *18*, 72–88.

Appetecchi G. B., Croce F., Romagnoli P., Scrosati B., Heider U., & Oesten R. (1999). High-performance gel-type lithium electrolyte membranes. *Electrochemistry Communications*, *1*(2), 83–86.

Aravindan V., Gnanaraj J., Madhavi S., & Liu H. K. (2011). Lithium-ion conducting electrolyte salts for lithium batteries. *Chemistry: A European Journal*, *17*(51), 14326–14346.

Armand M. B. (1994). The history of polymer composites. *Solid State Ionics*, *69*(3–4), 309–319.

Armand M. B., Bruce P. G., Forsyth M., Scrosati B., & Wieczorek W. (2011a). Polymer electrolytes. *Energy Materials*, 1–31.

Armand M. B., Chabagno J., & Duclot M. (1979). *Fast ion transport in solids*, ed. P. Vashishta, J.N. Mundy & G.K. Shenoy. Amsterdam, North Holland.

Armand M. B., Endres F., MacFarlane D. R., Ohno H., & Scrosati B. (2011b). Ionic-liquid materials for the electrochemical challenges of the future. *Materials for Sustainable Energy: A Collection of Peer-Reviewed Research and Review Articles from Nature Publishing Group*, 129–137.

Arof A. K., Amirudin S., Yusof S., & Noor I. (2014). A method based on impedance spectroscopy to determine transport properties of polymer electrolytes. *Physical Chemistry Chemical Physics*, *16*(5), 1856–1867.

Arya A., & Sharma A. L. (2017). Polymer electrolytes for lithium ion batteries: A critical study. *Ionics*, *23*(3), 497–540.

Arya A., & Sharma A. L. (2018). Structural, microstructural and electrochemical properties of dispersed-type polymer nanocomposite films. *Journal of Physics D: Applied Physics*, *51*(4), 045504.

Bae J., Li Y., Zhang J., Zhou X., Zhao F., Shi Y., . . . Yu G. (2018). A 3D nanostructured hydrogel-framework-derived high-performance composite polymer lithium-ion electrolyte. *Angewandte Chemie International Edition*, *57*(8), 2096–2100.

Bandara T., Dissanayake M., Albinsson I., & Mellander B.-E. (2011). Mobile charge carrier concentration and mobility of a polymer electrolyte containing PEO and Pr_4N^+ I^- using electrical and dielectric measurements. *Solid State Ionics*, *189*(1), 63–68.

Bao J., Shi G., Tao C., Wang C., Zhu C., Cheng L., . . . Chen C. (2018). Polycarbonate-based polyurethane as a polymer electrolyte matrix for all-solid-state lithium batteries. *Journal of Power Sources*, *389*, 84–92.

Bruce P. G. (1995). Structure and electrochemistry of polymer composites. *Electrochimica Acta*, *40*, 2077–2085.

Bruce P. G., & Gray F. M. (1995). Polymer electrolytes II: physical principles. *Solid State Electrochemistry*, 119–162.

Bruce P. G., & Vincent C. (1993). Polymer electrolytes. *Journal of the Chemical Society, Faraday Transactions*, *89*(17), 3187–3203.

Cameron G. (1988). *Polymer electrolyte reviews-l*, ed. J. R. MacCallum & C. A. Vincent. Elsevier Applied Science Publishers, London, 1987. pp. x+ 351, price£ 48.00. ISBN 1-85166-07 1-2: Wiley Online Library.

Chen L., Li Y., Li S.-P., Fan L.-Z., Nan C.-W., & Goodenough J. B. (2018). PEO/garnet composite electrolytes for solid-state lithium batteries: From "ceramic-in-polymer" to "polymer-in-ceramic". *Nano Energy*, *46*, 176–184.

Chen X., Tang H., Putzeys T., Sniekers J., Wübbenhorst M., Binnemans K., . . . Luo J. (2016). Guanidinium nonaflate as a solid-state proton conductor. *Journal of Materials Chemistry A*, *4*(31), 12241–12252.

Cheon S.-E., Ko K.-S., Cho J.-H., Kim S.-W., Chin E.-Y., & Kim H.-T. (2003). Rechargeable lithium sulfur battery: I. Structural change of sulfur cathode during discharge and charge. *Journal of the Electrochemical Society*, *150*(6), A796.

Choe H., Giaccai J., Alamgir M., & Abraham K. (1995). Preparation and characterization of poly (vinyl sulfone)-and poly (vinylidene fluoride)-based electrolytes. *Electrochimica Acta*, *40*(13–14), 2289–2293.

Choi B., Kim Y., & Shin H. (2000). Ionic conduction in PEO: PAN blend polymer electrolytes. *Electrochimica Acta*, *45*(8–9), 1371–1374.

Choi J.-W., Kim J.-K., Cheruvally G., Ahn J.-H., Ahn H.-J., & Kim K.-W. (2007). Rechargeable lithium/sulfur battery with suitable mixed liquid electrolytes. *Electrochimica Acta*, *52*(5), 2075–2082.

Choudhary S., & Sengwa R. J. (2015). Structural and dielectric studies of amorphous and semicrystalline polymers blend-based nanocomposite electrolytes. *Journal of Applied Polymer Science*, *132*(3).

Cohen M. H., & Turnbull D. (1959). Molecular transport in liquids and glasses. *The Journal of Chemical Physics*, *31*(5), 1164–1169.

Cong B., Song Y., Ren N., Xie G., Tao C., Huang Y., . . . Bao J. (2018). Polyethylene glycol-based waterborne polyurethane as solid polymer electrolyte for all-solid-state lithium ion batteries. *Materials & Design*, *142*, 221–228.

Croce F., Persi L., Scrosati B., Serraino-Fiory F., Plichta E., & Hendrickson M. (2001). Role of the ceramic fillers in enhancing the transport properties of composite polymer electrolytes. *Electrochimica Acta*, *46*(16), 2457–2461.

Dam T., Karan N., Thomas R., Pradhan D. K., & Katiyar R. (2015). Observation of ionic transport and ion-coordinated segmental motions in composite (polymer-salt-clay) solid polymer electrolyte. *Ionics*, *21*(2), 401–410.

Duan H., Yin Y.-X., Zeng X.-X., Li J.-Y., Shi J.-L., Shi Y., . . . Wan L.-J. (2018). In-situ plasticized polymer electrolyte with double-network for flexible solid-state lithium-metal batteries. *Energy Storage Materials*, *10*, 85–91.

Earle M. J., Esperança J. M., Gilea M. A., Canongia Lopes J. N., Rebelo L. P., Magee J. W., . . . Widegren J. A. (2006). The distillation and volatility of ionic liquids. *Nature*, *439*(7078), 831–834.

Eiamlamai P. (2015). *Polymer electrolytes based on ionic liquids for lithium batteries*. Université Grenoble Alpes, France.

Ericson H., Svanberg C., Brodin A., Grillone A., Panero S., Scrosati B., & Jacobsson P. (2000). Poly (methyl methacrylate)-based protonic gel electrolytes: A spectroscopic study. *Electrochimica Acta*, *45*(8–9), 1409–1414.

Esterly D. M. (2002). *Manufacturing of poly (vinylidene fluoride) and evaluation of its mechanical properties*. Virginia Tech, USA.

Fenton D. (1973). Complexes of alkali metal ions with poly (ethylene oxide). *Polymer*, *14*, 589.

Fu X., Jewel Y., Wang Y., Liu J., & Zhong W.-H. (2016). Decoupled ion transport in a protein-based solid ion conductor. *The Journal of Physical Chemistry Letters*, *7*(21), 4304–4310.

Fu X., Li C., Wang Y., Kovatch L. P., Scudiero L., Liu J., & Zhong W. (2018a). Building ion-conduction highways in polymeric electrolytes by manipulating protein configuration. *ACS Applied Materials & Interfaces*, *10*(5), 4726–4736.

Fu X., Wang Y., Fan X., Scudiero L., & Zhong W. H. (2018b). Core: Shell hybrid nanowires with protein enabling fast ion conduction for high-performance composite polymer electrolytes. *Small*, *14*(49), 1803564.

Garcia-Calvo O., Lago N., Devaraj S., & Armand M. (2016). Cross-linked solid polymer electrolyte for all-solid-state rechargeable lithium batteries. *Electrochimica Acta*, *220*, 587–594.

Gebbie M. A., Valtiner M., Banquy X., Fox E. T., Henderson W. A., & Israelachvili J. N. (2013). Ionic liquids behave as dilute electrolyte solutions. *Proceedings of the National Academy of Sciences*, *110*(24), 9674–9679.

Giacomelli C., Schmidt V., Aissou K., & Borsali R. (2010). Block copolymer systems: From single chain to self-assembled nanostructures. *Langmuir*, *26*(20), 15734–15744.

Gibbs J. H., & DiMarzio E. A. (1958). Nature of the glass transition and the glassy state. *The Journal of Chemical Physics*, *28*(3), 373–383.

Gong J., Gong Z.-L., Yan X.-L., Gao S., Zhang Z.-L., & Wang B. (2012). Investigation of the free volume and ionic conducting mechanism of poly (ethylene oxide)-$LiClO_4$ polymeric electrolyte by positron annihilating lifetime spectroscopy. *Chinese Physics B*, *21*(10), 107803.

Gopalan A. I., Santhosh P., Manesh K. M., Nho J. H., Kim S. H., Hwang C.-G., & Lee K.-P. (2008). Development of electrospun PVdF-PAN membrane-based polymer electrolytes for lithium batteries. *Journal of Membrane Science*, *325*(2), 683–690.

Gray F. M. (1997). *Polymer composites*. Royal Society of Chemistry, Cambridge, p. 175.

Grünebaum M., Hiller M. M., Jankowsky S., Jeschke S., Pohl B., Schürmann T., . . . Müller R. (2014). Synthesis and electrochemistry of polymer based electrolytes for lithium batteries. *Progress in Solid State Chemistry*, *42*(4), 85–105.

Hellio D., & Djabourov M. (2006). *Physically and chemically crosslinked gelatin gels.* Macromolecular Symposia, 241:23-27.

Jeddi K., Qazvini N. T., Jafari S. H., & Khonakdar H. A. (2010). Enhanced ionic conductivity in PEO/PMMA glassy miscible blends: Role of nano-confinement of minority component chains. *Journal of Polymer Science Part B: Polymer Physics*, *48*(19), 2065–2071.

Jeong H.-S., Choi E.-S., Lee S.-Y., & Kim J. H. (2012). Evaporation-induced, close-packed silica nanoparticle-embedded nonwoven composite separator membranes for high-voltage/high-rate lithium-ion batteries: Advantageous effect of highly percolated, electrolyte-philic microporous architecture. *Journal of Membrane Science*, *415*, 513–519.

Ji J., Li B., & Zhong W.-H. (2012). An ultraelastic poly (ethylene oxide)/soy protein film with fully amorphous structure. *Macromolecules*, *45*(1), 602–606.

Jönsson M., Welch K., Hamp S., & Strømme M. (2006). Bacteria counting with impedance spectroscopy in a micro probe station. *The Journal of Physical Chemistry B*, *110*(20), 10165–10169.

Karimi M. B., Khanbabaei G., & Sadeghi G. M. M. (2017). Vegetable oil-based polyurethane membrane for gas separation. *Journal of Membrane Science*, *527*, 198–206.

Kim G. T., Appetecchi G. B., Carewska M., Joost M., Balducci A., Winter M., & Passerini S. (2010). UV cross-linked, lithium-conducting ternary polymer electrolytes containing ionic liquids. *Journal of Power Sources*, *195*(18), 6130–6137.

Kokorin A. (2011). *Ionic liquids: Theory, properties, new approaches*. IntechOpen, London, UNITED KINGDOM.

Kuo, P. L., Tsao, C. H., Hsu, C. H., Chen, S. T., & Hsu, H. M. (2016). A new strategy for preparing oligomeric ionic liquid gel polymer electrolytes for high-performance and nonflammable lithium ion batteries. *Journal of Membrane Science,* 499, 462–469.

Landauer R. (1952). The electrical resistance of binary metallic mixtures. *Journal of Applied Physics*, *23*(7), 779–784.

Lee H., Yoo J. K., Park J. H., Kim J. H., Kang K., & Jung Y. S. (2012). A stretchable polymer: Carbon nanotube composite electrode for flexible lithium-ion batteries: Porosity engineering by controlled phase separation. *Advanced Energy Materials*, *2*(8), 976–982.

Lee Y. H., Kim J. S., Noh J., Lee I., Kim H. J., Choi S., . . . Lee J.-Y. (2013). Wearable textile battery rechargeable by solar energy. *Nano Letters*, *13*(11), 5753–5761.

Lewin M., Mey-Marom A., & Frank R. (2005). Surface free energies of polymeric materials, additives and minerals. *Polymers for Advanced Technologies*, *16*(6), 429–441.

Li H., Li M., Siyal S. H., Zhu M., Lan J.-L., Sui G., . . . Yang X. (2018). A sandwich structure polymer/polymer-ceramics/polymer gel electrolytes for the safe, stable cycling of lithium metal batteries. *Journal of Membrane Science*, *555*, 169–176.

Li T., & Balbuena P. B. (1999). Theoretical studies of lithium perchlorate in ethylene carbonate, propylene carbonate, and their mixtures. *Journal of the Electrochemical Society*, *146*(10), 3613.

Li Y., Zhang W., Dou Q., Wong K. W., & Ng K. M. (2019). Li_7 La_3 Zr_2 O_{12} ceramic nanofiber-incorporated composite polymer electrolytes for lithium metal batteries. *Journal of Materials Chemistry A*, *7*(7), 3391–3398.

Liang X., Han D., Wang Y., Lan L., & Mao J. (2018). Preparation and performance study of a PVdF-LATP ceramic composite polymer electrolyte membrane for solid-state batteries. *RSC Advances*, *8*(71), 40498–40504.

Liang X., Wen Z., Liu Y., Wu M., Jin J., Zhang H., & Wu X. (2011). Improved cycling performances of lithium sulfur batteries with $LiNO_3$-modified electrolyte. *Journal of Power Sources*, *196*(22), 9839–9843.

Lin Y.-C., Ito K., & Yokoyama H. (2018). Solid polymer electrolyte based on crosslinked polyrotaxane. *Polymer*, *136*, 121–127.

Liu L., Chu L., Jiang B., & Li M. (2019). $Li_{1.4}Al_{0.4}Ti_{1.6}(PO_4)_3$ nanoparticle-reinforced solid polymer electrolytes for all-solid-state lithium batteries. *Solid State Ionics*, *331*, 89–95.

Liu W., Lee S. W., Lin D., Shi F., Wang S., Sendek A. D., & Cui Y. (2017). Enhancing ionic conductivity in composite polymer electrolytes with well-aligned ceramic nanowires. *Nature Energy*, *2*(5), 1–7.

Liu W., Lin D., Sun J., Zhou G., & Cui Y. (2016). Improved lithium ionic conductivity in composite polymer electrolytes with oxide-ion conducting nanowires. *ACS* Nano, *10*(12), 11407–11413.

Lu Q., Fang J., Yang J., Yan G., Liu S., & Wang J. (2013). A novel solid composite polymer electrolyte based on poly (ethylene oxide) segmented polysulfone copolymers for rechargeable lithium batteries. *Journal of Membrane Science*, *425*, 105–112.

Lu Q., He Y. B., Yu Q., Li B., Kaneti Y. V., Yao Y., . . . Yang Q. H. (2017). Dendrite-free, high-rate, long-life lithium metal batteries with a 3D cross-linked network polymer electrolyte. *Advanced Materials*, *29*(13), 1604460.

Luo H., Zhou X., Ellingford C., Zhang Y., Chen S., Zhou K., . . . Wan C. (2019). Interface design for high energy density polymer nanocomposites. *Chemical Society Reviews*, *48*(16), 4424–4465.

Luo J., Conrad O., & Vankelecom I. F. (2013). Imidazolium methanesulfonate as a high temperature proton conductor. *Journal of Materials Chemistry A*, *1*(6), 2238–2247.

Luo J., Hu J., Saak W., Beckhaus R., Wittstock G., Vankelecom I. F., . . . Conrad O. (2011). Protic ionic liquid and ionic melts prepared from methanesulfonic acid and 1H-1, 2, 4-triazole as high temperature PEMFC electrolytes. *Journal of Materials Chemistry*, *21*(28), 10426–10436.

J.R. MacCallum, C.A. Vincent, *Polymer Electrolyte Reviews, Elsevier*, London, Vol. 2 (1989)

MacFarlane D. R., & Forsyth M. (2001). Plastic crystal electrolyte materials: New perspectives on solid state ionics. *Advanced Materials*, *13*(12–13), 957–966.

Marcinek M., Syzdek J., Marczewski M., Piszcz M., Niedzicki L., Kalita M., . . . Trzeciak T. (2015). Electrolytes for Li-ion transport: Review. *Solid State Ionics*, *276*, 107–126.

Maxwell J. C. (1881). *A treatise on electricity and magnetism: pt. III. Magnetism. pt. IV. Electromagnetism* (Vol. 2). Clarendon Press, United Kingdom.
Mindemark J., Lacey M. J., Bowden T., & Brandell D. (2018). Beyond PEO: Alternative host materials for Li^+-conducting solid polymer electrolytes. *Progress in Polymer Science, 81*, 114–143.
Moskwiak M., Giska I., Borkowska R., Zalewska A., Marczewski M., Marczewska H., & Wieczorek W. (2006). Physico-and electrochemistry of composite electrolytes based on PEODME-LiTFSI with TiO_2. *Journal of Power Sources, 159*(1), 443–448.
Muldoon J., Bucur C. B., Boaretto N., Gregory T., & Di Noto V. (2015). Polymers: Opening doors to future batteries. *Polymer Reviews, 55*(2), 208–246.
Müller-Plathe F., & van Gunsteren W. F. (1995). Computer simulation of a polymer electrolyte: Lithium iodide in amorphous poly(ethylene oxide). *The Journal of Chemical Physics, 103*(11), 4745–4756.
Norman, G. E., & Pisarev, V. V. (2012). Molecular dynamics analysis of the crystallization of an overcooled aluminum melt. *Russian Journal of Physical Chemistry A*, 86(9), 1447–1452.
Nan C.-W., & Smith D. M. (1991). Ac electrical properties of composite solid electrolytes. *Materials Science and Engineering: B, 10*(2), 99–106.
Nunes-Pereira J., Costa C., & Lanceros-Méndez S. (2015). Polymer composites and blends for battery separators: State of the art, challenges and future trends. *Journal of Power Sources, 281*, 378–398.
Papke B., Ratner M., & Shriver D. (1982). Conformation and ion-transport models for the structure and ionic conductivity in complexes of polyethers with alkali Metal Salts. *Journal of the Electrochemical Society, 129*(8), 1694.
Petrowsky M., & Frech R. (2008). Concentration dependence of ionic transport in dilute organic electrolyte solutions. *The Journal of Physical Chemistry B, 112*(28), 8285–8290.
Pitawala J., Navarra M. A., Scrosati B., Jacobsson P., & Matic A. (2014). Structure and properties of Li-ion conducting polymer gel electrolytes based on ionic liquids of the pyrrolidinium cation and the bis (trifluoromethanesulfonyl) imide anion. *Journal of Power Sources, 245*, 830–835.
Ponrouch A., Monti D., Boschin A., Steen B., Johansson P., Palacin M. R. (2015). Non-aqueous composites for sodium-ion batteries. *Journal of Materials Chemistry A, 3*(1), 22–42.
Ratner M. A., Johansson P., & Shriver D. F. (2000). Polymer electrolytes: Ionic transport mechanisms and relaxation coupling. *Mrs Bulletin, 25*(3), 31–37.
Ren S., Chang H., He L., Dang X., Fang Y., Zhang L., . . . Lin Y. (2013). Preparation and ionic conductive properties of all-solid polymer electrolytes based on multiarm star block polymers. *Journal of Applied Polymer Science, 129*(3), 1131–1142.
Sequeira, C. A. C., & Santos, D. M. F. (2010). Introduction to polymer electrolyte materials. In *Polymer electrolytes* (pp. 3–61). Woodhead Publishing Cambridge, UK.
Shanmukaraj D., Wang G., Murugan R., & Liu H.-K. (2008). Ionic conductivity and electrochemical stability of poly (methylmethacrylate): Poly (ethylene oxide) blend-ceramic fillers composites. *Journal of Physics and Chemistry of Solids, 69*(1), 243–248.
Sharma A. L., & Thakur A. K. (2011). Polymer matrix: Clay interaction mediated mechanism of electrical transport in exfoliated and intercalated polymer nanocomposites. *Journal of Materials Science, 46*(6), 1916–1931.
Shi J., Yang Y., & Shao H. (2018). Co-polymerization and blending based PEO/PMMA/P (VDF-HFP) gel polymer electrolyte for rechargeable lithium metal batteries. *Journal of Membrane Science, 547*, 1–10.
Shin W.-K., Cho J., Kannan A. G., Lee Y.-S., & Kim D.-W. (2016). Cross-linked composite gel polymer electrolyte using mesoporous methacrylate-functionalized SiO_2 nanoparticles for lithium-ion polymer batteries. *Scientific Reports, 6*(1), 1–10.
Shukla P., & Agrawal S. (1999). On the description of conductivity in PVA-based composite polymer electrolytes: EMT approach. *Physica Status Solidi (a), 172*(2), 329–339.

Sloop S. E., Pugh J. K., Kerr J. B., & Kinoshita K. (2001). Chemical Reactivity of PF_5 and $LiPF_6$ in Ethylene Carbonate/Dimethyl Carbonate. Electrochemical and Solid-State Letters, 4(4), A42.

Song M.-K., Kim Y.-T., Cho J.-Y., Cho B. W., Popov B. N., & Rhee H.-W. (2004). Composite polymer electrolytes reinforced by non-woven fabrics. *Journal of Power Sources*, *125*(1), 10–16.

Tamura T., Yoshida K., Hachida T., Tsuchiya M., Nakamura M., Kazue Y., . . . Watanabe M. (2010). Physicochemical properties of glyme: Li salt complexes as a new family of room-temperature ionic liquids. *Chemistry Letters*, *39*(7), 753–755.

Tang C., Hackenberg K., Fu Q., Ajayan P. M., & Ardebili H. (2012). High ion conducting polymer nanocomposite electrolytes using hybrid nanofillers. *Nano Letters*, *12*(3), 1152–1156.

Wang S., Jeung S., & Min K. (2010). The effects of anion structure of lithium salts on the properties of in-situ polymerized thermoplastic polyurethane electrolytes. *Polymer*, *51*(13), 2864–2871.

Wang X., Fu X., Wang Y., & Zhong W. (2016). A protein-reinforced adhesive composite electrolyte. *Polymer*, *106*, 43–52.

Wang X., Zhang Y., Zhang X., Liu T., Lin Y.-H., Li L., . . . Nan C.-W. (2018). Lithium-salt-rich $PEO/Li_{0.3}La_{0.557}TiO_3$ interpenetrating composite electrolyte with three-dimensional ceramic nano-backbone for all-solid-state lithium-ion batteries. *ACS Applied Materials & Interfaces*, *10*(29), 24791–24798.

Wieczorek W., Such K., Florjanczyk Z., & Stevens J. (1994). Polyether, polyacrylamide, $LiClO_4$ composite electrolytes with enhanced conductivity. *The Journal of Physical Chemistry*, *98*(27), 6840–6850.

Wieczorek W., Such K., Wyciślik H., & Płocharski J. (1989). Modifications of crystalline structure of PEO polymer electrolytes with ceramic additives. *Solid State Ionics*, *36*(3–4), 255–257.

Williams M. L., Landel R. F., & Ferry J. D. (1955). The temperature dependence of relaxation mechanisms in amorphous polymers and other glass-forming liquids. *Journal of the American Chemical Society*, *77*(14), 3701–3707.

Wright, P. V. Electrical conductivity in ionic complexes of poly(ethylene oxide). *Br. Polym. J. 1975*, 7, 319–327.

Xiao W., Li X., Wang Z., Guo H., Li Y., & Yang B. (2012). Performance of PVdF-HFP-based gel polymer electrolytes with different pore forming agents. *Iranian Polymer Journal*, *21*(11), 755–761.

Xiao Y., Jiang L., Liu Z., Yuan Y., Yan P., Zhou C., & Lei J. (2017). Effect of phase separation on the crystallization of soft segments of green waterborne polyurethanes. *Polymer Testing*, *60*, 160–165.

Xiao Z., Zhou B., Wang J., Zuo C., He D., Xie X., & Xue Z. (2019). PEO-based electrolytes blended with star polymers with precisely imprinted polymeric pseudo-crown ether cavities for alkali metal ion batteries. *Journal of Membrane Science*, *576*, 182–189.

Xu C., Huang Y., Tang L., & Hong Y. (2017). Low-initial-modulus biodegradable polyurethane elastomers for soft tissue regeneration. *ACS Applied Materials & Interfaces*, *9*(3), 2169–2180.

Xu H., Wang A., Liu X., Feng D., Wang S., Chen J., . . . Zhang L. (2018). A new fluorine-containing star-branched polymer as electrolyte for all-solid-state lithium-ion batteries. *Polymer*, *146*, 249–255.

Xu K. (2004). Nonaqueous liquid electrolytes for lithium-based rechargeable batteries. *Chemical Reviews*, *104*(10), 4303–4418.

Xue Z., He D., & Xie X. (2015). Poly (ethylene oxide)-based electrolytes for lithium-ion batteries. *Journal of Materials Chemistry A*, *3*(38), 19218–19253.

Youcef, HB., Garcia-Calvo, O., Lago, N., Devaraj, S., & Armand, M. (2016). Cross-linked solid polymer electrolyte for all-solid-state rechargeable lithium batteries. *Electrochimica Acta*, 220, 587–594.

Young W.-S., & Epps III T. H. (2012). Ionic conductivities of block copolymer electrolytes with various conducting pathways: Sample preparation and processing considerations. *Macromolecules*, *45*(11), 4689–4697.

Young W. S., Kuan W. F., & Epps III T. H. (2014). Block copolymer electrolytes for rechargeable lithium batteries. *Journal of Polymer Science Part B: Polymer Physics*, *52*(1), 1–16.

Zhang H., Zhang P., Li Z., Sun M., Wu Y., & Wu H. (2007). A novel sandwiched membrane as polymer electrolyte for lithium ion battery. *Electrochemistry Communications*, *9*(7), 1700–1703.

Zhang J., Ma C., Liu J., Chen L., Pan A., & Wei W. (2016). Solid polymer electrolyte membranes based on organic/inorganic nanocomposites with star-shaped structure for high performance lithium ion battery. *Journal of Membrane Science*, *509*, 138–148.

Zhang J., Yang J., Dong T., Zhang M., Chai J., Dong S., . . . Cui G. (2018). Aliphatic polycarbonate-based solid-state polymer electrolytes for advanced lithium batteries: Advances and perspective. *Small*, *14*(36), 1800821.

Zhang J., Zang X., Wen H., Dong T., Chai J., Li Y., . . . Ma J. (2017). High-voltage and free-standing poly (propylene carbonate)/$Li_{6.75}La_3Zr_{1.75}Ta_{0.25}O_{12}$ composite solid electrolyte for wide temperature range and flexible solid lithium ion battery. *Journal of Materials Chemistry A*, *5*(10), 4940–4948.

Zhang X., Ji L., Toprakci O., Liang Y., & Alcoutlabi M. (2011). Electrospun nanofiber-based anodes, cathodes, and separators for advanced lithium-ion batteries. *Polymer Reviews*, *51*(3), 239–264.

Zhang X., Xie J., Shi F., Lin D., Liu Y., Liu W., . . . Liu K. (2018). Vertically aligned and continuous nanoscale ceramic: Polymer interfaces in composite solid polymer electrolytes for enhanced ionic conductivity. *Nano Letters*, *18*(6), 3829–3838.

Zhang Y., Lu W., Cong L., Liu J., Sun L., Mauger A., . . . Liu J. (2019). Cross-linking network based on Poly (ethylene oxide): Solid polymer electrolyte for room temperature lithium battery. *Journal of Power Sources*, *420*, 63–72.

Zhou W., Wang S., Li Y., Xin S., Manthiram A., & Goodenough J. B. (2016). Plating a dendrite-free lithium anode with a polymer/ceramic/polymer sandwich electrolyte. *Journal of the American Chemical Society*, *138*(30), 9385–9388.

Zhu M., Tan C., Fang Q., Gao L., Sui G., & Yang X. (2016). High performance and biodegradable skeleton material based on soy protein isolate for gel polymer electrolyte. *ACS Sustainable Chemistry & Engineering*, *4*(9), 4498–4505.

Zhu P., Yan C., Dirican M., Zhu J., Zang J., Selvan R. K., . . . Kiyak Y. (2018). $Li_{0.33}La_{0.557}TiO_3$ ceramic nanofiber-enhanced polyethylene oxide-based composite polymer electrolytes for all-solid-state lithium batteries. *Journal of Materials Chemistry A*, *6*(10), 4279–4285.

2 Hybrid Organic-Inorganic Polymer Composites

Sujeet Kumar Chaurasia, Kunwar Vikram, Manish Pratap Singh, and Manoj K. Singh

CONTENTS

2.1 INTRODUCTION

Owing to the rapid development of modern societies and the enormous consumption of fossil-fuels-based energy resources, the scientist looks for the development of durable and efficient electrochemical devices like batteries, fuel cells, solar cells, supercapacitors and so on (Liang et al. 2019; Wang & Zhong, 2015). A continuous

DOI: 10.1201/9781003208662-3

improvement has to be made in the components of these energy storage devices over time to enhance their efficiency for fulfilling the industrial and technological needs. The electrolyte has been a crucial part in these electrochemical devices as it is responsible for the transfer of electric charge/ions between the electrodes of these electrochemical devices. Earlier, it could be seen that most of the traditional energy storage devices employed liquid electrolytes (both aqueous and non-aqueous) due to their high ionic conductivity values. However, the devices made up of liquid electrolytes suffer from the several limitations like (i) leakage, (ii) short shelf life, (iii) inflammability, (iv) corrosion reaction between electrodes and electrolytes, (v) difficulty in miniaturization and (vi) instability towards changing temperature (Shalu et al., 2021). These shortcomings of liquid electrolytes initiated the research in solid electrolytes which are a good alternate to liquid-electrolyte-based systems due to their desirable properties such as non-volatility, structural sturdiness and high electrochemical stability. Recently, material scientists have been looking for developing materials which have exceptionally high room temperature ionic conductivities in the range of 10^{-2} to 10^{-3} S cm^{-1} with negligible electronic conductivity. These materials are also known as "fast ion conductors" or "superionic solids". The science and technology based on ion conducting solid electrolytes is termed as "Solid State Ionics" (Agrawal & Pandey, 2008). On the basis of physical properties and microstructures the solid electrolytes can be classified into the following four groups: (i) crystalline/polycrystalline electrolytes, (ii) glassy/amorphous electrolytes, (iii) polymer electrolytes, (iv) gel electrolytes and (iv) composite electrolytes. Out of these solid state electrolytes, ion conducting polymers or polymer electrolytes are the materials of current research interst because of their flexibility and higher room temperature ionic conductivity. These polymer electrolytes are generally obtained by immobilizing mobile ionic species into soft polymer matrices to provide semicrystalline conducive framework which supports easily ionic movement within the polymeric matrix (Gupta et al. 2016; Singh et al. 2016). Wright (Wright, 1975) first reported the ion conduction property in solid polymer electrolytes (SPEs) based on polymer PEO in 1973, and Armand and his co-workers (Armand et al. 1979) discuss its potential application in rechargeable batteries. Since then, various synthetic polar polymers like PEO, PVA, PMMA, PVdF, PVdF-HFP, PVP, etc. are commonly used as semi-crystalline or amorphous host polymeric matrices for the preparation of SPEs by the immobilization of mono-, di- or tri-valiant alkali metal salts to give conducive polymeric backbone by increasing the segmental motion of the polymeric chain (Quartarone & Mustarelli, 2011; Saroj et al. 2016). Nowadays, solid-like electrolytes are the most extensively studied research topic in the field of condensed matter physics since it represents an intriguing approach to enhance simultaneously achieving high ionic conductivity and resolving safety issues. Nevertheless, these SPEs have low ambient ionic conductivity, and interfacial compatibility towards electrodes falls short of practical applications in high-performance electrochemical devices. In addition to improved safety, various approaches like cross-linking of polymers (Young et al. 2014), addition of ceramic filler particles (Yoon et al. 2014), inclusion of plasticizers (Sadiq et al. 2021), forming organic-inorganic hybrids (Chaurasia & Chandra, 2017) have been employed for enhancing the conductivity without sacrificing electrochemical and thermal stability of the polymer electrolyte membranes. Alternatively, in polymer electrolytes ionic conductivity occurs due to the efficiently conductive cations via hopping mechanism within rigid crystalline polymeric backbone structure (Chaurasia et al., 2011; Chaurasia et al. 2022). Furthermore, in typical polymer electrolytes increasing the salt content generally leads to an increase in polymer amorphous fraction. However, the reduction in the crystalline phases of the polymer electrolytes cause worsening of its mechanical properties and dimensional stability. In recent years, quasi-solid state electrolytes are termed as "organic-inorganic hybrid electrolytes" that have been prepared by combining the organic polymer and inorganic nanofiller particles at the molecular level with a variety of known and hitherto unknown properties (Nicole et al. 2014). The organic part of the hybrid electrolyte materials imparts high flexibility and mouldability while the inorganic network provides high thermal stability, structural integrity and chemical resistance. Generally, one of the components (either organic or inorganic) in the organic-inorganic hybrid electrolytes appear within the range of nanoscale and thus these materials are fall within the category of nano materials. The concept of organic-inorganic hybrids nanocomposites was exploded in the second half of the 20th century, leading to its applications in many diverse fields of science such as organic synthesis, biotechnology, catalysis and electronics. These hybrid materials gained importance in the '80s with the development of soft inorganic chemistry processes and recently a lot of research is developing on this topic. The growth and importance of this multidisciplinary field can be seen by research activity going on this field recently and a tremendous increasing number of publications, reports and patents in the last few years (Mir et al. 2018; Sanchez et al. 2011). Figure 2.1 shows the research activity going on world-wide in the field of organic-inorganic hybrid electrolytes.

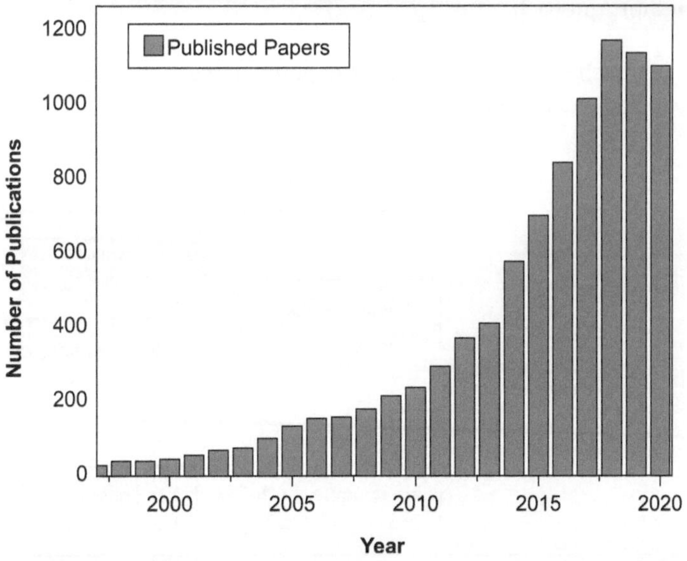

FIGURE 2.1 Progress of research papers published by the scientific community in the field of hybrid electrolytes.

Source: ISI web of KnowledgeTM.

These hybrid electrolytes have dual advantages of high ionic conductivity due to their organic components and high thermal stability as well as mechanical stability due to their inorganic components. Such materials not only merge the properties of both the phases but also exhibit new and superior features as a result of the hybridization process. The charge transport properties of these hybrid electrolytes are significantly dependent on the chemical nature of the organic and inorganic phases, their conformations, sizes, morphology and the nature of inter phase interactions. Thus, the hybridization of both the phases leads to the development of multifunctional materials with desired properties.

This approach of the synthesis of materials provides an opportunity to have a large set of materials with a variety of known and hitherto unknown properties. Besides the nature of the interfaces between the two phases and the kind of application for which the materials are to be designed, the final properties of these materials are strongly influenced by their method of processing. The aim of this chapter is to describe different classes of polymeric-based organic-inorganic hybrid electrolytes and discuss the most recent developments in these fields along with their possible applications in developing solid state electrochemical devices.

2.2 SYNTHESIS ROUTES FOR ORGANIC-INORGANIC HYBRID ELECTROLYTES

There are two main approaches adopted for the preparation of organic-inorganic hybrid electrolytes:

2.2.1 Building Block Approach
2.2.2 In-Situ Approach

These two approaches for the synthesis of organic-inorganic hybrid electrolytes are schematically shown in Figure 2.2.

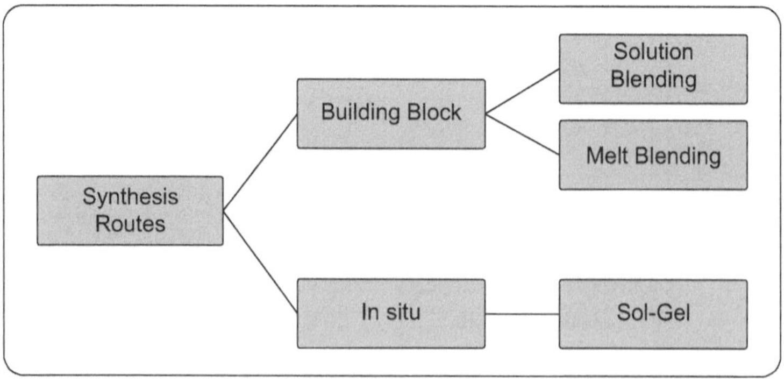

FIGURE 2.2 Synthesis approaches adopted for the preparation of organic-inorganic hybrid electrolytes.

2.2.1 Building Block Approach

In this approach, the building blocks (i.e. precursors) at least partially keep their original molecular integrity throughout the materials which permits the survival of their crucial properties within the organic matrix. A few examples of such well-defined building blocks are inorganic clusters which comprise at least one functional group that permits their interaction with the organic matrix and nanosized particles, nanorods etc., that can be used to form nanocomposites. The building blocks can be designed effectively to form materials with better structural integrity. This building approach of the synthesis of organic-inorganic hybrid electrolytes consists of two approaches as described next:

2.2.1.1 Solution Blending Method

The solution blending technique for the preparation of organic-inorganic hybrid electrolytes is an effective way to modify the properties of existing polymers and to develop new materials with the combination of properties of both organic and inorganic phases for the specific applications. In this approach, polymers or polymer-salt complexes are dissolved into a suitable solvent to form a homogeneous mixture, and then required amounts of ceramic filler particles (SiO_2, TiO_2, Al_2O_3, etc.) are added for getting the polymer-ceramic filler composite solutions. After the complete evaporation of the volatile organic solvents, thin composite polymer electrolyte films (which are also termed as ex-situ organic-inorganic hybrid electrolytes) are obtained.

2.2.1.2 Melt Blending Method

In this approach, there is no need to use any solvents for the preparation of organic-inorganic hybrids. In this method, first polymers are melted and then mixed with appropriate amounts of filler particles to form a homogeneous melted mixture of both the constituents, and then the obtained solution is hot-pressed between two stainless steel plates to get polymer composite films.

2.2.2 In-Situ Approach

This method is based on the chemical transformation of well-defined discrete molecules (i.e. precursors) into multidimensional networks (e.g. 3D silica, titania, etc. matrices) with completely different characteristics from the original molecules/precursors. The internal morphology of the final materials depends on both the composition of the precursors as well as the synthesis conditions. The alternation in any of these variables can lead to the formation of two very different materials. Moreover, the collective effect of various factors such as proper synthesis conditions, an optimized stoichiometric ratio of organic and inorganic components and nature of the interaction between the two phases decides the structural, morphological and electrochemical properties of the prepared organic-inorganic hybrids. This approach involves the in-situ formation of either one (i.e. organic or inorganic) or both the components simultaneously. For example, a change from acid to base catalyst in the in-situ formation of silica (inorganic) network within the polymer matrix makes a huge difference in terms of particle size, surface area, porosity, pore size distribution, etc.

2.2.2.1 Sol-Gel Method

Among the various processing techniques, the classical sol-gel route has been extensively utilized for the preparation of these organic-inorganic hybrid electrolytes. To fulfill the need of a specific degree of dispersion of the filler, avoiding unwanted agglomeration and overcoming its poor compatibility with the polymer matrix, an elegant way of processing hybrid electrolytes known as the sol-gel technique has been introduced. The sol-gel method involves five steps: (i) formation of sol—a colloidal suspension, (ii) gelation of sol—to form a network, (iii) aging of gel—polycondensation reaction continues until the gel transforms into a solid mass, (iv) drying of gel and (v) densification at high temperatures. The major advantage of this approach is that it employs mild processing conditions such as low temperature and pressure. This is a bottom-up approach by which in-situ production and dispersion of the filler particles is achieved within the host polymeric matrix from the inorganic precursors. This method has been found to be useful and effective for the reinforcement of the particles within the polymer matrix over the traditional solution or melt blending methods as it can subtly control the surface characteristics of the growing inorganic phase within the matrix. This approach also prevents the aggregation of the inorganic nanofiller particles and also enables their homogeneous distribution within the polymer matrix.

2.3 POLYMER-BASED ORGANIC-INORGANIC HYBRID ELECTROLYTES

A typical polymer composite is a combination of polymer and filler. The polymer composites are used to ameliorate the drawbacks of conventional polymers. In general, the weak intermolecular bonding between the fillers and polymer matrix exists at a micrometer level and least probable chemical bonding is involved in between polymer matrix and fillers. But the combination of filler at the nanometer level and functional polymers interacting at the atomic level leads to the formation of chemical bonding and consequently significant improvements in the mechanical property; especially strength, electrical conductivity, thermal stability, thermal conductivity and oxidation stability have been observed. Such materials come in an important class of composites known as organic-inorganic nanostructured materials (Gai et al. 2019; Judeinstein & Sanchez, 1996; Kaushik et al. 2015). Several polymers-based organic-inorganic hybrid electrolytes had been formed by hybridization of the organic polymers with various types of incorporating inorganic ceramic fillers; fast ion conducive ceramics, glasses, ionic liquids are discussed in the subsequent sections.

2.3.1 Polymer/Inorganic Hybrid Composite Electrolytes

The polymer/inorganic hybrid electrolytes are a unique new class of material prepared by combining polymers with inorganic ceramic filler particles at the molecular level. The ultrafine dispersion of inorganic filler particles within the polymeric matrix is considered to be an important route for developing multifunctional materials with unusual improved properties. It has recently gained a great deal of attention because of the superior properties in terms of increased strength and modulus, improved heat resistance, decreased crystallinity, increased electrical conductivity and improved dielectric properties. With these improved set of properties, these hybrids have become promising candidates for a wide spectrum of electrochemical applications (Chakrabarty et al. 2012; Lim et al. 2015). The inorganic ceramic filler particles added to the polymer matrix to form polymer/inorganic hybrid electrolytes can be classified into two categories. One is passive ceramic fillers that cannot take part in the transport process and the other is known as active inorganic ceramic fillers (or fast ion conductors) that couple with the ionic movements. These inert filler particles usually act as solid plasticizers to the polymer matrix and disrupt the polymer chain regularity (or crystallinity) since the incorporation of fillers create a large surface area within the matrix which inhibits the recrystallization of polymers (Chen et al. 2018; Zheng & Hu, 2018). The different types of inert ceramic fillers such as SiO_2 (Guyomard-Lack et al. 2014), TiO_2 (Pal & Ghosh, 2018), Al_2O_3 (Liang et al. 2015), zeolite (Zhang et al. 2021), etc. had been incorporated into polymers to solve the problems of low ionic conductivity as well as improve the mechanical properties of polymer electrolytes. However, the effect of the filler particles on the performances of the poly(ethylene oxide) (PEO)-based polymer electrolytes has been a subject of discussion for a long time. Croce et al. (Croce et al. 1998) studied the impact of the addition of TiO_2 and Al_2O_3 inorganic inert filler particles to the PEO-$LiClO_4$ polymer electrolytes and reported an enhanced ionic conductivity of the system ~10^{-5} S cm^{-1} at room temperature. It can be seen that filler plays a crucial role in determining the interfacial properties between an organic polymer matrix and fillers. For achieving good interfaces the contact surface area between the filler particles and polymeric matrix must be increased. Krawiec et al. (Krawiec et al. 1995) studied the impact of the addition of micro- and nanosized Al_2O_3 fillers to the PEO-$LiBF_4$ polymer electrolytes and found that nanometer-range-particle-containing systems give more conductivity because of good interfacial interactions. Lin et al. (Lin et al. 2018) prepared and studied a PEO-silica aerogel hybrid electrolyte system which gives high ionic conductivity 6.0×10^{-4} S cm^{-1} and high modulus 0.43 GPa. In addition to this, Chaurasia and Chandra (Chaurasia & Chandra, 2017) studied the incorporation of silica inorganic matrix to the PEO-$LiCF_3SO_3$ systems via in-situ approach, which gives high thermal and electrochemical stability (stability window 5.2 V), layered structure morphology and high ionic conductivity of the order of 10^{-5} S cm^{-1} at room temperature. On the other hand, other types of filler particles such as active ceramic filler particles are also added to the polymers and polymer electrolytes to form polymer/inorganic hybrids that can provide better ionic conductivity because of the existence of extra ion conduction pathways within the polymer matrix through the cross-linking. Sharma and Nair et al. (Nair et al. 2009) studied the addition of $Mg_2B_2O_5$ nanowire to the PEO-LiTFSI electrolyte system and obtained ionic conductivity 1.53×10^{-4} S cm^{-1} at 40 °C. Liu et al. (Liu et al. 2013) studied the addition of $LiAlO_2$ filler to the polymer electrolyte PEO-$LiClO_4$ and found the maximum ionic conductivity ~10^{-4} S cm^{-1} at 60 °C. Sharma and Hashmi (Sharma & Hashmi, 2019) studied the effect of the incorporation of Mg aluminate ($MgAl_2O_4$) nanofiller particles in the poly(vinylidenefluoride-hexafluoropropylene) (PVdF-HFP)-Mg-triflate-based system which offers maximum ionic conductivity value of 4.3×10^{-3} S cm^{-1} at room temperature. The incorporation of 20 wt % of active $MgAl_2O_4$ nanofiller significantly enhanced the Mg^{2+} ion transport number up to 0.66, indicating a substantial increase in Mg-ion conductivity in the polymeric system.

2.3.2 Polymer/Fast Ion Conducting Ceramic Hybrid Electrolytes

Fast ion conductors are one of the most important classes of solid state electrolytes employed as electrolyte materials at ambient temperature in several electrochemical applications. They possess very high ambient temperature ionic conductivity up to 10^{-2} S cm^{-1}, low electronic conductivity, and ions are the dominate charge carriers (nearly unity ionic transference number). They have wide temperature range of operations. These fast ion conductors have made their mark in the last few decades as they are leak free, environmentally safe, mechanically and electrochemically stable. On the basis of their microstructures, fast ion conductors are categorized into four types: (i) garnet-type, (ii) perovskite-type, (iii) NASICON-type and (iv) sulfide-type as displayed in Figure 2.3.

The poor rigid contacts and high interfacial resistance restricts their practical applications in high-performance electrochemical devices. Therefore, the combinations of fast ion conducive solid crystalline or amorphous framework network with flexible polymers are one of the good approaches to synthesize polymer/fast ion conducting hybrid electrolytes for utilization in high-performance electrochemical devices. The various types of polymeric/fast ion conducting hybrid electrolytes are discussed next.

2.3.2.1 Polymer/Garnet-Type Hybrid Electrolytes

The garnet-type solid electrolytes and their derivatives have attracted much attention because of their high ionic conductivity, superior electrochemical and chemical

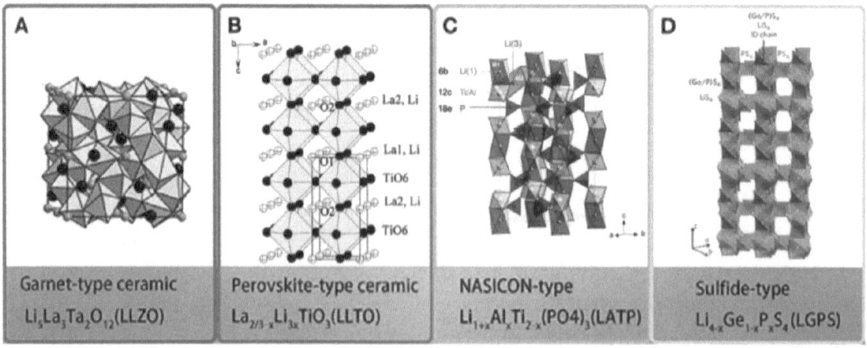

FIGURE 2.3 Structures of different types of fast ion conductors: (a) framework structure of garnet-type ceramic; (b) crystal structure of perovskite-type ceramic; (c) crystal structure of NASICON-type ceramics and (d) crystal structure of sulfide-type ceramic.

Source: This figure is reprinted with permission from Yao et al. (2019, Copyright 2019, Frontiers).

stabilities. The chemical formula for garnet-type solid state electrolytes is represented by $A_3B_2(XO_4)_3$ where *A*= Ca, Mg, Y, La or rare earth elements, and *B* = Al, Fe, Ga, Mn, Ni or V. The first report on garnet-type Li-ion conducting solid state electrolyte $Li_7La_3Zr_2O_{12}$ (LLZO) was reported in the year 2007. Since then various kinds of garnet-type solid state electrolytes and their derivatives have been developed (Murugan et al. 2007; Verduzco et al. 2021; Xie et al. 2018). The Li-ion-based garnet-type structure solid electrolyte $Li_5La_3M_2O_{12}$ (where M = Nb or Ta) is prepared which shows room temperature ionic conductivity of the order ~10^{-3} S cm^{-1} and also exhibits excellent electrochemical stability over wide temperature ranges (Verduzco et al. 2021). The garnet-type Li or Na-ion-based solid state electrolyte possesses high ionic conductivity and wide electrochemical window. These solid electrolytes are suffering from poor electrode/electrolyte contacts which enhance the interfacial resistances of the electrochemical devices. To resolve these issues and promote high charge/ion transport through interfaces, the garnet ceramic is mixed with polymers to provide polymer/garnet hybrid flexible electrolytes by physical or chemical blending methods, which shows both high flexibility due to polymer and high ionic conductivity due to presence of garnet-type ceramics/electrolytes. Choi et al. (Choi et al. 2015) prepared PEO-tetragonal LLZO organic-inorganic hybrid composite electrolytes with different amounts of garnet ceramic contents and showed maximum ion conductivity of 4.42×10^{-4} S cm^{-1} at 55 °C with 52.5 wt % LLZO. Chen et al. (Chen et al. 2017) have also prepared the polymer PEO/garnet LLZO hybrid composites with a small amount of ceramic powders which gives maximum ionic conductivity of 5.5×10^{-4} S cm^{-1} at 30 °C for the optimal concentration 7.5 wt % of filler LLZO. Morphologies and sizes of the ceramic fillers have seriously affected the ionic transport of polymer/garnet-type ceramic filler composites. Likewise ceramic particles and random nanowires reinforced polymer hybrid composites; the aligned nanowires combined with polymers provide continuous pathways for the conduction of dopant ions. Unlike 1D nanowires, the 3D ceramic $Li_{6.28}La_3Zr_2Al_{0.24}O_{12}$ (LLZO) network prepared by hydrothermal mixed with polymer to form solid electrolytes gives very high conductivity and chemical stability because of the appearance of a continuous 3D network for conduction (Bae et al. 2018a). Similarly, 3D porous LLZO structure is prepared and combined with PEO-LiTFSI polymer electrolyte to form 3D-garnet-polymer hybrid films which exhibit high conductivity of the order of 2.5×10^{-5} S cm^{-1} at 25 °C (Fu et al. 2016). Thus, this 3D-formed ion transport organic-inorganic hybrid network is considered a novel strategy for designing multifunctional materials with improved transport and electrochemical properties.

2.3.2.2 Polymer/Perovskite Hybrid Electrolytes

The perovskite-type solid electrolytes are represented by structural formula $Li_{3x}La_{2/3-x}TiO_3$ (LLTO), having cubic structure with P4/mmm and C-mmm space groups. Perovskites reveal high mechanical and chemical stabilities, higher oxidation potential and also stability at high voltages. They reveal high ionic conductivity but at the same time show larger electrode-electrolyte interfacial impedances, higher grain boundaries and instability towards lithium/sodium metal electrodes which restricts their electrochemical applications. To solve these issues as well as engineering challenges such as large-scale manufacturing of flexible sodium/lithium ion rechargeable batteries, improving interfacial impedances still hinder their immediate applications in solid state electrochemical devices (Inaguma et al. 1993; Yan et al. 2021). A possible approach to tackle these issues is the combination of inorganic perovskite electrolyte with the flexible polymeric which may provide many advantageous features like improved ionic conductivity and charge transfer at the electrode-electrolyte interfaces. However, preparatory conditions are very strict to tailor the ionic conductivity and interfacial properties of polymer/perovskite hybrid electrolytes. Zhu et al. (Zhu et al. 2018) studied the polymer-perovskite-type organic-inorganic electrolyte by incorporating nanowires $Li_{0.33}La_{0.557}TiO_3$ within PEO/$LiClO_4$ electrolytes and found that these hybrids show maximum ionic conductivity of 2.4×10^{-4} S cm^{-1} at 25 °C. Furthermore, Liu et al. (Liu et al. 2015) studied the impact of two different types of LLTO ceramics (nanowire and nanoparticle) to the polymer PAN which reveals that the introduction of LLTO nanowire gives more conducive polymer/perovskite hybrid network at room temperature as compared to pristine PAN films or nanoparticle imbedded hybrids. Moreover, the three-dimensional (3D) nanostructured $Li_{0.35}La_{0.55}TiO_3$ (LLTO) framework is designed by using the hydrogel method as 3D nanofiller for forming high-performance PEO/LLTO polymer composites as Li-ion conducting electrolytes for lithium battery application. This polymer-based composite hybrid electrolyte avoids the agglomeration of nanofillers as compared to the hybrid electrolytes prepared by simple dispersion process. Moreover, ultra-high specific surface area of LLTO ceramics provides a continuous interface network as lithium ion transport channels within the PEO/LLTO hybrid electrolytes. This synthesized polymer PEO/LLTO perovskite hybrid electrolyte displayed ionic conductivity of 8.8×10^{-5} S cm^{-1} at room temperature (Bae et al. 2018b). Thus, the formation of polymer-perovskite-type organic-inorganic hybrid flexible fast ion conducive electrolytes opens a new horizon in material science for developing high-performance electrochemical energy devices.

2.3.2.3 Polymer/NASICON-Type Hybrid Electrolytes

NASICON-type ceramics (particularly known as "sodium super ion conductors") are promising oxide-based solid electrolytes that are similar to garnet-type and perovskite-type and were first reported with composition $NaM_2(PO_4)_3$ (where M = Ge, Ti, Zr) in the year 1968 (Epp et al. 2015). However, the first report on NASICON-type solid state electrolyte material $LiZr_2(PO_4)_3$ was given in the year 1986; since then there has been growing interest on the use of Na and Li-ion-based NASICON-type ceramic electrolytes (i.e. $NaTi_2(PO_4)_3$ and $LiTi_2(PO_4)_3$) and their analogous structures in forming polymer hybrid composite electrolytes (Aono, 1990; Horowitz et al. 2020; Petit et al. 1986). Nairn et al., 1996 prepared the composite hybrid electrolytes by dispersing Li-ion conducting NASICON structure $Li_{1.3}Al_{0.3}Ti_{1.7}(PO_4)_3$ (LATP) to the PEG-$LiCF_3SO_3$ polymer electrolyte matrix and found maximum ionic conductivity of 1.90×10^{-4} S cm^{-1} at 40 °C with 66 wt % of ceramic fillers. Since then, a large number of works have been reported on Na and Li-ion-based NASICON-type ceramic structures as dispersoids to form organic-inorganic hybrid composite electrolytes. In another study, other NASICON-type ceramic filler particles, $Li_{1.5}Al_{0.5}Ge_{1.5}(PO_4)_3$ (LAGP) and $Li_{1.4}Al_{0.4}Ti_{1.6}(PO_4)_3$ (LATP), have been prepared and incorporated into PEA_{18}LiTFSI polymer electrolytes to fabricate an organic-inorganic composite solid electrolyte. The experimental results confirm that the obtained composite solid electrolytes had high ionic conductivity with weight ratio of 70% of 1.86×10^{-4} S cm^{-1} for LATP-PEO_{18}LiTFSI and 1.11×10^{-4} S cm^{-1} for LAGP-PEO_{18}LiTFSI along with superior chemical and electrochemical stabilities (Huang et al. 2012). In another report, the hybrid composite polymer electrolytes PEO-LAGP have been fabricated by dispersing different particle sizes of LAGP to the polymer PEO matrix, and experimental results showed that nanosized particles exhibited more favorable effects in enhancing transport and electrochemical properties. The polymer composite hybrid electrolyte PEO-LAGP, having 20 wt % LAGP nanoparticle concentration, had showed the highest ionic conductivity of 6.76×10^{-4} S cm^{-1} at 60 °C. This hybrid composite electrolyte was also used for the fabrication of solid state rechargeable batteries with assembly Li//PEO-LAGP//$LiFePO_4$ showing very high capacity retaining (~ 90% after 50 cycles at 1C) along with outstanding rate performances (Zhao et al. 2016).

2.3.2.4 Polymer/Sulfide-Type Hybrid Electrolytes

Sulfide-type inorganic solid electrolytes are another class of electrolytes showing supreme ionic conductivity of the order of 10^{-2} S cm^{-1} at room temperature and a wide electrochemical stability potential window (Kamaya et al. 2011). The sulfide-type electrolytes can be broadly classified into three categories: glasses, glass-ceramic and ceramic electrolytes. The various types of sulfide electrolytes such as $Li_{10}GeP_2S_{12}$ (LGPS), $Li_{10}SnP_2S_{12}$ (LSPS), Li_3PS_4, Li_2S-P_2S_5 sulfide glass and argyrodite-type Li_6PS_5X (where X = Cl, Br, I) are promising sulfide-type solid electrolytes showing very high ionic conductivity nearer to the liquid electrolytes. However, they demonstrated poor mechanical properties and undesirable high interfacial contact resistances between electrodes and electrolytes, which restricts their possible utilization in solid state electrochemical application (Yang et al. 2020). These inorganic solid electrolytes are also showing chemical and structural instability due to the reaction with moisture present in the air. The combination of polymer and high ionic conducive sulfide electrolytes is considered to be the designing of novel organic-inorganic composite electrolytes, which simultaneously includes the merits of flexible polymer and high conducting sulfide filler ceramics. Zhao et al. (Zhao et al. 2016) first prepared the hybrid composite electrolyte membranes by incorporation of the LGPS sulfide filler particles into the PEO-LiTFSI matrix. This PEO-LITFSI/sulfide composite electrolyte gives the maximum ionic conductivity of ~10^{-5} S cm^{-1} at room temperature and reaches 1.21×10^{-3} S cm^{-1} at 80 °C and electrochemical stability of these hybrids spans between 0 to 5.7 V. Recently, Zheng et al. (Zheng et al. 2019) examined the effect of sulfide LGPS and salt LiTFSI on the interface formed within the composite electrolyte and showed that the incorporation of 70 wt % LGPS to the PEO-LiTFSI electrolyte prepared by ball-milling exhibited ionic conductivity of 2.2×10^{-4} S cm^{-1} at room temperature. The hybrid composite electrolyte based on polymer poly(vinyl carbonate) and high conducting sulfide $Li_{10}SnP_2S_{12}$ (LSPS) was prepared by in-situ polymerization technique. The experimental results confirm that an integrated interface has been formed between the electrodes and hybrid composite (PVC-LSPS) which favors the decrease in the interfacial impedances. The hybrid composite electrolyte also shows a series of enhanced electrical and electrochemical properties like high conductivity of 2.0×10^{-4} S cm^{-1} at room temperature, wider span of electrochemical stability window of 4.5 V, improved Li-ion cationic transport number of 0.60 and good compatibility of hybrid composite electrolytes with Li metal anode. The fabricated all solid state battery cell $Li/LiFe_{0.2}Mn_{0.8}PO_4$ using this hybrid composite electrolyte shows high specific capacity of 130 $mAhg^{-1}$ and cycling stability of 140 cycles at 0.5 C (Ju et al. 2018). Thus, the results obtained from various studies showed that polymer-sulfide-type hybrid composites will be successfully used in high-performance electrochemical applications.

2.4 POLYMERIC/IONIC LIQUID QUASI-SOLID HYBRID GELS

The application of ionic liquids has been extensively used in several areas of science and technology. Ionic liquids are powerful solvents, and they have been used as electrolytes. In general, the melting point of ionic liquids (ILs) is lower than 100 °C. Ionic liquids are salt, which has shown properties such as ion conducting capability, non-flammability, good thermal stability, wide electrochemical window (up to 5–6 V vs. Li^+/Li), and non-volatility. It is possible to immobilize ionic liquid with various host polymeric matrices in the presence or absence of inorganic fillers to form IL-based hybrid quasi-solid electrolytes. The ionic liquid in the quasi-solid hybrid act as a supplier of charge carriers (mobile cations and anions) for conduction; a solvation media of ionic salts; and stable plasticizer for polymeric-based hybrid matrices (Chaurasia et al. 2022; Yang et al. 2020; S. Zhang et al. 2014). The ideal hybrid electrolyte characteristics are high conductivity, electronic insulation, physical stability, chemical stability, electrochemical stability, being inert, having excellent contact and compatibility with electrodes, being robust against abuse conditions, environmental friendliness. It has been believed that the ionic liquid incorporated hybrid electrolyte systems that wisely utilize the strength of their components and the interactions among the components can lead to real a breakthrough for next-generation large-scale battery applications. With the incorporation of ionic liquid, some of the polymeric quasi-solid hybrid networks show ionic conductivity within the range of 10^{-2} to 10^{-4} S cm^{-1} at room temperature (Guyomard-Lack et al. 2014; Tripathi, 2021). However, the ternary hybrids formed with polymers, ILs and inorganic materials have capabilities to achieve improved physical and chemical properties towards developing next-generation lithium ion rechargeable battery electrolytes. Such polymer-based quasi-solid hybrid electrolytes belong to a subcategory of organic/inorganic and have gained much attention because of their flexibility, high conductivity, wide temperature range of operation, wider span of electrochemical stable window and higher temperature operations. Huo et al. (Huo et al. 2017) prepared a hybrid membrane composed of PEO, 200 nm sized $Li_{6.4}La_3Zr_{1.4}Ta_{0.6}O_{12}$ particles and IL $[BMIM]TF_2N$ for interfacial wetting. The oxidation decomposition potential from the LSV curve increased from 4.65 V to 4.85 V when $[BMIM]TF_2N$ was added. Similarly, Zhan and co-workers (Zhan et al. 2019) grafted 1-vinyl-3 cyanopropyl imidazolium bis(trifluoromethyl sulfonyl) imide IL and oligomeric PEO chains to the backbone of polysiloxane due to the presence of strong electron withdrawing nitrile groups in the graft. IL, together with the imidazole ring, deceptively increased the anodic oxidation decomposition potential of the hybrid (0.7 V difference compared with the group without nitrile) as well; suppressing the growth of lithium dendrite is critical for ensuring battery safety and cycling stability. It has also been used a UV curing method to prepare cross-linked composite polymer electrolytes (PIL) with evenly dispersed boron nitride nanosheets and IL electrolytes. It has also been found that adding proper nanosized ceramic fillers and ILs at the electrolyte/electrode interface improved compatibility and was favorable for stable SEI formation (Ma et al. 2019; Yang et al. 2018). Yang et al. (Yang et al. 2020) have shown that a unique class of organic-inorganic hybrid electrolytes based on IL-grafted POSS exhibits many remarkable features, including non-flammability, superior room temperature ion conductivity and excellent electrochemical stability, for Li-metal batteries. Batteries with this electrolyte have been cycled at room temperature. However, considerable work remains to be done. We believe that this work provides a promising structural platform for the development of new solid state electrolytes for ambient-temperature lithium batteries.

2.5 APPLICATION OF ORGANIC-INORGANIC HYBRID ELECTROLYTES IN ELECTROCHEMICAL DEVICES

In general, organic-inorganic composite solid electrolytes (CSEs) form by the blending of polymer and inorganic ceramics to obtain enhancement in the performance solid electrolytes. Comparing to polymer and inorganic electrolytes, organic-inorganic hybrid electrolytes exhibit excellent mechanical characteristics, enhanced ionic conductivity and improved interfacial stability. Indeed, the consistent progress in the organic-inorganic composite electrolytes with remarkable properties is the need of future electrochemical devices. A schematic representation of organic-inorganic hybrid electrolytes which arises from combining the flexible polymer and inorganic fast ion conductors with their improved properties is shown in Figure 2.4.

Till now, various organic-inorganic CSEs have been reported. The development of organic-inorganic CSEs in various electrochemical devices have been evaluated in terms of the ion transport mechanism, electrochemical and mechanical properties, and their potential to overcome the challenges of future batteries and supercapacitors.

2.5.1 Application of Organic-Inorganic Hybrid Electrolytes in Batteries

Batteries have shown several advantages such as including flexibility, process-ability and compatibility with electrodes. Solid Polymer Electrolytes (SPEs), which have shown promising electrolytes for rechargeable batteries, contain polymer matrices and salt of lithium. A series of polymer hosts such as poly(ethylene oxide) (PEO), polycarbonate (PEC), polyvinylidene fluoride (PVdF), polyacrylonitrile (PAN) and their derivatives have been extensively used for practical polymer electrolytes in many electrochemical applications (Gupta et al. 2016; Yi et al. 2018). However, most of the SPEs show the ionic conductivity less than 10^{-8} to 10^{-5} S cm^{-1} at room temperature, but this doesn't meet the practical required conductivity of 10^{-3} S cm^{-1} (Inda et al. 2007). Inorganic solid electrolytes (ISEs) exhibit much higher ionic

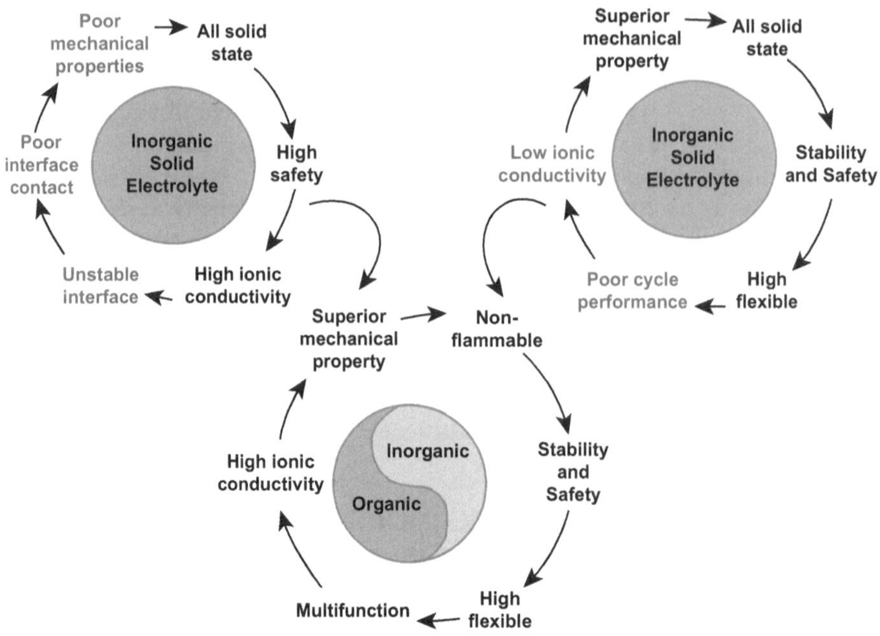

FIGURE 2.4 Advantage of organic-inorganic composite solid electrolytes.

Source: This figure is reprinted with permission from Zhang et al. (2020, Copyright 2019, Wiley-VCH Verlag GmbH & Co).

conductivity (range of 10^{-4} to 10^{-2} S cm^{-1}) at room temperature, and such higher ionic conductivity makes them suitable candidates for practical solid state batteries. The ion diffusion process of ISEs is mainly controlled by the ionic size and lattice volume. However, ISEs have limitations which hinder their applications. These limitations, due to some shortcomings like poor mechanical properties and higher rigidity, prohibit their application in flexible devices. Moreover, most of the ISEs are unstable with electrodes. A passivated interface layer between ISEs and electrodes can form after contact with electrodes; it reduces the cycling and rate performance of all solid state batteries. The unstable interface layer also restricts their practical application in Li-metal batteries. Researchers have made many attempts to solve the previously mentioned problems in past decades, and optimized properties of electrolytes are difficult for any single SSE (Zhou et al. 2018).

In the past few years, organic-inorganic CSEs have gained extensive attention due to their superior thermostability, mechanical properties, excellent processability, straightforward fabrication and improved safety (Dai et al. 2018). However, it is urgent to improve their electrochemical response for high-energy solid state batteries (Hu et al. 2017), such as room temperature ionic conductivity, interface stability with electrodes and electrochemical window. The recent advances of organic-inorganic hybrid electrolytes have been summarized in terms of their fabrication strategy, ion movement mechanism, electrochemical and mechanical properties, and device performances. It was found that the addition of inorganic ceramics into the polymer electrolytes can significantly enhance the ionic conductivity, expand the electrochemical window and improve mechanical strength (Yang et al. 2017). Therefore, the organic-inorganic hybrid electrolyte is the suitable candidate for high-energy lithium batteries, which has opened up a new path for solid electrolytes. Although various efforts have been carried out in this field, serious problems and challenges in the implementation of the practical application must be sorted out.

In order to obtain the better ionic conductivity of the CSEs, achieving improved comprehensive electrochemical properties are essential. The improvements in the ionic conductivity, the inorganic component which shows the high ionic conductivity and continuous positive ion transport channels, have been explored. Thus, porous one-dimensional or three-dimensional inorganic ceramic nanostructures design was found more effective. Electrolytes with high positive ion transportation can availably reduce concentration polarization in solid state batteries, thus selecting the polymer matrixes with single ionic conductivity might be beneficial to improving the positive ion transfer number. Stable and strong interfacial attachments with anodes as well as cathodes suppress the interfacial resistance, which assure the cycling stability of solid state batteries. Therefore, flexible polymer materials could be employed as the matrix of organic-inorganic hybrid electrolyte, which is an effective strategy to achieve a compact contact between the electrolyte and electrodes (Liu et al. 2018).

The conducting mechanism in organic-inorganic hybrid electrolytes embedded in solid state batteries still requires deep understanding. The origin of space charge region due to the interface between the polymer matrix and inorganic components plays a crucial role in ion transport. The mechanism can be understand well using in-situ techniques such as an in-situ transmission electron microscope (TEM), high-resolution nuclear magnetic resonance (NMR). Simultaneously, theoretical simulations would be favorable for understanding the ion migration behavior. Such fundamental research would serve as guidelines for the rational design of new organic-inorganic composite electrolytes.

The research area of hybrid electrolytes in the areas of solid state batteries needs to explore also thermal stability and cost of solid composite electrolytes. For example, electrolytes exhibiting low temperature ionic conductivity with superior non-flammability would be likely more reliable to apply in various fields. Apart from low cost and handy fabrication procedures, it should be considered for large-scale practical applications, which demand additional efforts in the future. However, few composite electrolytes have exhibited excellent comprehensive performances for practical solid state batteries. Future research would be based on improving the electrochemical performance along with the mechanical properties of the composite electrolytes. Some studies on the hybrid materials are elucidated. Diganta Saikia et al. have reported a enhancement in the conductive organic inorganic hybrid polymer electrolytes which is obtained by the reaction of Poly(propylene glycol)-block-poly(ethylene glycol)-block-poly(propylene glycol) bis(2-aminopropyl ether), 2,4,6-trichloro-1,3,5-triazine and alkoxysilane precursor 3-(glycidyloxypropyl) trimethoxysilane, and further doping of $LiClO_4$. This solid hybrid electrolyte exhibits a higher ionic conductivity of 1.6×10^{-4} S cm^{-1} at 300 K, which is the highest among the organic inorganic hybrid electrolytes. This hybrid was electrochemically stable up to 4.2 V

(Saikia et al. 2016). The results of lithium-ion battery tests show this organic inorganic hybrid electrolytes fulfill almost all conditions for the applications in electrochemical devices and lithium-ion batteries. In a recent report, Nataly Carolina Rosero-Navarro et al. proposed that organic-inorganic hybrid electrolyte obtained from an organic network of poly(ethylene oxide) chains inducted an inorganic Si-O-Si backbone system with lithium salt, as a novel interfacial material between a garnet-type oxide solid electrolyte and high potential cathodes. The low resistance solid-solid interface is studied between the solid electrolyte and $LiCoO_2$ which achieved by the hybrid solid electrolyte. They have shown that the hybrid material supports an ionic/electronic percolation of active material particles and resulting outstanding adherence properties (Rosero-Navarro et al. 2020).

2.5.2 Application of Organic-Inorganic Electrolytes in Supercapacitors

The supercapacitors and pseudo capacitors store charge through surface ion adsorption (non-faradaic) and surface redox reaction (faradaic), respectively. In order to achieve higher performance with stable electrochemical electrodes, architectural development in the electrode as well as electrocatalytical research is required. Electrochemical supercapacitors based on metal oxides and conducting polymers favor voltage-changing redox processes, where faradaic and non-faradaic mechanisms taking place concomitantly, are called hybrid supercapacitors (Gupta et al. 2017; Singh & Chaurasia, 2021; Zhang et al. 2020). The organic-inorganic hybrid supercapacitors are the basically nanoscale hybridization of super capacitive nanocarbons (activated carbon, nanotubes, mesoporous carbon and graphene-family nano materials) and pseudocapacitive or redox active transition metal oxides. In this context, graphene is the most suitable candidate of activated carbon because of its extraordinary physical-chemical properties. Moreover, functionalized graphene, such as graphene oxide (GO) and reduced form (rGO), is attracting a great deal of attention because of its tunable electrical conductivity, mechanical robustness and easy processability. Such materials with high surface area and higher functional groups can easily increase speedy faradaic reactions at the surface as well as storing electrical energy in transient chemical bonds possessing high specific power and energy density and excellent cycling stability. The continuous improvement in the new class of hybrid supercapacitors blurs the boundaries between batteries and supercapacitors. The significant progress toward hybrid supercapacitors requires deep understanding of storage mechanisms of transition metal oxides on graphene support with optimal loading and at the molecular level (Bagri et al. 2010; Loh et al., 2010; Novoselov et al. 2004).

The molecular clusters of large metal cluster anions are formed by transition metal oxides which possess an ability to form dynamic structures called Polyoxometalates (POMs). Polyoxometalates (POMs) can attain various shapes and sizes, and complex co-ordinations correspond to large inorganic molecules. POMs are classified into three broad subsets (Byrappa & Yoshimura, 2012). Heteropoly anions (HPAs) are metal oxide clusters containing hetero anions like SO_4^{2-} and PO_4^{3-} which have shown high reversible redox activities and are the most exploited subset for electrochemical and catalytic activity. They possess Keggin $[XM_{12}O_{40}]^{n-}$ and Well-Dawson $[X_2M_{18}O_{62}]n^-$ anion structures (where M = Mo(VI), W(VI), Ta, Nb or V(V); and X is a tetrahedral template) and have high charges and strongly acid oxygen surfaces. The second ones are isopolyanions, composed of a metal oxide framework; however without the internal heteroatom/heteroanion isopolyanions they are less stable. Lastly, reduced POM clusters are related to molybdenum blue species and usually their composition is largely unknown with substantial ongoing research. The POM molecules that are central to this work are mostly symmetric HPAs including Gupta & Price (2015).

The integration of POMs with graphene have been leading to the development of new concepts of hybrid materials where surface functional moieties on GO and rGO serve as chemical linkers and POMs clusters as molecular spacers for graphene sheets that allow excess large specific surface, enabling higher storage capacity otherwise inaccessible for ion adsorption due to self-aggregation. However, the syntheses of graphene-POM hybrids as high-performance electrochemical electrodes remain elusive and a fundamental understanding of interfacial capacitance with optimal loading and associated physicochemical properties are yet to be determined (Gupta & Carrizosa, 2016).

The development of a series of high-performance "organic-inorganic" hybrids consisting of pseudo-capacitive phosphomoylbdate (and phosphotungstic) molecular network anchored to or distributed on the super-capacitive reduced graphene oxide (r-GO) nanosheets have been explored. The nanosized polynuclear redox active molecular network with higher oxidation state transition metals are suitable to achieve high specific energy capacity due to multi-electron redox reactions. The interesting properties are obtained on the interaction of functionalized graphene, POW and POM nanodots. Such hybrid materials have elucidated extended better operating voltage, harvested enhanced specific energy and power density (Wen et al. 2012). The outstanding response is that hybrid nanoscale inorganic with organic materials, especially r-GOs, provide chemical functionality, synergistic coupled charge storage mechanisms. On the other hand, mesoporous and topologically interconnected morphology of r-GO nanosheets generates electronic as well as ionic conducting paths also. The mechanism can be explored where molecularly bridged and chemically anchored POM and POW nanoclusters create tailored interactions, provide high specific capacitance, interfacial contact area at electrode/electrolyte interfaces and accessibility for ion adsorption, which provides new conductive pathways for charge transfer. The redox active electrolyte provides extra protons and improved ionic conductivity and additional faradaic active sites for storage capacity.

2.6 CONCLUSIONS AND FUTURE PROSPECTS

In this chapter, an overview of the significance of organic-inorganic hybrid materials and their possible applications in the field of solid state electrochemical devices are discussed. It has been already reported that organic-inorganic hybrid materials have been used potentially in the development of nanoelectronics, drug delivery systems, bioimaging, hybrid solar cells and many more fields. In organic-inorganic hybrid materials, the polymers play a role as a stabilizer whereas an inorganic metal provides mechanical strength, transparency, conductivity, optical properties, etc. Scientists have shown a strong commitment to develop nano assemblies of organic-inorganic hybrid material species in the areas of energy conversion, storage fields.

Organic-inorganic hybrid materials as composite solid electrolytes (CSEs) exhibit superior thermostability, mechanical properties, excellent processability, easy fabrication and high safety relative to traditional solid electrolytes. But still many challenges remain to be overcome for large-scale applications. CSEs are yet to show a better response in their electrochemical performances, like ambient temperature ionic conductivity, stability at the interface electrodes, and the need for a broader electrochemical window. This chapter is dedicated to exploring the history of organic-inorganic hybrid electrolytes and evaluates the possibility of the broad level application of CSEs to fulfill the requirements in the field of energy. The advantages of organic-inorganic CSEs have been summarized in terms of their design principles, ion transport mechanisms, electrochemical and mechanical properties, and battery performances. The authors also discussed the impact of induction inorganic ceramics into polymer electrolytes to improve enhancement in their ionic conductivity, to extend the electrochemical window and mechanical strength. It is believed that organic-inorganic CSEs will be optimal candidates for high-energy batteries and high-power density supercapacitors/hybrid supercapacitors, which opens up a new avenue for solid electrolytes.

REFERENCES

Agrawal, R. C., & Pandey, G. P. (2008). Solid polymer electrolytes: Materials designing and all-solid-state battery applications: an overview. *Journal of Physics D: Applied Physics, 41*(22), 223001. https://doi.org/10.1088/0022-3727/41/22/223001

Aono, H. (1990). Ionic conductivity and sinterability of lithium titanium phosphate system. *Solid State Ionics, 40–41*, 38–42. https://doi.org/10.1016/0167-2738(90)90282-V

Armand, M. B., Chabagno, J. M., & Duclot, M. J. (1979). *Fast ion transport in solids*, ed. P. Vashishta, J. N. Mundy, & G. K. Shenoy. Amsterdam, North Holland.

Bae, J., Li, Y., Zhang, J., Zhou, X., Zhao, F., Shi, Y., Goodenough, J. B., & Yu, G. (2018a). A 3D nanostructured hydrogel-framework-derived high-performance composite polymer lithium-ion electrolyte. *Angewandte Chemie International Edition, 57*(8), 2096–2100. https://doi.org/10.1002/anie.201710841

Bae, J., Li, Y., Zhang, J., Zhou, X., Zhao, F., Shi, Y., Goodenough, J. B., & Yu, G. (2018b). A 3D nanostructured hydrogel-framework-derived high-performance composite polymer lithium-ion electrolyte. *Angewandte Chemie International Edition, 130*(8), 2118–2122. https://doi.org/10.1002/ange.201710841

Bagri, A., Mattevi, C., Acik, M., Chabal, Y. J., Chhowalla, M., & Shenoy, V. B. (2010). Structural evolution during the reduction of chemically derived graphene oxide. *Nature Chemistry, 2*(7), 581–587. https://doi.org/10.1038/nchem.686

Byrappa, K., & Yoshimura, M. (2012). *Handbook of hydrothermal technology*. William Andrew, Elsevier B. V., The Netherlands.

Chakrabarty, T., Singh, A. K., & Shahi, V. K. (2012). Zwitterionic silica copolymer based crosslinked organic: Inorganic hybrid polymer electrolyte membranes for fuel cell applications. *RSC Advances, 2*(5), 1949. https://doi.org/10.1039/c1ra00228g

Chaurasia, S. K., & Chandra, A. (2017). Organic-inorganic hybrid electrolytes by in-situ dispersion of silica nanospheres in polymer matrix. *Solid State Ionics, 307*, 35–43. https://doi.org/10.1016/j.ssi.2017.05.003

Chaurasia, S. K., Singh, M. P., Singh, M. K., Kumar, P., & Saroj, A. L. (2022). Impact of ionic liquid incorporation on ionic transport and dielectric properties of PEO-lithium salt-based quasi-solid-state electrolytes: Role of ion-pairing. *Journal of Materials Science: Materials in Electronics, 33*(3), 1641–1656. https://doi.org/10.1007/s10854-022-07706-y

Chaurasia, S. K., Singh, R. K., & Chandra, S. (2011). Structural and transport studies on polymeric membranes of PEO containing ionic liquid, EMIM-TY: Evidence of complexation. *Solid State Ionics, 183*(1), 32–39. https://doi.org/10.1016/j.ssi.2010.12.008

Chen, F., Yang, D., Zha, W., Zhu, B., Zhang, Y., Li, J., Gu, Y., Shen, Q., Zhang, L., & Sadoway, D. R. (2017). Solid polymer electrolytes incorporating cubic $Li_7La_3Zr_2O_{12}$ for all-solid-state lithium rechargeable batteries. *Electrochimica Acta, 258*, 1106–1114. https://doi.org/10.1016/j.electacta.2017.11.164

Chen, S., Wang, J., Zhang, Z., Wu, L., Yao, L., Wei, Z., Deng, Y., Xie, D., Yao, X., & Xu, X. (2018). In-situ preparation of poly(ethylene oxide)/Li_3PS_4 hybrid polymer electrolyte with good nanofiller distribution for rechargeable solid-state lithium batteries. *Journal of Power Sources, 387*, 72–80. https://doi.org/10.1016/j.jpowsour.2018.03.016

Choi, J. H., Lee, C. H., Yu, J. H., Doh, C. H., & Lee, S. M. (2015). Enhancement of ionic conductivity of composite membranes for all-solid-state lithium rechargeable batteries incorporating tetragonal $Li_7La_3Zr_2O_{12}$ into a polyethylene oxide matrix. *Journal of Power Sources, 274*, 458–463. https://doi.org/10.1016/j.jpowsour.2014.10.078

Croce, F., Appetecchi, G. B., Persi, L., & Scrosati, B. (1998). Nanocomposite polymer electrolytes for lithium batteries. *Nature, 394*(6692), 456–458. https://doi.org/10.1038/28818

Dai, J., Yang, C., Wang, C., Pastel, G., & Hu, L. (2018). Interface engineering for garnet-based solid-state lithium-metal batteries: Materials, structures, and characterization. *Advanced Materials, 30*(48), 1802068. https://doi.org/10.1002/adma.201802068

Epp, V., Ma, Q., Hammer, E.-M., Tietz, F., & Wilkening, M. (2015). Very fast bulk Li ion diffusivity in crystalline $Li_{1.5}Al_{0.5}Ti_{1.5}(PO_4)_3$ as seen using NMR relaxometry. *Physical Chemistry Chemical Physics, 17*(48), 32115–32121. https://doi.org/10.1039/C5CP05337D

Fu, K. (Kelvin), Gong, Y., Dai, J., Gong, A., Han, X., Yao, Y., Wang, C., Wang, Y., Chen, Y., Yan, C., Li, Y., Wachsman, E. D., & Hu, L. (2016). Flexible, solid-state, ion-conducting membrane with 3D garnet nanofiber networks for lithium batteries. *Proceedings of the National Academy of Sciences, 113*(26), 7094–7099. https://doi.org/10.1073/pnas.1600422113

Gai, J., Ma, F., Zhang, Z., Sun, D., Jin, Y., Guo, Y., & Kim, W. (2019). Flexible organic: Inorganic composite solid electrolyte with asymmetric structure for room temperature solid-state li-ion batteries. *ACS Sustainable Chemistry & Engineering, 7*(19), 15896–15903. https://doi.org/10.1021/acssuschemeng.9b01869

Gupta, H., Shalu, S., Balo, L., Singh, V. K., Chaurasia, S. K., & Singh, R. K. (2016). Effect of phosphonium based ionic liquid on structural, electrochemical and thermal behaviour of polymer poly(ethylene oxide) containing salt lithium bis(trifluoromethylsulfonyl) imide. *RSC Advances, 6*(91), 87878–87887. https://doi.org/10.1039/C6RA20393K

Gupta, S., Aberg, B., & Carrizosa, S. (2017). Functionalized graphene: Polyoxometalate nanodots assembly as "organic: Inorganic" hybrid supercapacitors and insights into electrode/electrolyte interfacial processes. *C, 3*(4), 24. https://doi.org/10.3390/c3030024

Gupta, S., & Carrizosa, S. B. (2016). Insights into electrode/electrolyte interfacial processes and the effect of nanostructured cobalt oxides loading on graphene-based hybrids by scanning electrochemical microscopy. *Applied Physics Letters, 109*(24), 243903. https://doi.org/10.1063/1.4972181

Gupta, S., & Price, C. (2015). Scanning electrochemical microscopy of graphene/polymer hybrid thin films as supercapacitors: Physical-chemical interfacial processes. *AIP Advances, 5*(10), 107113. https://doi.org/10.1063/1.4933190

Guyomard-Lack, A., Abusleme, J., Soudan, P., Lestriez, B., Guyomard, D., & Bideau, J. L. (2014). Hybrid silica-polymer ionogel solid electrolyte with tunable properties. *Advanced Energy Materials, 4*(8), 1301570. https://doi.org/10.1002/aenm.201301570

Horowitz, Y., Lifshitz, M., Greenbaum, A., Feldman, Y., Greenbaum, S., Sokolov, A. P., & Golodnitsky, D. (2020). Review-polymer/ceramic interface barriers: The fundamental challenge for advancing composite solid electrolytes for li-ion batteries. *Journal of the Electrochemical Society*, *167*(16), 160514. https://doi.org/10.1149/1945-7111/abcd12

Hu, J., Wang, W., Peng, H., Guo, M., Feng, Y., Xue, Z., Ye, Y., & Xie, X. (2017). Flexible organic: Inorganic hybrid solid electrolytes formed via thiol: Acrylate photopolymerization. *Macromolecules*, *50*(5), 1970–1980. https://doi.org/10.1021/acs.macromol.7b00035

Huang, L.-Z., Wen, Z.-Y., Jin, J., & Liu, Y. (2012). Preparation and characterization of PEO-LATP/LAGP ceramic composite electrolyte membrane for lithium batteries: Preparation and characterization of PEO-LATP/LAGP ceramic composite electrolyte membrane for lithium batteries. *Journal of Inorganic Materials*, *27*(3), 249–252. https://doi.org/10.3724/SP.J.1077.2012.00249

Huo, H., Zhao, N., Sun, J., Du, F., Li, Y., & Guo, X. (2017). Composite electrolytes of polyethylene oxides/garnets interfacially wetted by ionic liquid for room-temperature solid-state lithium battery. *Journal of Power Sources*, *372*, 1–7. https://doi.org/10.1016/j.jpowsour.2017.10.059

Inaguma, Y., Liquan, C., Itoh, M., Nakamura, T., Uchida, T., Ikuta, H., & Wakihara, M. (1993). High ionic conductivity in lithium lanthanum titanate. *Solid State Communications*, *86*(10), 689–693. https://doi.org/10.1016/0038-1098(93)90841-A

Inda, Y., Katoh, T., & Baba, M. (2007). Development of all-solid lithium-ion battery using Li-ion conducting glass-ceramics. *Journal of Power Sources*, *174*(2), 741–744. https://doi.org/10.1016/j.jpowsour.2007.06.234

Ju, J., Wang, Y., Chen, B., Ma, J., Dong, S., Chai, J., Qu, H., Cui, L., Wu, X., & Cui, G. (2018). Integrated interface strategy toward room temperature solid-state lithium batteries. *ACS Applied Materials & Interfaces*, *10*(16), 13588–13597. https://doi.org/10.1021/acsami.8b02240

Judeinstein, P., & Sanchez, C. (1996). Hybrid organic: Inorganic materials: A land of multidisciplinarity. *Journal of Materials Chemistry* *6*(4), 511–525. https://doi.org/10.1039/JM9960600511

Kamaya, N., Homma, K., Yamakawa, Y., Hirayama, M., Kanno, R., Yonemura, M., Kamiyama, T., Kato, Y., Hama, S., Kawamoto, K., & Mitsui, A. (2011). A lithium superionic conductor. *Nature Materials*, *10*(9), 682–686. https://doi.org/10.1038/nmat3066

Kaushik, A., Kumar, R., Arya, S. K., Nair, M., Malhotra, B. D., & Bhansali, S. (2015). Organic: Inorganic hybrid nanocomposite-based gas sensors for environmental monitoring. *Chemical Reviews*, *115*(11), 4571–4606. https://doi.org/10.1021/cr400659h

Krawiec, W., Scanlon, L. G., Fellner, J. P., Vaia, R. A., Vasudevan, S., & Giannelis, E. P. (1995). Polymer nanocomposites: A new strategy for synthesizing solid electrolytes for rechargeable lithium batteries. *Journal of Power Sources*, *54*(2), 310–315. https://doi.org/10.1016/0378-7753(94)02090-P

Liang, B., Tang, S., Jiang, Q., Chen, C., Chen, X., Li, S., & Yan, X. (2015). Preparation and characterization of PEO-PMMA polymer composite electrolytes doped with nano-Al2O3. *Electrochimica Acta*, *169*, 334–341. https://doi.org/10.1016/j.electacta.2015.04.039

Liang, J., Luo, J., Sun, Q., Yang, X., Li, R., & Sun, X. (2019). Recent progress on solid-state hybrid electrolytes for solid-state lithium batteries. *Energy Storage Materials*, *21*, 308–334. https://doi.org/10.1016/j.ensm.2019.06.021

Lim, Y. J., Kim, H. W., Lee, S. S., Kim, H. J., Kim, J.-K., Jung, Y.-G., & Kim, Y. (2015). Ceramic-based composite solid electrolyte for lithium-ion batteries. *ChemPlusChem*, *80*(7), 1100–1103. https://doi.org/10.1002/cplu.201500106

Lin, D., Yuen, P. Y., Liu, Y., Liu, W., Liu, N., Dauskardt, R. H., & Cui, Y. (2018). A silica-aerogel-reinforced composite polymer electrolyte with high ionic conductivity and high modulus. *Advanced Materials*, *30*(32), 1802661. https://doi.org/10.1002/adma.201802661

Liu, J., Xu, J., Lin, Y., Li, J., Lai, Y., Yuan, C., Zhang, J., & Zhu, K. (2013). All-solid-state lithium ion battery: Research and industrial prospects. *Acta Chimica Sinica*, *71*(06), 869. https://doi.org/10.6023/A13020170

Liu, W., Liu, N., Sun, J., Hsu, P.-C., Li, Y., Lee, H.-W., & Cui, Y. (2015). Ionic conductivity enhancement of polymer electrolytes with ceramic nanowire fillers. *Nano Letters*, *15*(4), 2740–2745. https://doi.org/10.1021/acs.nanolett.5b00600

Liu, X., Peng, S., Gao, S., Cao, Y., You, Q., Zhou, L., Jin, Y., Liu, Z., & Liu, J. (2018). Electric-field-directed parallel alignment architecting 3D lithium-ion pathways within solid composite electrolyte. *ACS Applied Materials & Interfaces*, *10*(18), 15691–15696. https://doi.org/10.1021/acsami.8b01631

Loh, K. P., Bao, Q., Eda, G., & Chhowalla, M. (2010). Graphene oxide as a chemically tunable platform for optical applications. *Nature Chemistry*, *2*(12), 1015–1024. https://doi.org/10.1038/nchem.907

Ma, F., Zhang, Z., Yan, W., Ma, X., Sun, D., Jin, Y., Chen, X., & He, K. (2019). Solid polymer electrolyte based on polymerized ionic liquid for high performance all-solid-state lithium-ion batteries. *ACS Sustainable Chemistry & Engineering*, *7*(5), 4675–4683. https://doi.org/10.1021/acssuschemeng.8b04076

Mir, S. H., Nagahara, L. A., Thundat, T., Mokarian-Tabari, P., Furukawa, H., & Khosla, A. (2018). Review-organic-inorganic hybrid functional materials: An integrated platform for applied technologies. *Journal of the Electrochemical Society*, *165*(8), B3137–B3156. https://doi.org/10.1149/2.0191808jes

Murugan, R., Thangadurai, V., & Weppner, W. (2007). Fast lithium ion conduction in garnet-type Li7La3Zr2O12. *Angewandte Chemie International Edition*, *46*(41), 7778–7781. https://doi.org/10.1002/anie.200701144

Nair, J. R., Gerbaldi, C., Chiappone, A., Zeno, E., Bongiovanni, R., Bodoardo, S., & Penazzi, N. (2009). UV-cured polymer electrolyte membranes for Li-cells: Improved mechanical properties by a novel cellulose reinforcement. *Electrochemistry Communications*, *11*(9), 1796–1798. https://doi.org/10.1016/j.elecom.2009.07.021

Nairn, K., Forsyth, M., Every, H., Greville, M., & MacFarlane, D. R. (1996). Polymer-ceramic ion-conducting composites. *Solid State Ionics*, *86–88*, 589–593. https://doi.org/10.1016/0167-2738(96)00212-3

Nicole, L., Laberty-Robert, C., Rozes, L., & Sanchez, C. (2014). Hybrid materials science: A promised land for the integrative design of multifunctional materials. *Nanoscale*, *6*(12), 6267–6292. https://doi.org/10.1039/C4NR01788A

Novoselov, K. S., Geim, A. K., Morozov, S. V., Jiang, D., Zhang, Y., Dubonos, S. V., Grigorieva, I. V., & Firsov, A. A. (2004). Electric field effect in atomically thin carbon films. *Science*, *306*(5696), 666–669. https://doi.org/10.1126/science.1102896

Pal, P., & Ghosh, A. (2018). Influence of TiO_2 nano-particles on charge carrier transport and cell performance of PMMA-$LiClO_4$ based nano-composite electrolytes. *Electrochimica Acta*, *260*, 157–167. https://doi.org/10.1016/j.electacta.2017.11.070

Petit, D., Colomban, Ph., Collin, G., & Boilot, J. P. (1986). Fast ion transport in $LiZr_2(PO_4)_3$: Structure and conductivity. *Materials Research Bulletin*, *21*(3), 365–371. https://doi.org/10.1016/0025-5408(86)90194-7

Quartarone, E., & Mustarelli, P. (2011). Electrolytes for solid-state lithium rechargeable batteries: Recent advances and perspectives. *Chemical Society Reviews*, *40*(5), 2525. https://doi.org/10.1039/c0cs00081g

Rosero-Navarro, N. C., Kajiura, R., Miura, A., & Tadanaga, K. (2020). Organic-inorganic hybrid materials for interface design in all-solid-state batteries with a garnet-type solid electrolyte. *ACS Applied Energy Materials*, *3*(11), 11260–11268. https://doi.org/10.1021/acsaem.0c02164

Sadiq, M., Raza, M. M. H., Chaurasia, S. K., Zulfequar, M., & Ali, J. (2021). Studies on flexible and highly stretchable sodium ion conducting blend polymer electrolytes with enhanced structural, thermal, optical, and electrochemical properties. *Journal of Materials Science: Materials in Electronics*, *32*(14), 19390–19411. https://doi.org/10.1007/s10854-021-06456-7

Saikia, D., Ho, S.-Y., Chang, Y.-J., Fang, J., Tsai, L. D., & Kao, H. M. (2016). Blending of hard and soft organic: Inorganic hybrids for use as an effective electrolyte membrane in lithium-ion batteries. *Journal of Membrane Science*, *503*, 59–68. https://doi.org/10.1016/j.memsci.2015.12.024

Sanchez, C., Belleville, P., Popall, M., & Nicole, L. (2011). Applications of advanced hybrid organic: Inorganic nanomaterials: From laboratory to market. *Chemical Society Reviews*, *40*(2), 696. https://doi.org/10.1039/c0cs00136h

Saroj, A. L., Chaurasia, S. K., Kataria, S., & Singh, R. K. (2016). Isothermal and non-isothermal crystallization kinetics of PVA+ionic liquid [BDMIM][BF_4]-based polymeric films. *Phase Transitions*, *89*(6), 578–597. https://doi.org/10.1080/01411594.2015.1080260

Shalu, S., Singh, R. K., & Dhar, R. (2021). Momentous past and key advancements in ionic liquid mediated polymer electrolyte for application in energy storage. *International Journal of Energy Research*, *45*(11), 15646–15675. https://doi.org/10.1002/er.6833

Sharma, J., & Hashmi, S. (2019). Magnesium ion-conducting gel polymer electrolyte nanocomposites: Effect of active and passive nanofillers. *Polymer Composites*, *40*(4), 1295–1306. https://doi.org/10.1002/pc.24853

Singh, M. K., & Chaurasia, S. K. (2021). Performance of ionic liquid: Based quasi-solid-state hybrid battery supercapacitor fabricated with porous carbon capacitive cathode and proton battery anode. *Energy Storage*. https://doi.org/10.1002/est2.310

Singh, V. K., Shalu, S., Chaurasia, S. K., & Singh, R. K. (2016). Development of ionic liquid mediated novel polymer electrolyte membranes for application in Na-ion batteries. *RSC Advances*, *6*(46), 40199–40210. https://doi.org/10.1039/C6RA06047A

Tripathi, A. K. (2021). Ionic liquid: Based solid electrolytes (ionogels) for application in rechargeable lithium battery. *Materials Today Energy*, *20*, 100643. https://doi.org/10.1016/j.mtener.2021.100643

Verduzco, J. C., Vergados, J. N., Strachan, A., & Marinero, E. E. (2021). Hybrid polymer-garnet materials for all-solid-state energy storage devices. *ACS Omega*, *6*(24), 15551–15558. https://doi.org/10.1021/acsomega.1c01368

Wang, Y., & Zhong, W. H. (2015). Development of electrolytes towards achieving safe and high-performance energy-storage devices: A review. *ChemElectroChem*, *2*(1), 22–36. https://doi.org/10.1002/celc.201402277

Wen, S., Guan, W., Wang, J., Lang, Z., Yan, L., & Su, Z. (2012). Theoretical investigation of structural and electronic propertyies of $[PW_{12}O_{40}]^{3-}$ on graphene layer. *Dalton Transactions*, *41*(15), 4602. https://doi.org/10.1039/c2dt12465c

Wright, P. V. (1975). Electrical conductivity in ionic complexes of poly(ethylene oxide). *British Polymer Journal*, *7*(5), 319–327. https://doi.org/10.1002/pi.4980070505

Xie, H., Yang, C., Fu, K. K., Yao, Y., Jiang, F., Hitz, E., Liu, B., Wang, S., & Hu, L. (2018). Flexible, scalable, and highly conductive garnet-polymer solid electrolyte templated by bacterial cellulose. *Advanced Energy Materials*, *8*(18), 1703474. https://doi.org/10.1002/aenm.201703474

Yan, S., Yim, C.-H., Pankov, V., Bauer, M., Baranova, E., Weck, A., Merati, A., & Abu-Lebdeh, Y. (2021). Perovskite solid-state electrolytes for lithium metal batteries. *Batteries*, *7*(4), 75. https://doi.org/10.3390/batteries7040075

Yang, G., Chanthad, C., Oh, H., Ayhan, I. A., & Wang, Q. (2017). Organic: Inorganic hybrid electrolytes from ionic liquid-functionalized octasilsesquioxane for lithium metal batteries. *Journal of Materials Chemistry A*, *5*(34), 18012–18019. https://doi.org/10.1039/C7TA04599A

Yang, G., Fan, B., Liu, F., Yao, F., & Wang, Q. (2018). Ion pair integrated organic-inorganic hybrid electrolyte network for solid-state lithium ion batteries. *Energy Technology*, *6*(12), 2319–2325. https://doi.org/10.1002/ente.201800431

Yang, G., Song, Y., Wang, Q., Zhang, L., & Deng, L. (2020). Review of ionic liquids containing, polymer/inorganic hybrid electrolytes for lithium metal batteries. *Materials & Design*, *190*, 108563. https://doi.org/10.1016/j.matdes.2020.108563

Yao, P., Yu, H., Ding, Z., Liu, Y., Lu, J., Lavorgna, M., Wu, J., & Liu, X. (2019). Review on polymer-based composite electrolytes for lithium batteries. *Frontiers in Chemistry*, *7*, 522. https://doi.org/10.3389/fchem.2019.00522

Yi, Q., Zhang, W., Li, S., Li, X., & Sun, C. (2018). Durable sodium battery with a flexible $Na_3Zr_2Si_2PO_{12}$—PVdF—HFP composite electrolyte and sodium/carbon cloth anode. *ACS Applied Materials & Interfaces*, *10*(41), 35039–35046. https://doi.org/10.1021/acsami.8b09991

Yoon, I. N., Song, H., Won, J., & Kang, Y. S. (2014). Shape dependence of SiO_2 nanomaterials in a quasi-solid electrolyte for application in dye-sensitized solar cells. *The Journal of Physical Chemistry C*, *118*(8), 3918–3924. https://doi.org/10.1021/jp4104454

Young, W. S., Kuan, W. F., & Epps, T. H. (2014). Block copolymer electrolytes for rechargeable lithium batteries. *Journal of Polymer Science Part B: Polymer Physics*, *52*(1), 1–16. https://doi.org/10.1002/polb.23404

Zhan, X., Zhang, J., Liu, M., Lu, J., Zhang, Q., & Chen, F. (2019). Advanced polymer electrolyte with enhanced electrochemical performance for lithium-ion batteries: Effect of nitrile-functionalized ionic liquid. *ACS Applied Energy Materials*, *2*(3), 1685–1694. https://doi.org/10.1021/acsaem.8b01733

Zhang, D., Xu, X., Qin, Y., Ji, S., Huo, Y., Wang, Z., Liu, Z., Shen, J., & Liu, J. (2020). Recent progress in organic: Inorganic composite solid electrolytes for all-solid-state lithium batteries. *Chemistry: A European Journal*, *26*(8), 1720–1736. https://doi.org/10.1002/chem.201904461

Zhang, S., Sun, J., Zhang, X., Xin, J., Miao, Q., & Wang, J. (2014). Ionic liquid-based green processes for energy production. *Chemical Society Reviews 43*(22), 7838–7869. https://doi.org/10.1039/C3CS60409H

Zhang, Y., Josien, L., Salomon, J.-P., Simon-Masseron, A., & Lalevée, J. (2021). Photopolymerization of zeolite/polymer-based composites: Toward 3D and 4D printing applications. *ACS Applied Polymer Materials*, *3*(1), 400–409. https://doi.org/10.1021/acsapm.0c01170

Zhao, Y., Huang, Z., Chen, S., Chen, B., Yang, J., Zhang, Q., Ding, F., Chen, Y., & Xu, X. (2016). A promising PEO/LAGP hybrid electrolyte prepared by a simple method for all-solid-state lithium batteries. *Solid State Ionics*, *295*, 65–71. https://doi.org/10.1016/j.ssi.2016.07.013

Zhao, Y., Wu, C., Peng, G., Chen, X., Yao, X., Bai, Y., Wu, F., Chen, S., & Xu, X. (2016). A new solid polymer electrolyte incorporating $Li_{10}GeP_2S_{12}$ into a polyethylene oxide matrix for all-solid-state lithium batteries. *Journal of Power Sources*, *301*, 47–53. https://doi.org/10.1016/j.jpowsour.2015.09.111

Zheng, J., & Hu, Y. Y. (2018). New insights into the compositional dependence of li-ion transport in polymer: Ceramic composite electrolytes. *ACS Applied Materials & Interfaces*, *10*(4), 4113–4120. https://doi.org/10.1021/acsami.7b17301

Zheng, J., Wang, P., Liu, H., & Hu, Y.-Y. (2019). Interface-enabled ion conduction in li $_{10}GeP_2S_{12}$: Poly(ethylene Oxide) hybrid electrolytes. *ACS Applied Energy Materials*, *2*(2), 1452–1459. https://doi.org/10.1021/acsaem.8b02008

Zhou, Q., Zhang, J., & Cui, G. (2018). Rigid: Flexible coupling polymer electrolytes toward high-energy lithium batteries. *Macromolecular Materials and Engineering*, *303*(11), 1800337. https://doi.org/10.1002/mame.201800337

Zhu, P., Yan, C., Dirican, M., Zhu, J., Zang, J., Selvan, R. K., Chung, C.-C., Jia, H., Li, Y., Kiyak, Y., Wu, N., & Zhang, X. (2018). $Li_{0.33}La_{0.557}TiO_3$ ceramic nanofiber-enhanced polyethylene oxide-based composite polymer electrolytes for all-solid-state lithium batteries. *Journal of Materials Chemistry A*, *6*(10), 4279–4285. https://doi.org/10.1039/C7TA10517G

3 Ion Dynamics and Dielectric Relaxation in Polymer Composites

Anil Arya, Annu Sharma, A. L. Sharma, and Vijay Kumar

CONTENTS

3.1 INTRODUCTION

Dielectric spectroscopy (DES) is an extremely effective technique for characterizing the molecular dynamics of polymers, monomers and other insulating materials over a very broad frequency range from kilohertz to gigahertz. The ion dynamics and the relaxation in the polymer matrix at the microscopic level can be investigated in the polymer composites by dielectric spectroscopy. To understand the charge storage and ion transport

DOI: 10.1201/9781003208662-4

perfectly, deep insights into dielectric parameters (complex permittivity, complex conductivity, modulus, loss tangent, relaxation time) are vital parameters for quantifying the energy storage ability and for determining the exact electrical transport mechanism of the charge. A polymer composites matrix comprises a polymer host with salt dissolved in it. This salt gets dissociated into ions on interaction with polymer chains. Cations, due to their smaller size, contribute to conduction, while anions due to their bulky size remain immobilized in the matrix. Overall, the charge migration occurs via the coordinating sites provided by the electron-rich group of the polymer chains. Polymer chain relaxation promotes the forward ion migration via the segmental mechanism and hopping mechanism. For faster ion dynamics, shorter relaxation time is favorable and in relation to conductivity, relaxation time is given as $\sigma\tau T = constant$, in any polymer electrolyte system. Conductivity is a crucial parameter for any device and decides the overall performance of the device. Conductivity is linked to the dielectric constant (ε'), dielectric loss (ε''), complex conductivity (σ^*), relaxation time (τ), and the segmental motion of the polymer chain role [Ratner, and Shriver 1988; Wang et al. 2012; Anantha and Hariharan 2005].

This chapter starts with the basics of the dielectric and some characteristic terms essential for a proper understanding of ion dynamics. Subsequently, various physical models proposed by different researchers have been summarized and then some key results have been presented.

3.1.1 Complex Dielectric Permittivity

The complex dielectric permittivity is represented as $\varepsilon^* = \varepsilon' - j\varepsilon''$. The real part of dielectric permittivity (ε') indicates the alignment of dipoles or polarization, and the imaginary part of (ε'') is termed as a dielectric loss. The real and imaginary parts are expressed in terms of the real and imaginary part of impedance as given in equation 3.1.

$$\varepsilon' = \frac{-Z''}{\omega C_o(Z'^2 + Z''^2)} \text{ and } \varepsilon'' = \frac{Z'}{\omega C_o(Z'^2 + Z''^2)} \tag{3.1}$$

The dielectric constant is also linked to the extent of salt dissociation, and in polymer electrolytes, the number of charge carriers depends on the dissociation energy and the dielectric constant. For a high dielectric constant, more free ions are required as expressed by equation 3.2 (Woo et al. 2011).

$$n = n_o \exp\left(-\frac{U}{\varepsilon' kT}\right) \tag{3.2}$$

For the PEO-PVP blend, Arya and Sharma (Arya and Sharma, 2019) evidenced the decrease of the dielectric constant at high frequency. At high frequency, ion dipoles fail to respond to the applied electric field. Now, the ion-ion interaction decreases in transient dipoles (cation-ether group), and only long-range ion charge carriers contribute to conduction (Ngai et al. 2018; Awadhia et al. 2006; Ravi et al. 2011).

In polymer electrolytes, the origin of dielectric loss is linked to dipole response to the field. When the field is applied, then ions migrate along the field direction but on field reversal, all dipoles fail to respond and so half the ion diffusion occurs in the field direction. As a result, heat is generated in the dielectric and is termed dielectric energy loss (i.e. $\varepsilon'' = 0$ for $\omega\tau = 0$). The dielectric loss is explained by a three-step process during a high periodic reversal field. In the first step, the ion is de-accelerated on a change of direction of the field, followed by a steady-state in the second step. Then the ion is again accelerated in the opposite direction and for the zero relaxation time, ε'' is zero (i.e. for $\omega, \rightarrow$ "0) (Ravi et al. 2011). The absence of a relaxation peak in the dielectric loss plot is due to the dominance of the electrode polarization effect which masks the relaxation behavior of polymer composites. Cole-Cole proposed the distribution of relaxation time for Debye processes (Cole and Cole, 1941), given by equation 3.3:

$$\varepsilon_* = \epsilon_\infty + \frac{\Delta\varepsilon}{1+(jx)^{1-\alpha}} \quad 0 \le \alpha < 1 \tag{3.3}$$

Where, α is distribution parameter and $x = \omega\tau$; ω is the angular frequency of the applied field and τ is Debye relaxation time. The real and imaginary parts of the dielectric constant can be obtained by separating equation 3.3 and expressing it as (Cao and Gerhardt, 1990):

$$\varepsilon' = \epsilon_\infty + \frac{\Delta\varepsilon\left(1+x^{1-\alpha}\sin\pi/2\alpha\right)}{1+2x^{1-\alpha}\sin\pi/2\alpha + x^{2(1-\alpha)}} \tag{3.3a}$$

$$\varepsilon'' = \Delta\varepsilon\frac{x^{1-\alpha}\cos\pi/2\alpha)}{1+2x^{1-\alpha}\sin\pi/2\alpha + x^{2(1-\alpha)}} \tag{3.3b}$$

These equations can be written in another form by replacing α with $1-\alpha$ in equation 3.3a, b (Sharma and Thakur, 2015).

The real and imaginary parts of the dielectric constant are also expressed as equations 3.3c and d.

$$\varepsilon' = \epsilon_\infty + \frac{\Delta\varepsilon\left(1+x^{\alpha}\cos\frac{\alpha\pi}{2}\right)}{1+2x^{\alpha}\cos\frac{\alpha\pi}{2} + x^{2\alpha}} \tag{3.3c}$$

$$\varepsilon'' = \Delta\varepsilon\frac{x^{\alpha}\sin\frac{\alpha\pi}{2}}{1+2x^{\alpha}\cos\frac{\alpha\pi}{2} + x^{2\alpha}} \tag{3.3d}$$

Here, ε_s is static dielectric constant ($x \rightarrow 0$), ϵ_∞ is dielectric constant ($x \rightarrow \infty$), $x = \omega\tau$; ω is the angular frequency of an applied field and τ is Debye relaxation time. The relaxation time ($\tau_{\varepsilon'}$) can be extracted from the fitting of experimental data with these equations. Further, molecular relaxation time (τ_M) can be obtained using the equation 3.4:

$$\tau_M = \frac{(2\varepsilon_s + \varepsilon_\infty)}{3\varepsilon_s} \times \tau_{\varepsilon'} \tag{3.4}$$

Here, ε_s and ε_∞ are obtained from the dielectric strength ($\Delta\varepsilon = \varepsilon_s - \varepsilon_\infty$).

3.1.2 Loss Tangent

Loss tangent (tan δ) is the ratio of dielectric loss to dielectric permittivity (= ε″/ ε′), and it consists of a single peak. The information obtained from the loss tangent curve is correlated with ion mobility and ionic conductivity. The area under the loss tangent plot reflects the number of ions contributing to the relaxation process (Ngai et al. 2018). Relaxation peak gives an idea about the relaxation time. A molecule comprises translational, vibrational and rotational/orientation motion in the absence of a field (E = 0). But, with the application of a field (E ≠ 0), molecules or dipoles fail to get aligned along the field direction. When the frequency of the applied external electric field and frequency of molecule rotation matches properly, then proper alignment is achieved. At this instant maximum power transfer to the molecular dipoles from the applied field occurs and heat is produced in the system which is termed loss (Singh, 2012). The relaxation time can be obtained by observing the relaxation peak frequency. For deep analysis, fitting the tan δ vs. frequency plot with equation 3.5 can be done:

$$\tan\delta = \frac{(r-1)}{r+x^2}x \tag{3.5}$$

Where r is the relaxation ratio ($\varepsilon_s / \varepsilon_\infty$), ε_s is static dielectric constant ($x \rightarrow 0$), ε_∞ is dielectric constant ($x \rightarrow \infty$), $x = \omega\tau$; ω is the angular frequency of the applied field and τ is Debye relaxation time (reciprocal of jump frequency in the absence of an external electric field).

This equation proposed by Debye provides satisfactory fitting for a single particle and non-interacting system (null interaction between dipoles). But, in the low-frequency window, the Debye model is not in good agreement owing to the presence of multi-type dipole polarization. To examine the broad peak, the ideal Debye equation is modified for better simulation of experimental results. Therefore, to meet the experimental needs it becomes essential to do certain empirical modifications by adding some parameters, as one parameter is used in Cole-Cole, Davidson-Cole, Williams-Wats, and two parameters in Havriliak-Negami fluctuations (Hill and Jonscher, 1983). A modified equation was proposed and was in good agreement with the experimental results in the whole frequency window (Arya and Sharma, 2018). So, in this ideal Debye equation shape, parameter α is added as the power law exponent with value $0 \leq \alpha \leq 1$ to fit the broad tangent delta plot (equation 3.6). This proposed empirical equation and the presence of this factor confirm the presence of more strong interaction in our system. The modified equation is expressed as equation 3.6:

$$\tan\delta = \left(\frac{(r-1)}{r+x^2}x\right)^\alpha \tag{3.6}$$

Figure 3.1 displays a close agreement of measured and fitted results. The fitted curve is represented by the solid red lines and describes the observed results with accuracy. The negligible slight deviation on the low-frequency side probably may be due to the electrode polarization. One can return to the Debye model for α =1 in equation 3.6. Figure 3.1 shows the loss tangent plot for the PEO and PVP blend complexed with LiBOB salt (Arya and Sharma, 2019).

It may be noted that all samples exhibit the same trend with a single relaxation peak at a particular frequency. For better analysis, this plot is divided into three regions: (i) lower frequency region increase of loss tangent is attributed to the dominance of the Ohmic component over the capacitive component, (ii) the peak in the plot is associated with the maximum transfer of energy for the particular frequency ($\omega\tau$ = 1) on the application of the field and (iii) at high frequency, decrease is attributed to the independent nature of the Ohmic part and growing nature of the reactive component (Chopra et al. 2003). With the addition of salt, relaxation peak shifts towards high frequency, and it suggests a decrease in relaxation time. This is also in correlation with the conductivity value which demonstrates maxima at the lowest relaxation time (Arya and Sharma, 2018). The modified equation fits the loss tangent plot (solid line) in the whole frequency window, in contrast to the ideal Debye equation which is not reliable at the low-frequency window.

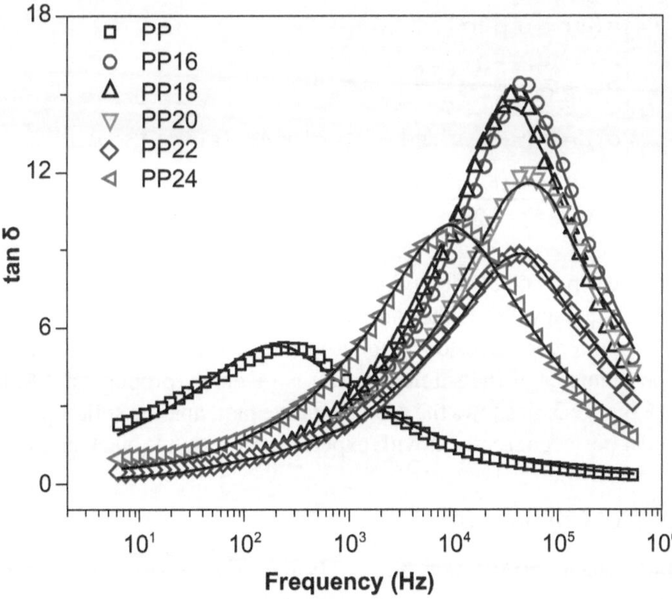

FIGURE 3.1 Frequency dependence of the tangent delta loss (tan δ) for blend polymer electrolyte.

Source: Reproduced with permission from Arya and Sharma [27(4), 334–345, 2019, Copyright Springer].

3.1.3 Sigma Representation

The suppressed features of the Cole-Cole plot at high frequencies are explored further with a new representation termed as Sigma representation. The Cole-Cole plot (ε'' vs. ε') is very useful for the materials possessing relaxation processes and following the Debye equation's Cole-Cole plot (ε'' vs. ε') provides significant information. But for high ionic conductivity polymer composites, Cole-Cole representation becomes less useful owing to the presence of dc conductivity which leads to a divergence of ε'' lower frequencies. σ-representation (σ'' vs. σ') successfully describes the divergence in the Cole-Cole plot (Wei and Sridhar, 1993). As both σ'' and σ' involve the multiplication of frequency with real and imaginary dielectric parameters, so high-frequency features suppressed in the Cole-Cole plot become noticeable in a σ-representation plot. It comprises a plot between the real (σ') and imaginary part (σ'') of complex conductivity (σ^*) with varying frequency. In the plot, a low-frequency x-intercept gives dc conductivity (σ_o) and a high-frequency x-intercept gives σ_∞ for $\sigma'' = 0$. The diameter of the semicircle ($D = \sigma_o - \sigma_\infty$) is associated with the relaxation time by the equation $D = [\varepsilon_v(\varepsilon_o - \varepsilon_\infty)]/(2\tau)$ and the large value of the diameter suggests slower relaxation time, hence faster segmental motion of the cation coordinated polymer chain (Figure 3.2a, b). The complex electrical conductivity can be written using the following expression (equations 3.7–3.10):

$$\begin{cases} \sigma(\omega) = \sigma' + i\sigma'' & (3.7) \\ \sigma_\infty = \sigma_0 + \dfrac{\varepsilon_v(\varepsilon_0 - \varepsilon_\infty)}{\tau} = \sigma_0 + \delta & (3.8) \\ \sigma_{ac} = \sigma' = \omega\varepsilon_v\varepsilon'' \text{and } \sigma_{dc} = \sigma'' = \omega\varepsilon_v(\varepsilon' - \varepsilon_\infty) = \omega\varepsilon_v\varepsilon' & (3.9) \\ r = \dfrac{\delta}{2} = \dfrac{\varepsilon_v(\varepsilon_0 - \varepsilon_\infty)}{2\tau} & (3.10) \end{cases}$$

Here, σ' is the real part of conductivity, σ'' is the imaginary part of conductivity, ω is the angular frequency, "r" is the radius of the semicircle. And, when $\sigma'' = 0$ then low-frequency x-intercept gives dc conductivity (σ_o), and high-frequency x-intercept gives σ_∞. The diameter of the semicircle (D) is inversely proportional to the relaxation time (τ). Figure 3.3b shows the representative plot, and the solid line is the fitted plot which is in perfect agreement with experimental results (black circle).

3.1.4 Complex Conductivity

The complex conductivity of the polymer composites is expressed as $\sigma^*(\omega) = \sigma'(\omega) + i\sigma''(\omega)$, $[\sigma' = \sigma_{ac} = \omega\varepsilon_0\varepsilon'' = \omega\varepsilon_0\varepsilon' \tan\delta, \sigma'' = \omega\varepsilon_0\varepsilon' = \omega\varepsilon_0\varepsilon'' \tan\delta]$. Terms have meaning as given earlier. From the frequency-dependent real part of the conductivity plot, dc conductivity (σ_{dc}) can be extracted by drawing the intercept corresponding to frequency-independent conductivity. The high-frequency region follows the well-known Jonscher's power law (JPL) which is the general characteristic of a polymer composite, given by equation 3.11.

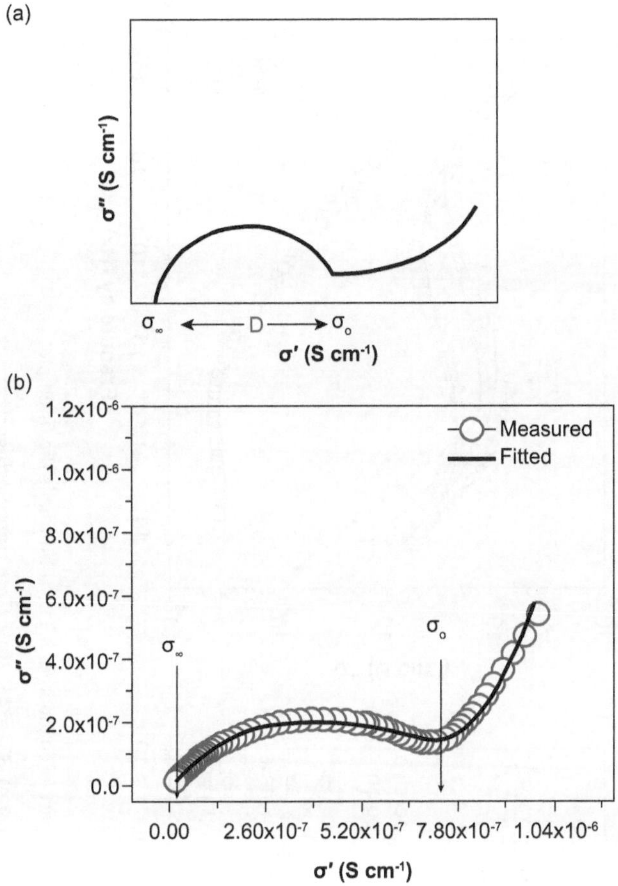

FIGURE 3.2 Sigma representation, (a) ideal case, and (b) PEO-PVC, and x % MMMT.

$$\begin{cases} \sigma_{ac} = \sigma_{dc}\left(1+\left(\omega/\omega_h\right)^n\right) & (3.11) \\ \qquad\qquad\qquad \textit{and} & \\ \sigma_{ac} = 2\sigma_{dc}\ \textit{when}\ \omega = \omega_h & (3.12) \end{cases}$$

Here, σ_{ac} and σ_{dc} are the ac and dc conductivities of electrolyte, while A and n are the frequency-independent Arrhenius constant and the power law exponent is a dimensionless frequency exponent that represents the degree of interaction between mobile ions and their surrounding, where $0 < n < 1$ (the solid line is for the fitting of JPL). For ideal Debye dielectric dipolar-type, value of n is zero and 1 for an ideal ionic-type crystal. At ω_h (hopping frequency) ac conductivity becomes double dc conductivity (eq. 3.12). For some polymers, n may be greater than unity also (Nasri

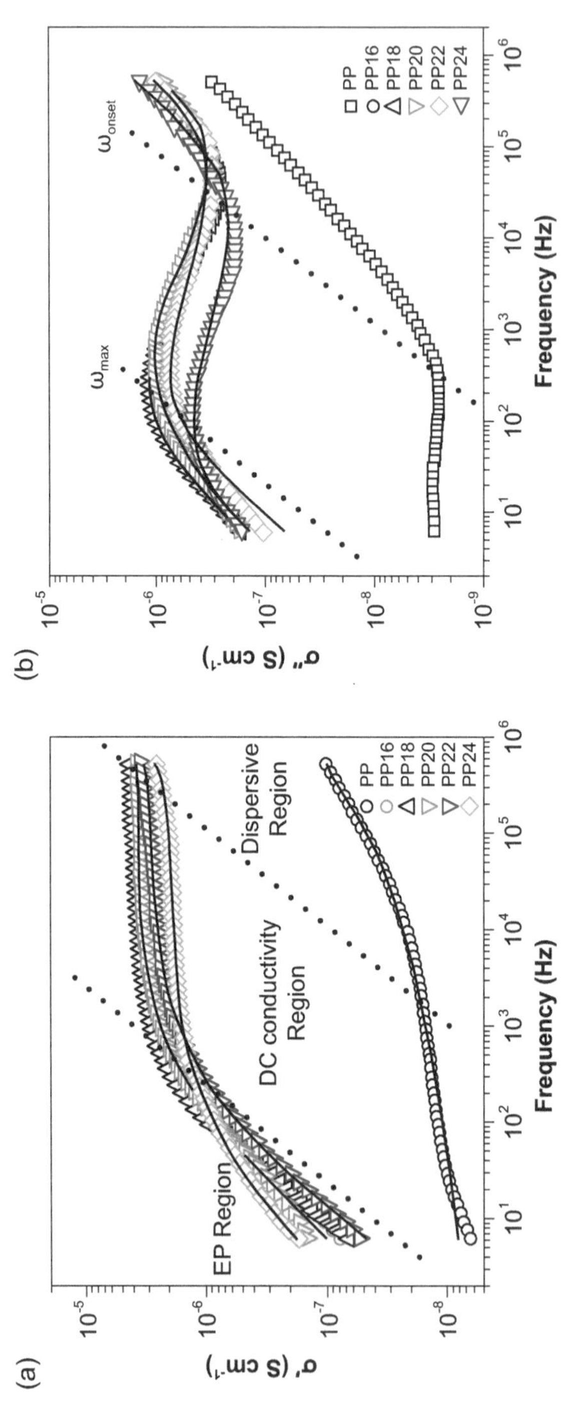

FIGURE 3.3 Representative plot for (a) real and (b) imaginary part of conductivity. The solid line is the fitted plot.

Source: Reproduced with permission from Arya and Sharma [27(4), 334–345, 2019, Copyright Springer].

et al. 2016; Jonscher, 1978; Papathanassiou et al. 2007). Limitation of JPL is that it is valid only at the high-frequency window and it does not consider the contribution of the universal electrode polarization (EP) region dominating in the low-frequency window. So, a modified equation proposed by Roy et al. (Roy et al. 2016) is considered for a better understanding of the ion dynamics parameters. One key feature is that it fits the experimental data in the whole frequency window. The effective complex conductivity can be written as equation 3.13:

$$\sigma^*_{eff} = \left(\frac{1}{\sigma_b} + \frac{1}{i\omega C_{dl}}\right)^{-1} + i\omega C_b \tag{3.13}$$

Now, considering equation 3.13, the real and imaginary part of the conductivity is written as

$$\sigma'(\omega) = \frac{\sigma_b^2 C_{dl}\omega^\alpha \cos\left(\frac{\alpha\pi}{2}\right) + \sigma_b (C_{dl}\omega^\alpha)^2}{\sigma_b^2 + 2\sigma_b C_{dl}\omega^\alpha \cos\left(\frac{\alpha\pi}{2}\right) + (C_{dl}\omega^\alpha)^2} \tag{3.13a}$$

$$\sigma''(\omega) = \frac{\sigma_b^2 C_{dl}\omega^\alpha \sin\left(\frac{\alpha\pi}{2}\right)}{\sigma_b^2 + 2\sigma_b C_{dl}\omega^\alpha \cos\left(\frac{\alpha\pi}{2}\right) + (C_{dl}\omega^\alpha)^2} + \omega C_b \tag{3.13b}$$

The real and imaginary part of conductivity in the high-frequency region has been expressed as

$$\begin{cases} \sigma'(\omega) = \sigma_b\left[1 + \left(\frac{\omega}{\omega_h}\right)^n\right] & (3.14\text{a}) \\ \sigma''(\omega) = A\omega^s & (3.14\text{b}) \end{cases}$$

Here, all parameters have the same meaning as earlier and both "n" and "s" have values less than unity. Now, to investigate the complete frequency response we replace the σ_b in equation 3.13a by equation 3.14a and equation 3.13b by equation 3.14b. Where, C_{dl} is frequency-independent double-layer capacitance, ω is the angular frequency, s and α are exponent terms with value <1 and C_b is the bulk capacitance of solid polymer electrolytes (Wei and Sridhar, 1993).

PEO-PVP polymer blend doped with LiBOB salt was prepared (Arya and Sharma, 2018). Figure 3.3a shows the profile of the real part of conductivity and solid lines are corresponding fits. At low frequency($\omega \rightarrow 0$), ions jump at a faster rate from one coordinating site (–Ö–) to another. For $\omega < \omega_h$ relaxation time increases due to long-range ion transport. But, at high frequency ($\omega \rightarrow \infty$) two competing hopping mechanisms are known; one is unsuccessful hopping when the ion jumps back to its initial position (correlated forward–backward–forward) and another is successful hopping when the neighborhood ions become relaxed concerning the ion's position (the ions stay in the new site; i.e., successful hopping). For, $\omega > \omega_h$ the number of successful hopping is more and indicates a more dispersive ac conductivity (Shukla et al. 2014; Choudhary and Sengwa, 2015). When the frequency exceeds ω_h, σ_{ac} increases proportionally, where $n < 1$. Figure 3.3b displays the variation of the imaginary part (σ'') of conductivity against the frequency and gradual increase in frequency at high frequency for pure polymer electrolyte. With addition of salt, frequency decreases only up to a certain frequency, termed onset frequency (ω_{on}), which corresponds to the start of an electrode polarization mechanism. With further decrease in frequency, maxima in (σ'') is observed for a particular frequency (i.e. maximum frequency; ω_{max}), indicating the complete build-up of polarization (Arya and Sharma, 2018; García-Bernabé et al. 2016; Das and Ghosh, 2017). The σ'' again decreases with a decrease of the frequency. It is important to note that both onset and maximum frequency peak lies toward high frequency as compared to other polymer composites. This suggests the increase in the number of free charge carriers and the effective electrode polarization effect.

Kamboj et al. (Kamboj et al. 2021) investigated the effect of ZrO_2 nanofiller on dielectric properties of polymer nanocomposite electrolyte based on the PEO-$NaPF_6$ matrix prepared via the standard solution cast method. Figure 3.4a shows the plot of the frequency-dependent imaginary part of the conductivity (σ''). In moving from right to left (towards low frequency) in the plot, σ'' decreases rapidly. At a particular frequency (ω_{max}) maximum polarization is achieved and maxima in conductivity are noticed, followed by a decrease in conductivity (Fuentes et al. 2017; Popov et al., 2016). The solid line in the plot is fitted plot, and relaxation time corresponding to ω_{max} and ω_{on} has been evaluated. Figure 3.4b shows the variation in relaxation time and it decreases with increase in temperature. With an increase in temperature, the electrode polarization region also increases. So, temperature plays an effective role in the enhancement of ion dynamics (Namikawa, 1975). It was concluded that ion

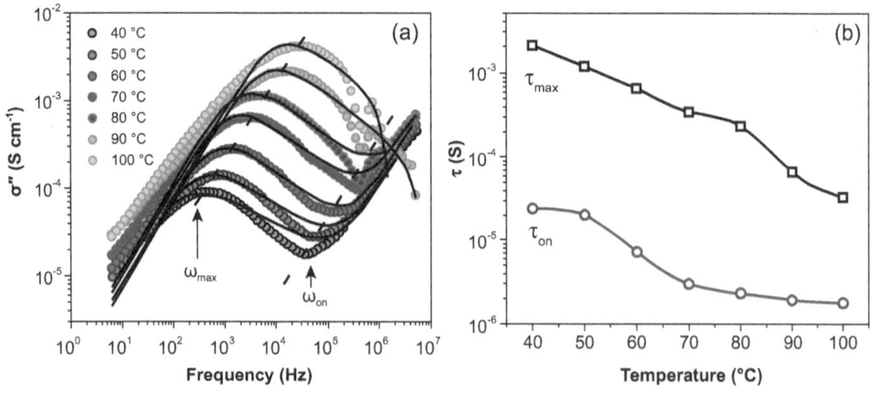

FIGURE 3.4 (a) Temperature-dependent imaginary part and (b) variation of relaxation time.

Source: Reproduced with permission from Kamboj et al. [2021, Copyright Springer].

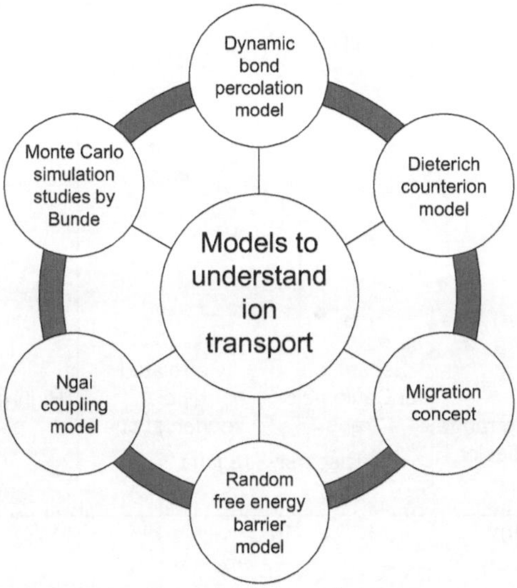

FIGURE 3.5 Proposed models to understand the ion dynamics.

dynamics are influenced by nanofiller and provide easier conduction path access to cations, which is reflected in the enhancement of the conductivity and dielectric properties.

3.2 PHYSICAL MODELS

The ionic transport in the polymer composites is explained with physical models for better understanding. All formalism is based on some key parameters such as the amorphous phase, glass transition temperature, ionic conductivity, type of ion migration mechanism and mobility of polymer host. Figure 3.5 summarizes some key models proposed to get insights into ion dynamics (Rodríguez et al. 2013; Ngai and Kanert, 1992; Knödler et al. 1996; Murch and Dyre, 1989; Ngai, 2015; Funke and Banhatti, 2006).

3.2.1 Dynamic Bond Percolation Theory

This formalism is used to explain ions/electrons diffusion in a disordered system that is going through dynamic rearrangement in a period of time (i.e. smaller than observation time). This model is applicable for polymer electrolytes, as the dynamic motion of polymer chains favors ion diffusion by changing the environment. Three characteristic parameters of this model are (i) average hopping rate, (ii) percentage of available bonds and (iii) mean renewal time (Druger et al. 1983).

3.2.2 Counter-Ion Model

The counter-ion model (CM) was proposed by Dieterich and co-workers in the 1990s. Figure 3.6 shows the conductivity spectra obtained by the CM model (Pendzig and

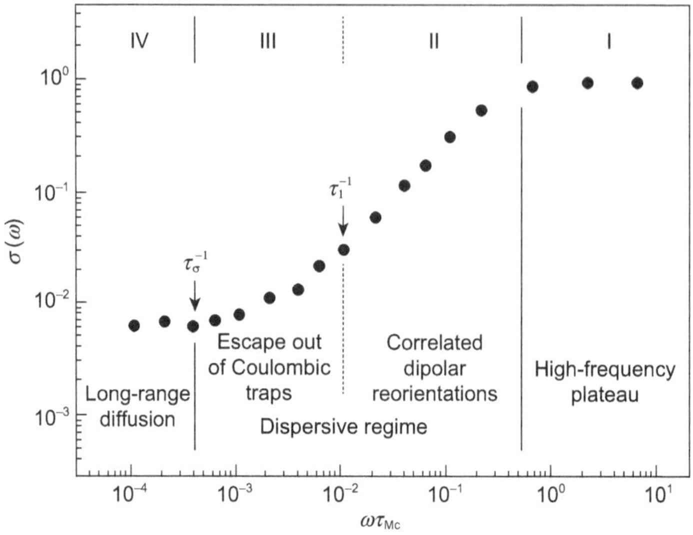

FIGURE 3.6 Illustration of different frequency regimes in the dynamic of the counter-ion model. The two intermediate regimes II and III pertain to "nearly constant loss" and Jonscher-type behavior, respectively.

Source: Reproduced with permission from Pendzig and Dieterich [1998, Copyright Elsevier].

Dieterich, 1998). In a high-frequency window, every hop contributes to conductivity, and a plateau is evidenced. While, at a lower frequency, the dispersive region occurs which comprises two regions: (i) the NCL region, where only correlated dipolar reorientations are observed (n equal to or even slightly larger than 1) and (ii) the Jonscher region, where ions escape out of a Coulombic trap ($n \sim 0.5$–0.6). For long range ion diffusion, dc plateau occurs (Pradel and Ribes, 2014).

3.2.3 Jump Relaxation Model

The jump relaxation model was proposed by Funke and Riess (Funke and Riess, 1984). Here conductivity dispersion is attributed to strong forward-backward jump correlations in the motion of an ion. When an ion migrates from one site to another, equilibrium is restored by two competitive relaxation mechanisms, (i) the surrounding environment can relax due to hopping of other ions, and (ii) the ion can move back to its former position. The longer the ion stays in the new position, the smaller the probability becomes for a backward jump to occur (Funke, 1993).

3.2.4 Random Free Energy Barrier Hopping Model

This model is based on the assumption that ion migration occurs via hopping, and hopping charge carriers are subject to spatially randomly varying energy barriers (Murti et al. 2016). Here, frequency-dependent conductivity is determined by the dc conductivity and the dielectric loss strength. Dyre (Dyre, 1988) solved this model for disordered solids in the continuous time random walk and in the effective medium approximation, and it was concluded that two solutions are almost indistinguishable.

3.3 RELAXATION TYPES

Under an ac field, frequency dispersion or dielectric relaxation is observed due to a number of different polarization mechanisms. The presence of any dielectric relaxation then corresponds to one or more of the possible polarization mechanisms that occur on a microscopic scale. Each relaxation process may be characterized by a relaxation time which describes the decay of its polarization with time in a periodic field. Experimentally, more than one relaxation process can be observed in the frequency range of interest but multiple relaxation times will not be addressed here. Cao and Gerhardt (Cao and Gerhardt, 1990) in 1990 proposed three models to calculate various relaxation times: (i) Debye, (ii) Cole-Cole and (iii) ideal conduction.

3.3.1 Debye Relaxation

Debye relaxation provides the information about the non-interacting population of freely rotating dipoles. The expression for the ideal frequency response of localized oscillation or motion in a condensed medium was given by Debye (Smyth, 1955) as expressed in Figure 3.7a, b. The complex permittivity is expressed as

$$\varepsilon^* = \epsilon_\infty + \frac{\Delta\varepsilon}{1 + jx} \tag{3.15a}$$

$$\varepsilon' = \epsilon_\infty + \frac{\Delta\varepsilon}{1 + x^2} \tag{3.15b}$$

$$\varepsilon'' = \frac{\Delta\varepsilon}{1 + x^2} \tag{3.15c}$$

Here, $x = \omega\tau$, τ is Debye relaxation time denoted as $\tau_{\varepsilon'}$, ε_s and ε_∞ are linked to dielectric strength as $\Delta\varepsilon = \varepsilon_s - \varepsilon_\infty$. We can also calculate $\tau_{\tan\delta}$ (tangent delta plot) and τ_m

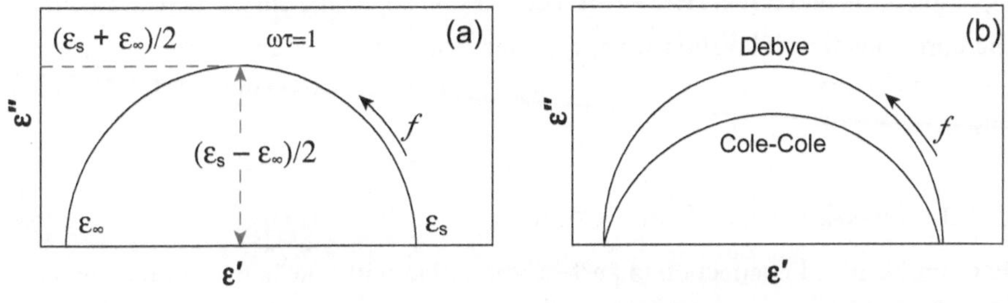

FIGURE 3.7 (a) Cole-Cole diagram; (b) comparison of Debye and Cole-Cole plot.

(modulus time) from using equation $\tau_{\tan\delta} = \tau / \sqrt{r}$ and, $\tau_m = \tau / r$, where r (relaxation ratio) = $\varepsilon_s / \varepsilon_\infty$.

Two other relaxation times can be calculated, τ_z an τ_y. Here, $\tau_z = \tau / r$, and $\tau_y = \tau$. Relation between three relaxation times is $\tau_{\varepsilon'} \geq \tau_y > \tau_{tan\delta} > \tau_z \geq \tau_m$. It suggests that the impedance and modulus plots place emphasis on the high-frequency data, while dielectric constant and admittance plots provide information in low-frequency data. The previously mentioned relation was also verified (Arya and Sharma, 2018). Authors concluded that this analysis of variation of various relaxation times is in absolute agreement with the ionic conductivity value and follows the same trend. Also, one remarkable point to be noted here is that all relaxation times of the same relaxation process follow the order $\tau_{\varepsilon'} > \tau_{tan\delta} > \tau_z > \tau_m$ (Cao and Gerhardt, 1990). It means that the dielectric constant and tangent delta loss peak lies at a lower frequency while the impedance and modulus peak is located at high frequency. So, due to their different relaxation peak positions a relaxation peak may be seen in one process and be absent in another as in modulus spectra.

3.3.2 Cole-Cole Relaxation

Polar dielectrics that have more than one relaxation time do not satisfy Debye equations. Cole and Cole (Cole and Cole, 1941) suggested the distribution of relaxation time when a depressed semicircle (center below x-axis) is observed in plot. The real and imaginary parts are also expressed as given:

$$\varepsilon' = \epsilon_\infty + \frac{\Delta\varepsilon\left(1 + x^{1-\alpha}\sin\frac{\pi\alpha}{2}\right)}{1 + 2x^{1-\alpha}\sin\frac{\pi\alpha}{2} + x^{2(1-\alpha)}} \quad (3.16)$$

$$\varepsilon' = \frac{\Delta\varepsilon\left(1 + x^{1-\alpha}\cos\frac{\pi\alpha}{2}\right)}{1 + 2x^{1-\alpha}\sin\frac{\pi\alpha}{2} + x^{2(1-\alpha)}} \quad (3.17)$$

Here α is distribution parameter; its values lie between 0.2 to 0.5 for most of materials. For single relaxation time, α = 0, then the previous equation becomes same as equation 3.15b and 3.15c. Here $\frac{\pi\alpha}{2}$ is a depression angle of a curve. We can calculate the corresponding relaxation time, $\tau_{tan\delta}$ (tangent delta plot) and τ_m (modulus time) from using equation $\tau_{tan\delta} = \tau / r^{/2}$ and, $\tau_m = \tau / r$, where r (relaxation ratio) = $\varepsilon_s / \varepsilon_\infty$, and = 1 / (1 – α).

3.3.3 Non-Localized Conduction

For non-localized conduction, a perfect semicircle will now appear in the complex impedance plane rather than in the dielectric constant plane (Grant, 1958). Now the dielectric function will be $\varepsilon^* = \epsilon_\infty(1 - jx)$. The relaxation time will be calculated as $\tau = \tau_z = \tau_m$. Now, $\tau_{\varepsilon'}$, τ_y, $\tau_{\tan\delta}$ do not exist since their dielectric functions have no maxima at any x value. The non-localized process can be considered as localized process when $\varepsilon_s \to \infty$, and $\tau_{\varepsilon'} \to \infty$.

3.3.4 Havriliak-Negami Model

This is most popular model till now to analyze the dielectric spectra for broad relaxations, and is a mixture of the Cole-Cole and the Cole-Davidson equations (Havriliak and Negami, 1967; Redondo-Foj et al. 2014). Here, asymmetry and broadness in dielectric curves are demonstrated by a and b values. Although no exact physical meaning is given to α and β.

$$\varepsilon^* = \epsilon_\infty + \frac{\Delta\varepsilon}{[1+(\mathrm{j}x)^\alpha]^\beta} \tag{3.18}$$

Here $\Delta\varepsilon = \varepsilon_s - \varepsilon_\infty$, and $x = \omega\tau$. ε_s and ε_∞ are the relaxed ($\omega = 0$) and unrelaxed ($\omega = \infty$) dielectric permittivities, and τ is the relaxation time of the process. Here, α and β are shape parameters related to distribution of relaxation time which satisfy conditions such as $0 < \alpha \leq 1$, and $\alpha = \beta = 1$ for the Debye process. The parameter α is related to the departure of the Cole-Cole plot from a semi-circumference at low frequencies, while β is related to the skewness of the plot along a straight line, at high frequency. Using the HN equation, relaxation time can be evaluated. This equation can be further modified for n relaxations (Brochier et al. 2010).

$$\varepsilon^* = \epsilon_\infty + \sum_{j=0}^{n} \frac{\Delta\epsilon_j}{[1+(\mathrm{j}x)^{\alpha_j}]^j} \tag{3.19}$$

The general relation comprised of modulus is in inverse relation with complex dielectric permittivity, real and imaginary parts are expressed as $M^* = M' + jM'' = M^* = j\omega C_0 Z^* = \omega C_0 Z'' + j\omega C_0 Z'$. The real and imaginary part are expressed by equation 3.20.

$$M' = \frac{\varepsilon'}{\varepsilon'^2 + \varepsilon''^2} \text{ and } M'' = \frac{\varepsilon''}{\varepsilon'^2 + \varepsilon''^2} \tag{3.20}$$

3.4 FREQUENCY AND TEMPERATURE ON POLARIZATION OF DIELECTRICS

Both frequency and temperature are important parameters that influence the polarization, and hence dielectric constant. This section discusses the ion dynamics influenced by frequency and temperature.

3.4.1 Effect of Frequency

Electronic polarization is the fastest polarization which will complete at the instant the field is applied. The reason is that the electrons are lighter elementary particles

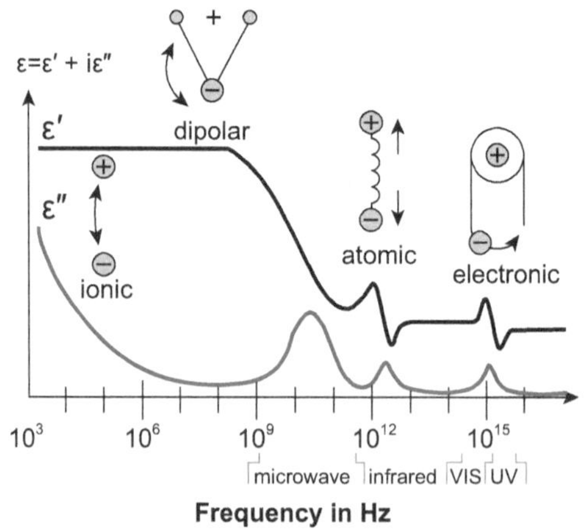

FIGURE 3.8 Schematic diagram of the complex permittivity spectrum vs. frequency, showing the several types of relaxation processes.

than ions. Therefore even for very high frequency (10^{16} Hz), applied electronic polarization occurs during every cycle of the applied field. Ionic polarization is a little slower than electronic polarization as ions are heavier than the electron cloud, and the time taken for displacement is large and occurs up to 10^{13} Hz. For higher frequencies ions do not respond. Orientational polarization is even slower than ionic polarization. This type of polarization occurs only at electrical frequency range (=10^{10} Hz) below microwave region. Space charge polarization is the slowest because it has to diffuse over several atomic distances. This process occurs at very low frequencies (10^4 Hz). Figure 3.8 explains the four types of polarization at different frequency ranges.

It is noted that at low frequencies, all the four types of polarizations occur and total polarization is very high. Total polarization decreases with increase in frequency and becomes minimum at optical frequency range. Dielectric constant also decreases with increase of frequency due to elimination of polarization contribution.

3.4.2 Effect of Temperature

With increase of temperature, electronic and ionic polarization displays no change. Orientational polarization is affected by temperature, and it decreases with increase of temperature due to enhanced disorder which hinders the dipoles orientation along the field, while the space charge region gets enhanced due to faster ion diffusion with increase of temperature. So, orientational and space charge polarization contribution can be altered by temperature. Figure 3.9a, b shows the energy barrier (ΔE) that needs to be crossed for orientation by any molecule. However, with application of field there is a change in the potential energy of the barrier. The potential energy of dipoles aligned with field reduces, and alignment against field increases. It means that small energy is required to cross barriers to align along the field, as compared to alignment against the field.

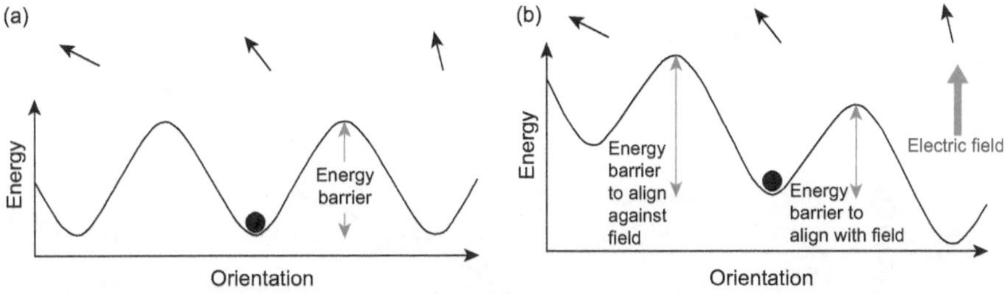

FIGURE 3.9 Orientation of dipoles without and with field.

When the energy of molecules is smaller than the energy barrier, then it fails to orient due to inability to cross the energy barrier. So, dielectric constant decreases owing to no contribution from orientation.

Three important effects of temperature are (i) increase of dielectric constant due to ease of dipole rotation, (ii) faster segmental motion of polymer chain and (iii) improved salt dissociation due to thermal activation (Arya and Sharma, 2019). The salt dissociation is linked with the lowering of the activation energy ($E_a = q^2 / 4\pi\varepsilon_o\varepsilon_s r$) (Fragiadakis et al. 2009). At ambient temperature, the energy barrier is high due to coordination of the cation with the host polymer and the cation migration is linked with the polymer chain motion. While, with increase of temperature polymer flexibility increases which lowers the time taken by cations to migrate from one coordinating site to another. As ion migration occurs via ion hopping, so the hopping distance alters due to thermal activation. Generally, with increase of the temperature average hopping length $\sqrt{\langle\ \rangle}$ decreases.

Another important parameter is hopping potential barrier (ΔE) that the ion must overcome for a successful jump to the forthcoming coordination site. Now the average hopping length (distance between the neighboring ion pairs) decreases with the increase of temperature. The lowering of the hopping length leads to overlap of Coulomb potential wells and ΔE decreases which results in the improved mobility of ions, hence the dc conductivity (Roy et al. 2016; Bruce and Gray, 1995). In brief, the probability of successful hopping for the cation increases that leads to the enhancement of the ionic conductivity which is also supported by lowering of relaxation time (Dam et al. 2016).

3.5 CORRELATION OF HOPPING FREQUENCY AND SEGMENTAL MOTION OF POLYMER CHAIN

Ion dynamics in the polymer composites are linked to hopping mechanism and segmental motion. Two important factors that determine the dynamics are amorphous phase and polymer chain flexibility. Enhancement of amorphous phase and segmental motion lowers the average hopping length (D) $\sqrt{\langle\ \rangle}$. Correlation between the ion hopping and the segmental motion was examined by Debye-Stokes-Einstein (DSE) plot as shown in Figure 3.10 for PEO-PVP blend (Arya and Sharma, 2019). The

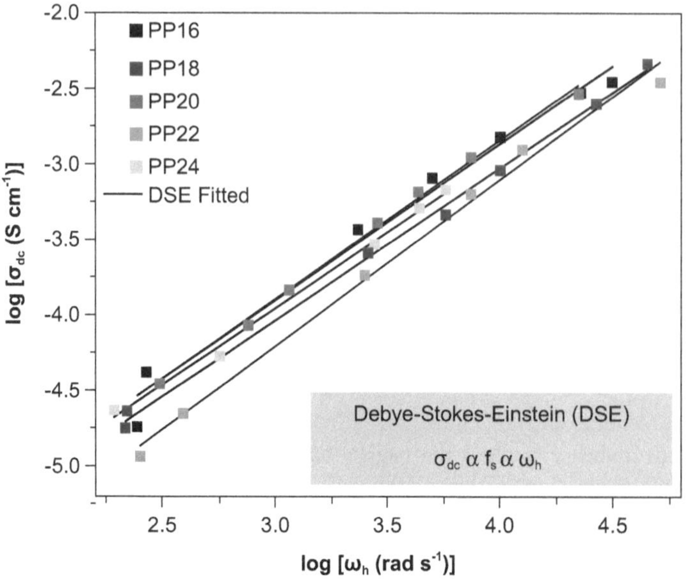

FIGURE 3.10 The plot of dc conductivity against the hopping frequency at different temperatures. Solid lines are the best fit for the DSE equation.

Source: Reproduced with permission from Arya and Sharma [27(4), 334–345, 2019, Copyright Springer].

DSE equation is used for polymer composites with weak dependency to temperature and is expressed as $\sigma_{dc} \propto f_s \propto \omega_h$ (f_s is the egmental frequency) (Fragiadakis et al. 2009). It may be noticed that the plot shows dc conductivity behavior according to the DSE equation for all samples with almost negligible deviation and evidence that the only free cation is contributing to conductivity with negligible contribution from the ion pairs (LaFemina et al. 2016; Yu et al. 2014). The authors concluded that at particular frequency all systems depict the coupling of the dc conductivity with the hopping frequency and only the cation is playing the key role in the enhancement of ion dynamics.

For further insights of ion dynamics, an interaction scheme based on the experimental results was proposed in two ways: (i) polymer segmental motion and (ii) hopping barrier (Arya and Sharma, 2019). In starting, ion migration occurs via the coordinating sites of the host polymer (Figure 3.11a). Distance between two coordinating sites is known as hopping length (D_1) and corresponding relaxation time of the polymer chains is τ_1. With increase of temperature, the polymer matrix switches from a rigid to a flexible nature, and hopping length of charge carriers decreases ($D_2 < D_1$) due to more coordinating sites' availability (Figure 3.11a). Hence ion migration is rapid and is confirmed by decrease of the relaxation time ($\tau_2 < \tau_1$). Hopping a potential barrier is the potential that must be overcome by the cation for successful hopping, and is a critical parameter. For lower temperature hopping, potential is E_1 and ions have to cross this barrier for successful hopping (Figure 3.11b). With

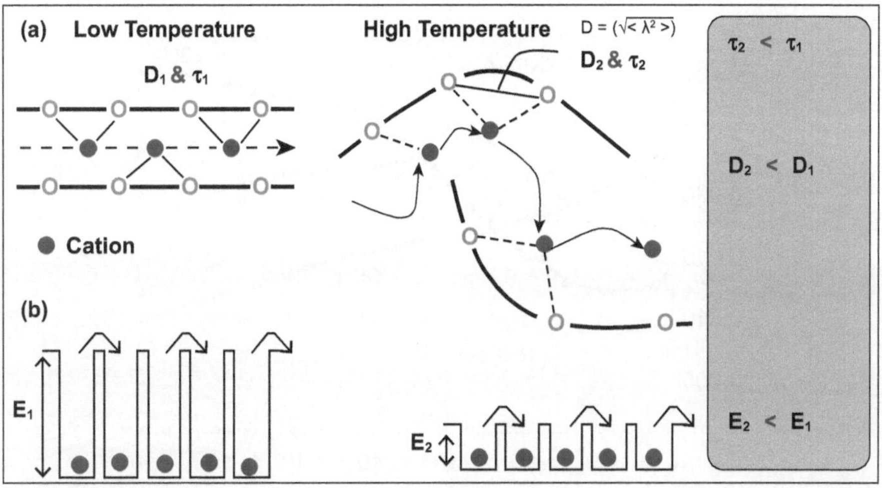

FIGURE 3.11 Temperature-dependent ion transport mechanism.

Source: Reproduced with permission from Arya and Sharma [27(4), 334–345, 2019, Copyright Springer].

increase of temperature, decrease in average hopping length results in overlapping of the potential barriers and hence overall barrier potential decreases as indicated by E_2. This reduction in the potential barrier ($E_2 < E_1$) suggests enhanced ion mobility, and hence high ionic conductivity. Temperature enhances the dielectric properties and is a combination of the three mechanisms, (i) increase of the polymer flexibility, (ii) decrease in hopping length and (iii) lowering of hopping potential barrier.

3.6 SEGMENTAL DYNAMICS AND CONDUCTIVITY MECHANISM

Polymers confined in pore systems show two phenomena: (i) slowing down of dynamics and (ii) faster relaxation time below glass transition temperature. As per IUPAC recommendations, micro, meso and macroporous materials are those characterized by pore diameters <2, 2–50, and >50 nm, respectively. The properties of polymer under confinement are influenced by the pore morphology, pore size and pore chemistry. So, keeping this in mind, the molecular confinement was examined by Bujan et al. (Barroso-Bujans et al. 2016). Authors examined the effect of Resorcinol-Formaldehyde Resin Nanoparticles (RNP) pore size on the dynamics and structure of poly(ethylene oxide) intercalated in the porous structure of this organic gel. RNPs exhibit a globular morphology, and intercalation of PEO was confirmed in the interstices between RNPs.

The dielectric response was examined by Cole-Cole function via broadband dielectric spectroscopy. Figure 3.12a shows the variation of imaginary part (ε'') of the dielectric permittivity of dry neat RNPs (as empty circles), and depicts temperature-independent

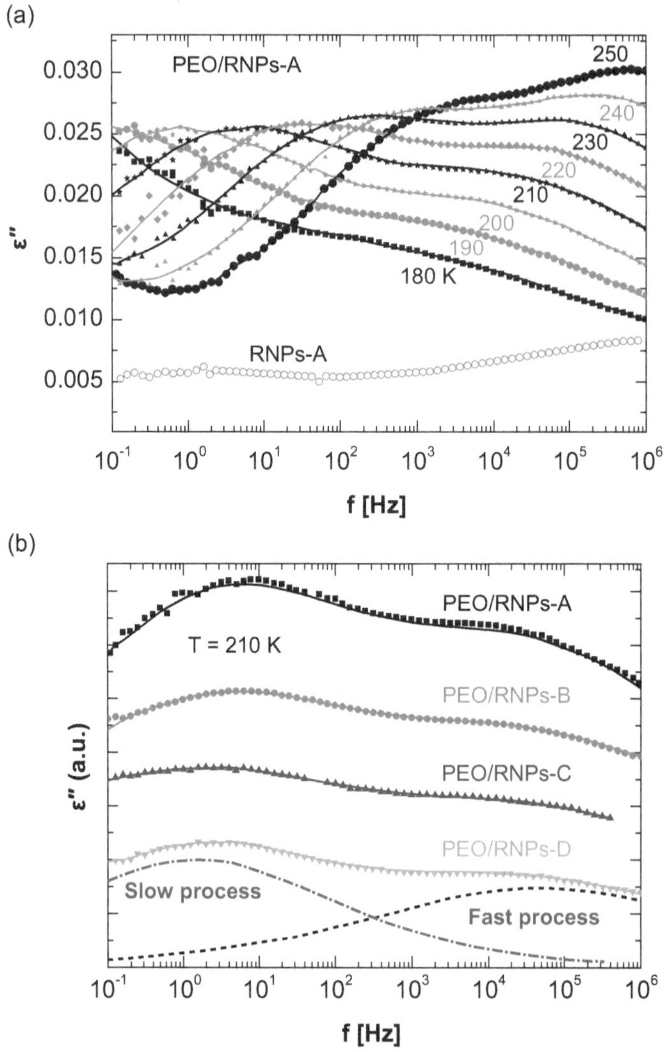

FIGURE 3.12 (a) Imaginary part of the complex dielectric permittivity of PEO/RNPs-A at different temperatures. Solid lines through the data points represent the fits to the experimental data. Data of neat RNPs-A are included for comparison. (b) Imaginary part of the complex dielectric permittivity and their corresponding fitting curves at 210 K for all the PEO/RNPs samples. Slow and fast processes of PEO/RNPs-D are shown by dash and dash-dotted lines, respectively.

Source: Reproduced with permission from Barroso-Bujans, et al. [2016, Copyright American Chemical Society, 49(15), 5704–5713].

low and almost flat permittivity losses. After incorporation of PEO into RNPs, two relaxation processes are noticed, and a shift towards a high-frequency window with increase of temperature. The relaxation behavior for different PEO/RNPs (at T = 210 K) is shown in Figure 3.12b, and fast and slow relaxation process are observed. The

α-parameter in the Cole-Cole function corresponding to slow relaxation is about 0.4, and 0.23 corresponding to fast component at 200 K. Relaxation time for PEO/RNPs is 10 times smaller than bulk PEO.

Choi et al. (Choi et al. 2016) examined the segmental dynamics of a series of siloxane-based polar copolymers combining pendant cyclic carbonates and short poly(ethylene oxide) (PEO) chains via dielectric relaxation spectroscopy (DRS). The cyclic carbonate (CECA, having four oxygens) monomer exhibits slightly higher static dielectric constant (ε_s) than propylene carbonate (PC, having three oxygens) and much higher ε_s than oligomeric PEG13. Incorporation of more PEO3 side chain lowers the static dielectric constant ε_s and the glass transition temperature T_g. Dielectric contestant decreases with increase of temperature as evaluated by the Onsager equation, and is attributed to the thermal dipole randomization. Then dielectric relaxation process were examined by fitting dielectric loss plot with the Havriliak-Negami equation. All samples (activation energy = ~ 33 kJ/mol) exhibit a single secondary β relaxation at lower temperatures and a single segmental α relaxation at higher temperatures. Three dielectric relaxation processes, α (linked to segmental motions of the amorphous parts of PEO), β (more local segmental motions of polymer chains in amorphous regions) and γ (local intramolecular twisting motion of ethylene ($-CH_2-CH_2-$) parts or local motions of the chain ends of the polymer) are observed in PEO (Ishida et al. 1965; Heaton et al. 1996). Money et al. (Money et al. 2012) examined the effect of δ-Al_2O_3 nanofillers in tuning the dc conductivity, glass transition and dielectric relaxations in the polymer electrolyte $(PEO)_4$:$LiClO_4$. DSC analysis confirmed the decrease of T_g with nanofiller content, and is attributed to the speed up of the structural (α) relaxation of the polymer chains owing to filler-polymer interaction dominating glass transition temperature. Both dc conductivity (σ_{dc}), and hopping rate (f_h) increase with nanofiller addition. Also, increase in mobile concentration factor (K) suggests increase in free charge carriers with nanofiller addition owing to dissociation of ion clusters via nanofiller surface groups (–OH). Modulus formalism confirmed that the conductivity relaxation is a temperature independent dynamical process for polymer electrolyte with and without Al_2O_3 nanofillers. Addition of nanofiller alters the local environment within polymer chains, rather than in absence of nanofiller. The conductivity as well as relaxation processes in polymer electrolytes, the imaginary part of the dielectric permittivity (ε''), was analyzed using a sum of a conductivity term, a Havriliak-Negami function for the α-process and Cole-Cole functions for the β- and γ-processes. Finally, fitting function is given as

$$\varepsilon'' = \frac{\sigma_{dc}}{\omega\epsilon_{o\,o}} + \frac{\Delta\varepsilon_\alpha}{\left[1+\left(i\omega\tau_\alpha\right)^a\right]^b} + \frac{\Delta\varepsilon_\beta}{1+(i\omega\tau_\beta)^a} + \frac{\Delta\varepsilon_\gamma}{1+(i\omega\tau_\gamma)^a} \tag{3.21}$$

Here σ_{dc} is dc conductivity, τ_α, τ_β, τ_γ are relaxation times for α, β and γ-processes. The shape parameter "a" determines the symmetric broadening of the process, and the parameter "b" controls the asymmetric broadening. The analysis of dielectric permittivity data confirmed the presence of three relaxations. Except PEO, all polymer electrolytes show absence of α-relaxation. Increase in intensity of the β-relaxation, and shift toward high frequency with addition of δ-Al_2O_3 concentration was noticed. This indicates the speed up of the relaxation process of polymer chains in amorphous phase, while the γ-relaxation depicts no change in peak position and intensity. Figure 3.13 shows the temperature dependences of α-, β- and γ-relaxation times in PEO and PEO with fillers.

The α-relaxation of both samples shows a non-Arrhenius temperature dependence, and with the addition of the nanofiller relaxation process speeds up what is the signature of increased free volume around polymer chains. While, both β- and γ-relaxation times depict Arrhenius temperature dependences, and do not show any significant change with addition of nanofillers. From the temperature dependence of the conductivity relaxation, "strength parameter" D was evaluated from VTF equation (reciprocally related to fragility). This is related to the "strong" and "fragile" corresponding to change in dynamic properties (viscosity, structural relaxation time and diffusion constant) above T_g. The increase in D value from 6.6 (nanofiller free) to 9.6 (4 wt % nanofiller) indicates the decrease in fragility of the polymer electrolyte system, which indicates that the polymer-filler interaction is nonattractive (Agapov and Sokolov, 2011; Starr and Douglas, 2011).

The conductivity behavior of polymer composites was also examined by exploring the dielectric properties (Chilaka and Ghosh, 2014). They synthesized the semi-Interpenetrating Polymer Network (IPN) of [poly(ethylene glycol)-polyurethanepolymethylmethacrylate] [60:40]-montmorillonite (MMT) nanocomposites. With addition of MMT, a shift in T_g toward low temperature suggested the enhanced amorphous content. The real and imaginary part of a dielectric permittivity

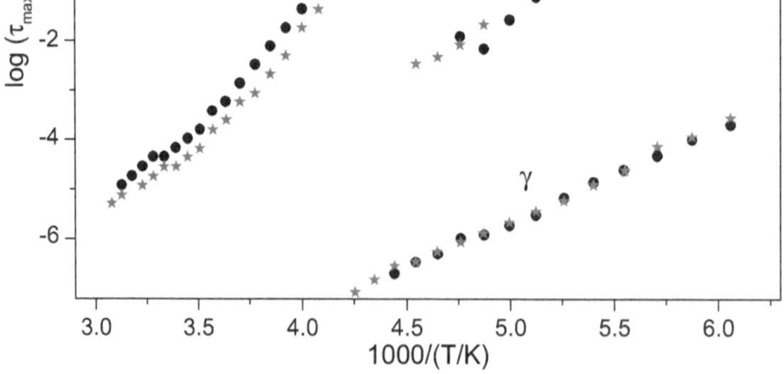

FIGURE 3.13 Temperature dependences of the α-, β- and γ-relaxation times in PEO and PEO with 4 wt % δ-Al2O3.

Source: Reproduced with permission from Money et al. [2012, Copyright American Chemical Society, 116(26), 7762–7770].

plot displayed the decrease with increase of frequency. High value of dielectric permittivity at low frequency is attributed to the presence of electrode polarization and space charge polarization owing to charge accumulation at electrode-electrolyte interface. It confirms the non-Debye-type dependence while decrease in high frequency is due to reduction of charge polarization (Choudhary and Sengwa, 2011). Further, the peak corresponding to relaxation in loss tangent (tan d) plot shifts towards high frequency with addition of MMT, which indicates the reduction of relaxation time. The sample with 5% MMT shows the lowest relaxation time (1.60×10^{-4} s), and highest dc conductivity (1.09×10^{-5} S cm^{-1}), which suggests increase of free charge carriers. Figure 3.14 shows the plot of the real part of complex conductivity against frequency. Increase of conductivity in the low-frequency window is due to a polarization effect, while at mid-frequency a frequency-independent dispersion region is due to true bulk (dc) conductivity of the samples. While at high frequency, conductivity again increases due to fast reversal of applied field.

Dam et al. (Dam et al. 2016) investigated a series of ion conducting polymer-clay composites (PEO_{20}-$LiCF_3SO_3$) prepared via solution casting technique. Further, relaxation dynamics and the ionic transport mechanism are examined via broadband dielectric spectroscopy over a wide frequency and temperature range. Sometimes, the high value of ionic conductivity masks the relaxation processes observed in

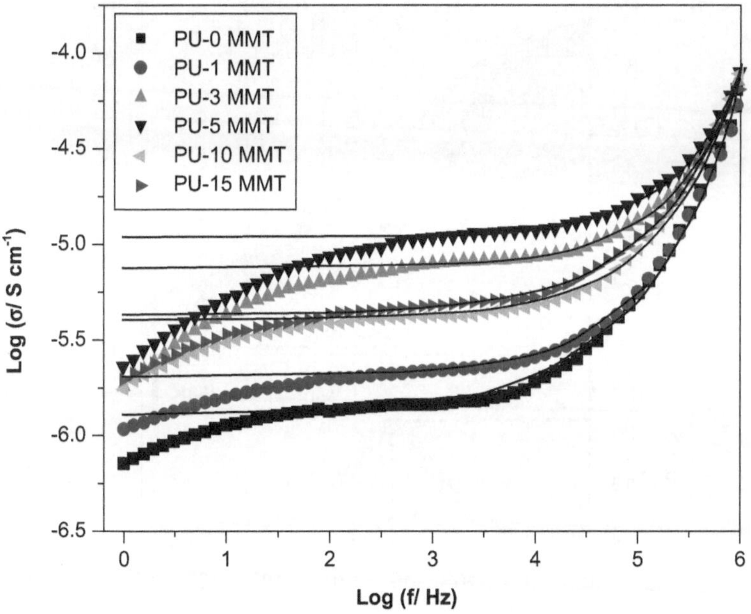

FIGURE 3.14 Conductivity versus log frequency graph for PU-xMMT with different MMT weight percent (x is 0, 1, 3, 5, 10 and 15 wt %) simulated with NLSF method using the UPL equation. The solid line represents the fit of experimental data to the Jonscher power law.

Source: Reproduced with permission from Chilaka and Ghosh [2014, Copyright Elsevier, 134, 232–241].

complex dielectric spectra so to examine the accurate relaxations the dc conduction free dielectric loss formalism is adopted and is expressed as

$$\varepsilon''_{\text{der}} = -\frac{\pi}{2} \times \frac{\partial \varepsilon'(\omega)}{\partial(\ln \omega)} \approx \varepsilon'' \tag{3.22}$$

So, the plot is fitted with the HN equation to extract information about relaxation phenomena due to the EP effect and segmental motion of the host polymer (Figure 3.15a). The value of shape parameters for EP relaxation is near unity, and is the signature of non-interacting Debye-type relaxation. While, for segmental relaxation shape parameters, values around 0.7 to 0.75 confirm the non-Debye-type relaxation. Figure 3.15b shows the loss tangent plot, and ion diffusivity is obtained by using the MacDonald-Trukhan model along with the Nernst-Einstein (NE) relation as given in equation 3.23.

$$D = \frac{\pi f_{max} L^2}{(16 \tan \partial_{max})^3} \tag{3.23}$$

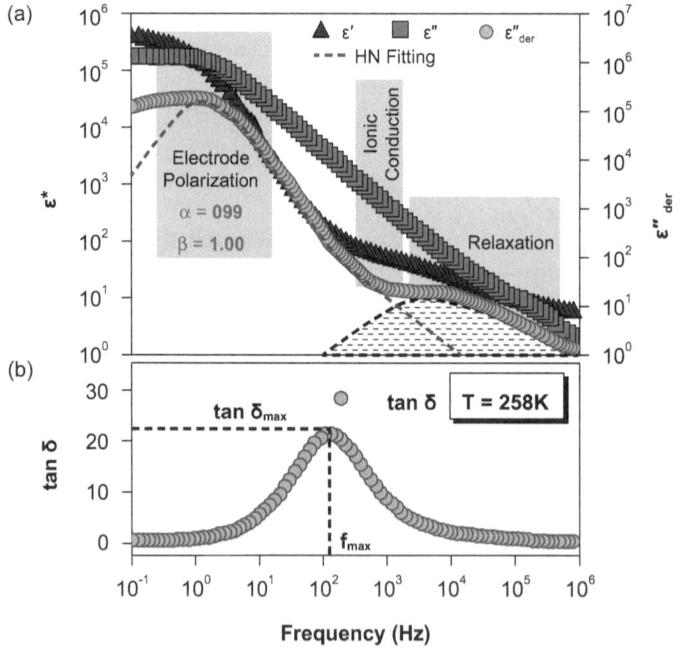

FIGURE 3.15 (a) Complex dielectric spectra with dc conduction free dielectric loss, (b) loss tangent (tan δ) of PEO_20-$LiCF_3SO_3$–3 wt % hMMT as a function of frequency at $T = 258$ K.

Source: Reproduced with permission from Dam et al. [2016, Copyright Royal Society of Chemistry, *18*(29), 19955–19965].

Here, L is polymer film thickness.

Further, the ac conductivity spectra was analyzed using the random free energy barrier model (RFEBM) and the phenomenological modified Almond-West (AW) approach. The modified Almond-West (AW) equation is expressed as (Singh et al. 2013)

$$\sigma' = \sigma_{dc}\left[\left(1+\left(\frac{f}{f_c}\right)^n\right]+Af^m\right. \tag{3.24}$$

Here, σ_{dc}, f_c and n represent the dc conductivity, crossover frequency and power law exponent, respectively. At crossover frequency transition from the dc ionic to the dispersive conductivity occurs. The value of n should be between 0 and 1. A is a weakly temperature-dependent parameter and m is another power law exponent whose value should be very close to unity. With addition of hMMT clay, increase in free charge carriers was confirmed, and hence high dc conductivty. The conductvity can be expressed by RFEBM formalism also as given in equation 3.25.

$$\sigma' = \frac{\sigma_{dc}\tau_e\, tan^{-1}(\tau_e)}{\frac{1}{4}\left[\ln\left(1+^2\tau_e{}^2\right)\right]^2+\left[tan^{-1}(\tau_e)\right]^2} \tag{3.25}$$

Here, σ_{dc} *and* τ_e represent the dc conductivity and relaxation time, respectively.

Three possible pathways for ion trasnport are shown in Figure 3.16a. First, ions jump from one favorable site to another along the PEO backbone, while in second pathway ion jumps from one favorable site to another favorable site of a different PEO chain. At ambient temperature PEO chains have thermal motion, which can create a new favorable site and destroy an existing favorable site. The clay-polymer and clay-ion interactions also contribute to ion transport, and motion of ions is coupled to host dynamics. Figure 3.16b shows the comparison of both formalism via energy barrier potential with different hopping distance. Value of dc conductivity evaluated from both formalism is almost equal, which suggests equal probability of cation hopping in both.

The small difference between the experimental and theoretical results is due to the assumption that hopping distance is equal in RFEBM formalism. This assumption may be true at low temperature, but at high temperature increased polymer flexibility results in varied hopping distance. So, AW formalism is better for explanation of ac conductivity spectra. Both relaxation time (conductivity relaxation time, τ_c, and segmental relaxation time, τ_s, follows the VTF behavior and increases with decrease of temperature ($D\tau_s = constant$). Further, the coupling of the ionic transport mechanism and segmental relaxation is confirmed by the Ratner approach (Wang et al. 2012).

The effect of different plasticizers—Polyethylene glycol, propylene carbonate, ethylene carbonate and dimethyl carbonate—on the ionic conductivity and dielectric relaxation of PEO-$LiClO_4$ solid polymer electrolytes was examined (Das and Ghosh, 2015). It was concluded from ac conductivity analysis that the PEG-based polymer electrolytes demonstrate the highest dc conductivity and high crossover frequency ($n < 1$). At high frequency dielectric permittivity is constant and is attributed to the

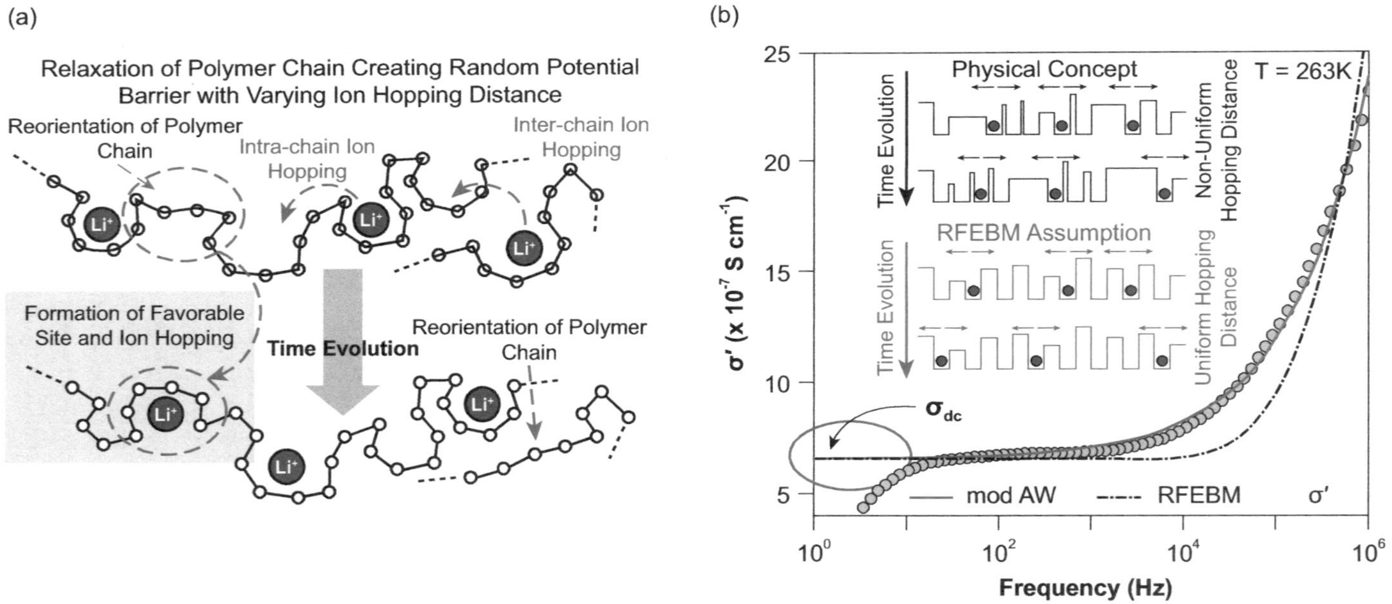

FIGURE 3.16 (a) Schematic of ion transport (b) between the modified Almond-West and the random free energy barrier model approach at $T =$ 263 K.

Source: Reproduced with permission from Dam et al. [2016, Copyright Royal Society of Chemistry, *18*(29), 19955–19965].

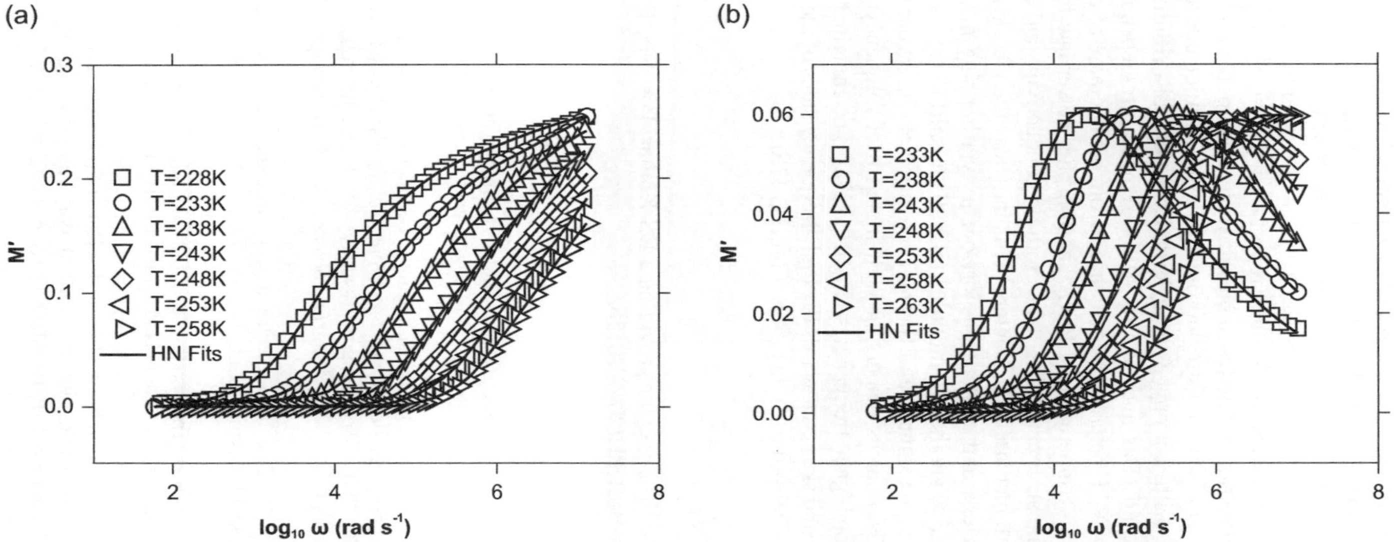

FIGURE 3.17 Frequency dependence of the (a) real part (M0) and (b) imaginary part (M00) of complex electric modulus M* for PEO/(PVdF-HFP)-LITFSI-30 wt % PMIMTFSI electrolytes.

Source: Reproduced with permission from Das and Ghosh [2016, Copyright AIP publishing, 119(9), 095101].

rapid polarization of atoms and electrons due to the applied field. Increase of dielectric permittivity with plasticizer addition is due to enhanced localization of charge carriers. Dielectric strength is in the order of $(\Delta\varepsilon)_{PEG} > (\Delta\varepsilon)_{EC} > (\Delta\varepsilon)_{PC} > (\Delta\varepsilon)_0 > (\Delta\varepsilon)_{DMC}$ [$(\Delta\varepsilon)_0$ is for plasticizer-free electrolyte]. Relaxation time obtained after fitting the HN equation is lowest for PEG-based polymer electrolyte and temperature dependence of relaxation time follows VTF nature. Another report from the same group reported the ion transport and relaxation dynamics in polyethylene oxide/poly(vinylidene fluoride-hexafluoropropylene)-lithium bis(trifluoromethane)sulfonimide blend polymer electrolytes embedded with 1-propyl-3-methyleimidazoliuum bis(trifluromethylesulfonyl)imide ionic liquid (Das and Ghosh, 2016). Temperature-dependent ionic conductivity follows the VTF behavior as evidenced from the decrease of activation energy above the melting temperature of PEO. The Jonscher power law formalism was used to examine the ac conductivity data. Both dc conductivity and crossover frequency increases with increase of IL content. This indicates the increase in number of free charge carriers, and cation transport occurs via hopping meachanism from one conducting site to another. In modulus formalism, shift of relaxation peak toward high frequency is the signature of thermal activation of charge carriers. The real part of modulus (M,) shows dispersing with increase of frequency, and getting saturated at M_∞ for higher frequencies (Figure 3.17a). The imaginary part of modulus is asymmetric and skewed towards the high-frequency sides of the maxima, which suggests the non-Debye-type relaxation (Figure 3.17b).

ACKNOWLEDGMENT

Anil Arya is thankful to the UGC for providing a DS Kothari Post Doc Fellowship via award letter number BSR/PH/20–21/0156.

REFERENCES

Awadhia, A., Patel, S. K., & Agrawal, S. L. (2006). Dielectric investigations in PVA based gel electrolytes. *Progress in Crystal Growth and Characterization of Materials*, 52(1–2), 61–68.

Agapov A. L., & Sokolov A. P. (2011). Decoupling ionic conductivity from structural relaxation: A way to solid polymer electrolytes? *Macromolecules*, *44*(11), 4410–4414.

Anantha P., & Hariharan K. (2005). Ac Conductivity analysis and dielectric relaxation behaviour of $NaNO_3$—Al_2O_3 composites. *Materials Science and Engineering: B*, *121*(1–2), 12–19.

Arya A., & Sharma A. (2018). Structural, electrical properties and dielectric relaxations in Na^+-ion-conducting solid polymer electrolyte. *Journal of Physics: Condensed Matter*, *30*(16), 165402.

Arya A., & Sharma A. L. (2019). Temperature and salt-dependent dielectric properties of blend solid polymer electrolyte complexed with LiBOB. *Macromolecular Research*, *27*(4), 334–345.

Barroso-Bujans F., Cerveny S., Palomino P., Enciso E., Rudić S., Fernandez-Alonso F., . . . Colmenero J. (2016). Dynamics and structure of poly (ethylene oxide) intercalated in the nanopores of resorcinol: Formaldehyde resin nanoparticles. *Macromolecules*, *49*(15), 5704–5713.

Brochier A., Aufray M., & Possart W. (2010). Dielectric spectra analysis: Reliable parameter estimation using interval analysis. In *Materials with Complex Behaviour* (pp. 99–123). Springer Nature, Switzerland.

Bruce P., & Gray F. (1995). Polymer electrolytes. II. General principles. In *Solid State Electrochemistry*, P. G. Bruce, Editor (p. 119). Cambridge University Press, Cambridge.

Cao W., & Gerhardt R. (1990). Calculation of various relaxation times and conductivity for a single dielectric relaxation process. *Solid State Ionics*, *42*(3–4), 213–221.

Chilaka N., & Ghosh S. (2014). Dielectric studies of poly (ethylene glycol)-polyurethane/poly (methylmethacrylate)/montmorillonite composite. *Electrochimica Acta*, *134*, 232–241.

Choi U. H., Liang S., Chen Q., Runt J., & Colby R. H. (2016). Segmental dynamics and dielectric constant of polysiloxane polar copolymers as plasticizers for polymer electrolytes. *ACS Applied Materials & Interfaces*, *8*(5), 3215–3225.

Chopra S., Sharma S., Goel T., & Mendiratta R. (2003). Structural, dielectric and pyroelectric studies of $Pb_{1-X}CaXTiO_3$ thin films. *Solid State Communications*, *127*(4), 299–304.

Choudhary S., & Sengwa R. J. (2011). Dielectric spectroscopy and confirmation of ion conduction mechanism in direct melt compounded hot-press polymer nanocomposite electrolytes. *Ionics*, *17*(9), 811–819.

Choudhary S., & Sengwa R. J. (2015). Structural and dielectric studies of amorphous and semicrystalline polymers blend-based nanocomposite electrolytes. *Journal of Applied Polymer Science*, *132*(3).

Cole K. S., & Cole R. H. (1941). Dispersion and absorption in dielectrics I. Alternating current characteristics. *The Journal of Chemical Physics*, *9*(4), 341–351.

Dam T., Jena S. S., & Pradhan D. K. (2016). The ionic transport mechanism and coupling between the ion conduction and segmental relaxation processes of PEO_{20}-$LiCF_3$ SO_3 based ion conducting polymer clay composites. *Physical Chemistry Chemical Physics*, *18*(29), 19955–19965.

Das S., & Ghosh A. (2015). Effect of plasticizers on ionic conductivity and dielectric relaxation of PEO-$LiClO_4$ polymer electrolyte. *Electrochimica Acta*, *171*, 59–65.

Das S., & Ghosh A. (2016). Structure, ion transport, and relaxation dynamics of polyethylene oxide/poly (vinylidene fluoride co-hexafluoropropylene): Lithium bis (trifluoromethane sulfonyl) imide blend polymer electrolyte embedded with ionic liquid. *Journal of Applied Physics*, *119*(9), 095101.

Das S., & Ghosh A. (2017). Charge carrier relaxation in different plasticized PEO/PVdF-HFP blend solid polymer electrolytes. *The Journal of Physical Chemistry B*, *121*(21), 5422–5432.

Druger S. D., Nitzan A., & Ratner M. A. (1983). Dynamic bond percolation theory: A microscopic model for diffusion in dynamically disordered systems. I. Definition and one-dimensional case. *The Journal of Chemical Physics*, *79*(6), 3133–3142.

Dyre J. C. (1988). The random free-energy barrier model for ac conduction in disordered solids. *Journal of Applied Physics*, *64*(5), 2456–2468.

Fragiadakis D., Dou S., Colby R. H., & Runt J. (2009). Molecular mobility and Li^+ conduction in polyester copolymer ionomers based on poly (ethylene oxide). *The Journal of Chemical Physics*, *130*(6), 064907.

Fuentes I., Andrio A., Teixidor F., Viñas C., & Compañ V. (2017). Enhanced conductivity of sodium versus lithium salts measured by impedance spectroscopy: Sodium cobaltacarboranes as electrolytes of choice. *Physical Chemistry Chemical Physics*, *19*(23), 15177–15186.

Funke K. (1993). Jump relaxation in solid electrolytes. *Progress in Solid State Chemistry*, *22*(2), 111–195.

Funke K., & Banhatti R. D. (2006). Ionic motion in materials with disordered structures. *Solid State Ionics*, *177*(19–25), 1551–1557.

Funke K., & Riess I. (1984). Debye-Hückel-type relaxation processes in solid ionic conductors. *Zeitschrift für Physikalische Chemie*, *140*(2), 217–232.

García-Bernabé A., Rivera A., Granados A., Luis S. V., & Compañ V. (2016). Ionic transport on composite polymers containing covalently attached and absorbed ionic liquid fragments. *Electrochimica Acta*, *213*, 887–897.

Grant F. (1958). Use of complex conductivity in the representation of dielectric phenomena. *Journal of Applied Physics*, *29*(1), 76–80.

Havriliak, S., & Negami, S. (1967). A complex plane representation of dielectric and mechanical relaxation processes in some polymers. *Polymer*, 8, 161–210.

Heaton N., Benavente R., Pérez E., Bello A., & Perena J. (1996). The γ relaxation in polymers containing ether linkages: conformational dynamics in the amorphous phase for a series of polybibenzoates containing oxyethylene spacers. *Polymer*, *37*(17), 3791–3798.

Hill R., & Jonscher A. (1983). The dielectric behaviour of condensed matter and its many-body interpretation. *Contemporary Physics*, *24*(1), 75–110.

Ishida, Y., Matsuo, M., & Takayanagi, M. (1965). Dielectric behavior of single crystals of poly (ethylene oxide). *Journal of Polymer Science Part B: Polymer Letters*, *3*(4), 321–324.

Jonscher A. (1978). Analysis of the alternating current properties of ionic conductors. *Journal of Materials Science*, *13*(3), 553–562.

Kamboj V., Arya A., Tanwar S., Kumar V., & Sharma A. (2021). Nanofiller-assisted Na+-conducting polymer nanocomposite for ultracapacitor: Structural, dielectric and electrochemical properties. *Journal of Materials Science*, *56*(10), 6167–6187.

Knödler D., Pendzig P., & Dieterich W. (1996). Ion dynamics in structurally disordered materials: Effects of random Coulombic traps. *Solid State Ionics*, *86*, 29–39.

LaFemina N. H., Chen Q., Colby R. H., & Mueller K. T. (2016). The diffusion and conduction of lithium in poly (ethylene oxide)-based sulfonate ionomers. *The Journal of Chemical Physics*, *145*(11), 114903.

Money B. K., Hariharan K., & Swenson J. (2012). Glass transition and relaxation processes of nanocomposite polymer electrolytes. *The Journal of Physical Chemistry B*, *116*(26), 7762–7770.

Murch G. E., & Dyre J. C. (1989). Correlation effects in ionic conductivity. *Critical Reviews in Solid State and Material Sciences*, *15*(4), 345–365.

Murti R., Tripathi S., Goyal N., & Prakash S. (2016). Random free energy barrier hopping model for ac conduction in chalcogenide glasses. *AIP Advances*, *6*(3), 035010.

Namikawa H. (1975). Characterization of the diffusion process in oxide glasses based on the correlation between electric conduction and dielectric relaxation. *Journal of Non-Crystalline Solids*, *18*(2), 173–195.

Nasri S., Hafsia A. B., Tabellout M., & Megdiche M. (2016). Complex impedance, dielectric properties and electrical conduction mechanism of La 0.5 Ba 0.5 FeO 3− δ perovskite oxides. *RSC Advances*, *6*(80), 76659–76665.

Ngai K. S. (2015). Interpreting the nonlinear dielectric response of glass-formers in terms of the coupling model. *The Journal of Chemical Physics*, *142*(11), 114502.

Ngai K. S., & Kanert O. (1992). Comparisons between the coupling model predictions, Monte Carlo simulations and some recent experimental data of conductivity relaxations in glassy ionics. *Solid State Ionics*, *53*, 936–946.

Ngai K. S., Ramesh S., Ramesh K., & Juan J. C. (2018). Electrical, dielectric and electrochemical characterization of novel poly (acrylic acid)-based polymer electrolytes complexed with lithium tetrafluoroborate. *Chemical Physics Letters*, *692*, 19–27.

Papathanassiou A., Sakellis I., & Grammatikakis J. (2007). Universal frequency-dependent ac conductivity of conducting polymer networks. *Applied Physics Letters*, *91*(12), 122911.

Pendzig P., & Dieterich W. (1998). Dispersive transport and dipolar effects in ionic glasses. *Solid State Ionics*, *105*(1–4), 209–216.

Popov I., Ishai P. B., Khamzin A., & Feldman Y. (2016). The mechanism of the dielectric relaxation in water. *Physical Chemistry Chemical Physics*, *18*(20), 13941–13953.
Pradel A., & Ribes M. (2014). Ionic conductivity of chalcogenide glasses. In *Chalcogenide Glasses* (pp. 169–208). Elsevier B.V., The Netherlands.
Ratner M. A., & Shriver D. F. (1988). Ion transport in solvent-free polymers. *Chemical Reviews*, *88*(1), 109–124.
Ravi M., Pavani Y., Kumar K. K., Bhavani S., Sharma A., & Rao V. N. (2011). Studies on electrical and dielectric properties of PVP: KBrO4 complexed polymer electrolyte films. *Materials Chemistry and Physics*, *130*(1–2), 442–448.
Redondo-Foj B., Carsi M., Ortiz-Serna P., Sanchis M., Vallejos S., García F., & García J. (2014). Effect of the dipole: Dipole interactions in the molecular dynamics of poly (vinylpyrrolidone)-based copolymers. *Macromolecules*, *47*(15), 5334–5346.
Rodríguez J., Navarrete E., Dalchiele E. A., Sánchez L., Ramos-Barrado J. R., & Martín F. (2013). Polyvinylpyrrolidone—$LiClO_4$ solid polymer electrolyte and its application in transparent thin film supercapacitors. *Journal of Power Sources*, *237*, 270–276.
Roy A., Dutta B., & Bhattacharya S. (2016). Correlation of the average hopping length to the ion conductivity and ion diffusivity obtained from the space charge polarization in solid polymer electrolytes. *RSC Advances*, *6*(70), 65434–65442.
Sharma A., & Thakur A. K. (2015). Relaxation behavior in clay-reinforced polymer nanocomposites. *Ionics*, *21*(6), 1561–1575.
Shukla, N., Thakur, A. K., Shukla, A., & Marx, D. T. (2014). Ion conduction mechanism in solid polymer electrolyte: an applicability of almond-west formalism. *Int J Electrochem Sci*, 9(12), 7644–7659.
Singh D. P., Shahi K., & Kar K. K. (2013). Scaling behavior and nearly constant loss effect in AgI—$LiPO_3$ composite glasses. *Solid State Ionics*, *231*, 102–108.
Singh R. J. (2012). *Solid State Physics*. Pearson Education, India
Smyth C. P. (1955). *Dielectric Behavior and Structure.* Toronto; London; McGraw-Hill.
Starr, F. W., & Douglas, J. F. (2011). Modifying fragility and collective motion in polymer melts with nanoparticles. *Physical review letters*, 106(11), 115702.
Wang Y., Agapov A. L., Fan F., Hong K., Yu X., Mays J., & Sokolov A. P. (2012). Decoupling of ionic transport from segmental relaxation in polymer electrolytes. *Physical Review Letters*, *108*(8), 088303.
Wei Y. Z., & Sridhar S. (1993). A new graphical representation for dielectric data. *The Journal of Chemical Physics*, *99*(4), 3119–3124.
Woo H., Majid S. R., & Arof A. K. (2011). Conduction and thermal properties of a proton conducting polymer electrolyte based on poly (ε-caprolactone). *Solid State Ionics*, *199*, 14–20.
Yu J., Wu W., Dai D., Song Y., Li C., & Jiang N. (2014). Crystal structure transformation and dielectric properties of polymer composites incorporating zinc oxide nanorods. *Macromolecular Research*, *22*(1), 19–25.

4 Synthesis Methods and Characterization Techniques for Polymer Composites

Avirup Das and Atma Rai

CONTENTS

DOI: 10.1201/9781003208662-5

4.1 INTRODUCTION

Polymer nanocomposite (PNC) represents a special class of polymeric matrix where an organic-inorganic filler network is embedded inside a polymer matrix. These organic/inorganic fillers generally have a dimension of 10–100 Å. Dispersion of this nanofiller into the polymer matrix drastically changes inherent properties like (i) thermal stability, (ii) mechanical stability, (iii) ionic conductivity of the pristine polymer. Further, because of the smaller (~ nanoscale) size, these advancement has been achieved with only a small vol % addition of nanofiller in the polymeric matrix compared to micro fillers. These improvements make nanocomposites an emerging solution to replace liquid electrolytes in Li-ion batteries. PNC is mainly fabricated by using different mechanical mixing or via a chemical route. However, the filler agglomeration inside the PNC matrix is a serious issue. To overcome this issue and to obtain a highly homogeneous filler network inside the polymer matrix, researchers have synthesized polymer nanocomposites (PNC) by different methods. These methods have been shown in Figure 4.1 and are described in the following sections.

In the present chapter different synthesis methods and characterization techniques have been discussed. The advantage and disadvantages of different synthesis methods have been described. A special focus has been given to the feasible techniques for large-scale implementation so that technology can successfully transform from the lab to the fabrication stage. Further, different characterization techniques

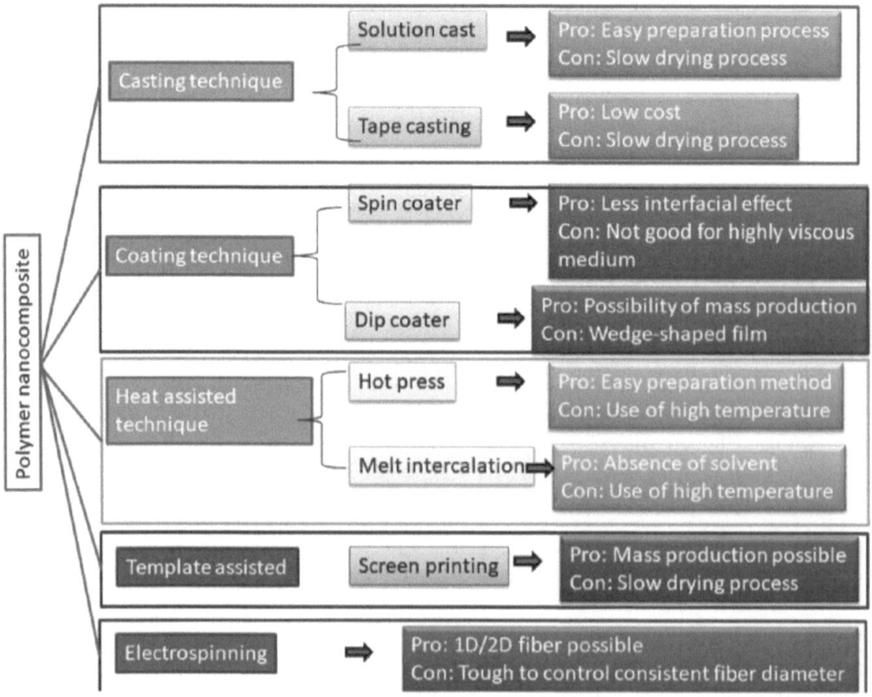

FIGURE 4.1 PNC preparation methods.

have been described and different information and analytical equations related to each characterization technique have been summarized.

4.2 SYNTHESIS METHODS OF POLYMER COMPOSITE

4.2.1 Solution Cast

The solution cast technique or solvent casting technique is one of the most popular techniques in polymer nanocomposite formation. This synthesis process is mostly performed at room temperature and in an inert atmosphere. However, depending upon the polymer matrix property sometimes elevated temperature can be used for proper polymerization. Using this technique, a thin polymer film in μm order can be obtained with homogeneously distributed nanofillers. This process has been used to prepare dispersed and intercalated phase nanocomposites. Different layered structure fillers like Montmorillonite, Hectorite, Graphite, and different metal oxides have been used for this synthesis method. Montmorillonite was a filler used by Ratna et al. for PNC preparation. PEO-$LiBF_4$-MMT composite was prepared by stirring the mixture at 75 °C following the solution casting method. For better hominization, the sonication method was used (Ratna et al. 2007). Das et al. have used PEO-PDMS polymer blend + $LiCF_3SO_3$ + Hectorite clay for PNC preparation. In this process, an adequate amount of polymer, salt, and filler have been mixed properly by using a magnetic stirrer. For better complexation initially, polymer salt complex was prepared then filler was added and stirred for 24 hours (Das and Thakur 2021). Sharma et al. have used MMT for PNC preparation using PAN and $LiCF_3SO_3$ as polymer salt complex using this process (Sharma & Thakur, 2010a). Apart from layered clay structure, Das et al. have used MWCNT for PNC preparation (Das et al. 2017), Padmaraj et al. have used ZnO to prepare a (PVdF-HFP)-$LiCLO_4$-ZnO PNC following a simple solution cast technique (Padmaraj et al. 2013). Typical steps for this synthesis method are described next and shown in Figure 4.2.

1. Initially the polymer matrix will be mixed in an appropriate solvent. For a proper homogeneous mixture, the solution will be stirred for an optimum time at room temperature or elevated temperature.
2. Then stoichiometric amounts of Li salt will be mixed and stirred for better polymer salt complex preparation. The stoichiometry of Li^+ will be calculated based on the available cation coordination site of the polymer monomer (Das et al. 2013; Das et al. 2014a).
3. For PNC preparation appropriate amount of filler will be mixed and stirred for the optimum time. The filler will be calculated based on the total polymer wt %.
4. For homogeneous distribution of filler among the polymer matrix, in some cases sonication method can be used (Ratna et al. 2007).
5. Finally the solution can be cast in a petri dish and kept for evaporation in an inert atmosphere.

Typical process parameters and their conductivity of PNC prepared via this technique have been described in Table 4.1. The advantage of this technique is its ease

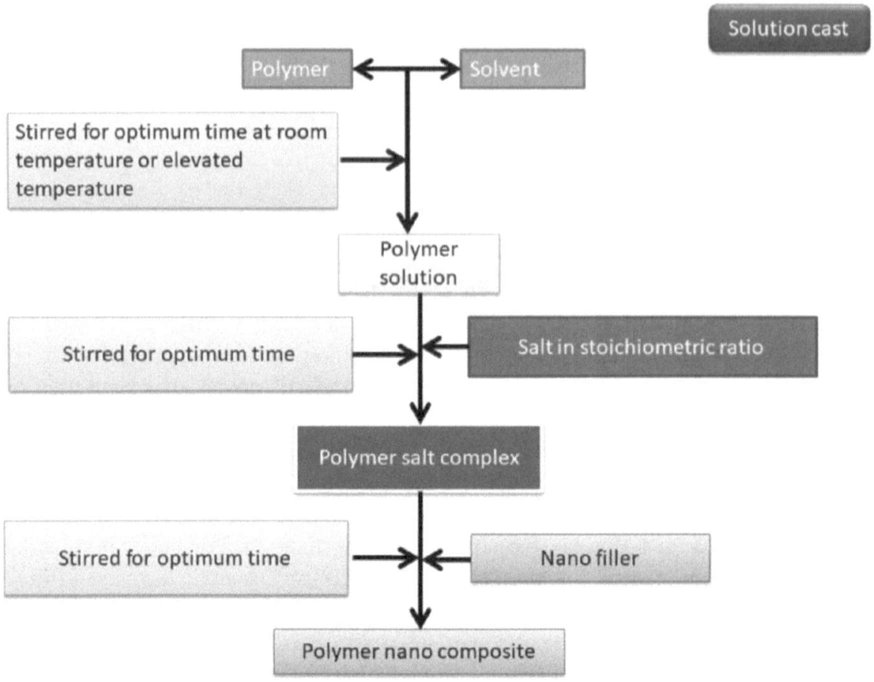

FIGURE 4.2 Preparation method of PNC via solution cast technique.

TABLE 4.1
PNC Prepared via Solution Casting.

Polymer	Salt	Solvent	Filler	Conductivity	Reference
PEO	$LiBF_4$	Acetonitrile	Organically modified clay	0.9×10^{-3}	Ratna et al. 2007
PAN	$LiCF_3SO_3$	N,N-dimethylformamide	Na-montmorillonite	$2.5 \times 10 - 2$ at 80°C	Sharma & Thakur, 2010a
PEO+PDMS	$LiCF_3SO_3$	THF	MWCNT	30 °C $8.04 \times 10 - 5$ S cm^{-1}	Das & Thakur 2017
PVdF-HFP	$LiClO_4$	THF	ZnO	1.043×10^{-3} S/cm	Padmaraj et al., 2013

of preparation method. However, depending upon the solvent and polymerization process the solvent evaporation method can take a much longer time. Since using this process a solid polymer electrolyte (SPE) will be produced, during cell formation there might exist an interfacial effect due to the uneven surface of SPE and electrode material.

4.2.2 Spin Coating

This synthesis method has been adopted for polymer electrolytes to reduce the electrode-electrolyte interfacial effect (Park et al. 2006). As depicted in Figure 4.3, in this method a solution of polymer nanocomposite is directly spin-coated onto an electrode. The PNC slurry is directly coated on the electrode surface so solid electrolyte interfacial (SEI) effect can be reduced. During this process, the thickness of PNC film can be controlled by controlling the viscosity, rotation duration, and speed of the coater. Park et al. have obtained a thickness of 25–30 μm polymer-salt (PEO-$LiClO_4$) complex using a 3000–5000 rpm. Sharanappa Chapi et al. have prepared a PNC by using a polymer blend (PEO-PVP) and nanofiller (ZnO) at 3000 and 6000 rpm

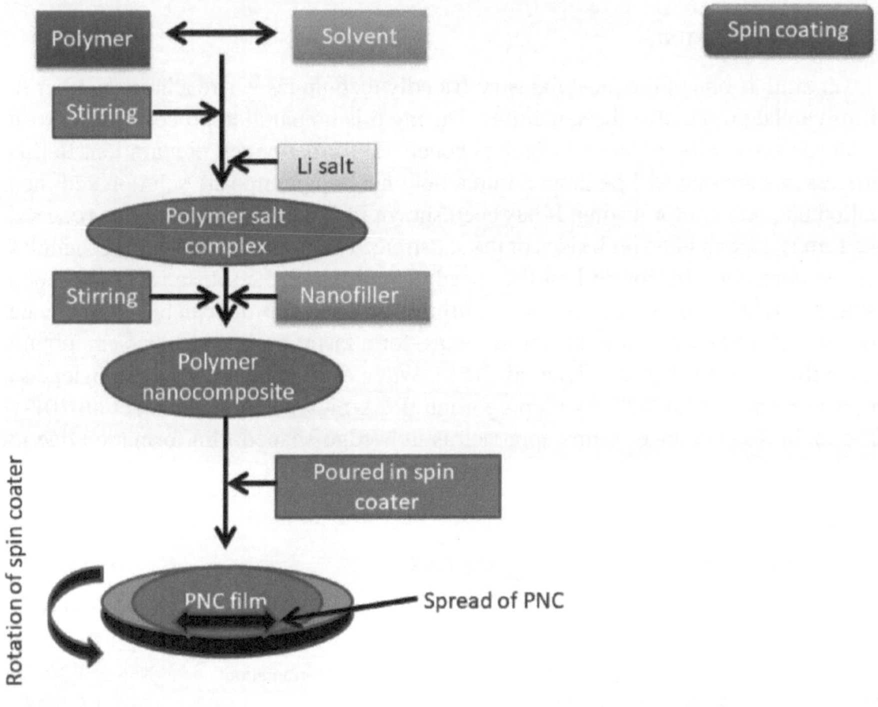

FIGURE 4.3 Preparation method of PNC via spin coating.

TABLE 4.2
PNC Prepared via Spin Coating Technique

Polymer	Solvent	Filler	Spin Speed (RPM)	Other Parameter	Ref
PEO+ $LiClO_4$	acetonitrile		3000–5000	25–30 μm	Park et al. 2006
PEO and PVP+	methanol	ZnO NPs	3000 for 40 s 6000 for 50 s		Chapi et al. 2016
PEO	Methanol	$CoCl_2 \cdot 6H_2O$	3500 rpm for the 40 s	1.21×10^{-3} S cm^{-1} at 343 K	Chapi, 2020

for 40s and 50s subsequently (Chapi et al. 2016) with a thickness of 3.9 ~ 4.2 μm. In this process initially, a solution of polymer nanocomposite will be prepared similar to the solution cast technique. The solution will be cast over the electrode or any other substrate kept inside a spin coater. The viscosity of the solution, rotation speed, and rotation duration control the film thickness and homogeneity. The main advantage of this kind of approach is its possibility of reduced interfacial effect. However, for a highly viscous medium, the spreading of slurry over the electrode surface may not be uniform due to high surface tension and moment of inertia of different constituents of the PNC. So a careful optimization of speed and rotation duration is needed for proper spreading of slurry (Norrman et al. 2005; Krebs, 2009; Wilson & Gottesfeld, 1992). Typical conductivity of PNC obtained via this route has been tabulated in Table 4.2.

4.2.3 Dip Coating

Dip coating is one of the most industry-friendly techniques where a large number of films can be prepared at the same time. During this preparation process, a film with a flat surface can be produced which is beneficial for better cell preparation. In this process, a substrate will be dipped into a polymer nanocomposite solution and then pulled up and kept for drying. It has been shown in Figure 4.4. During this process, the film is deposited on both sides of the substrate. The thickness of the PNC samples can be controlled by controlling the speed of pulling up the substance. As Phuong Nguyen-Tri et al. described, for a slow pulling rate a thicker film can be obtained due to faster evaporation of the solvent, whereas, for a faster-pulling rate one can obtain much thinner film [Nguyen-Tri et al. 2018]. Wang et al. have used polypropylene as a substrate and modified it by dipping it into PVA-SiO_2 solution (Wang et al. 2019). The main disadvantage of this approach is its wedge-shaped film formation due to

TABLE 4.3
PNC Preparation via Dip Coating Method.

Substrate	PNC	Other Parameter	Reference
PP	PVA-nano SiO_2	Standing for 3h	Wang et al. 2019
PE	γ-Al_2O_3/PVdF-HFP/TTT	electron beam irradiation	Nho et al. 2017
PE	PVdF-HFP-(Z-SiO_2)		Wua et al. 2019

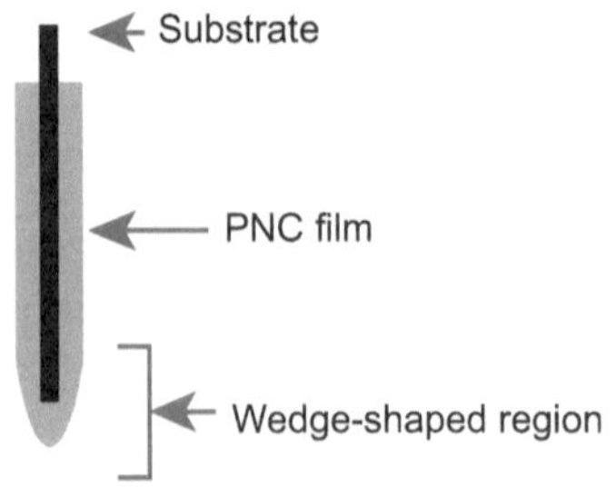

FIGURE 4.4 Thin film formation by dip coating.

solvent evaporation and draining (Brinker & Hurd, 1994). This type of film formation can create a huge obstacle for a better-performing cell. Typical conductivity of polymer composites obtained via this route has been summarized in Table 4.3.

4.2.4 Tape Casting

This is another low-cost synthesis method suitable for mass production of PNC samples at the industry level. Different steps involved in this process have been shown in Figure 4.5. Using this process a PNC film with a thickness of 10–1000 μm can be formed (Uchino, 2010; Kwon & Han, 2020). PNC preparation using the tap casting method needs a proper optimization of filler and polymer matrix. An excessive filler might improve the ion migration through the PNC but eventually increase the viscosity of the system. For a high viscous system removing air bubbles is very difficult, which can be fatal at the time of cell preparation. So, careful optimization of all the constituents is very necessary for the casting technique (Uchino, 2010). In this process, a drum filled with PNC slurry will be placed on top of a doctor's blade instrument. The slurry flow rate will be controlled by a pump connected with the slurry. Further, following the doctor's blade technique, a thin film will be cast and kept for drying in an inert atmosphere. To control the thickness of the thin film the micrometer attached with the blade can be used (Arya & Sharma, 2017). Pradhan et al. have prepared a sample using the standard tape casting technique. They have Montmorillonite as filler along with PEO as polymer matrix and $NaClO_4$ as salt. Initially, polymer, salt, and filler have been stirred and properly mixed to obtain homogeneous slurry and then it has been cast on a conventional tape casting machine to obtain a thin film (Pradhan et al. 2011). Though this process is suitable for the mass production of thin PNC film, the main drawback for this kind of synthesis method is its slow drying process. It hinders its large-scale application. The conductivity of PNC obtained via this process has been shown in Table 4.4.

4.2.5 Template-Assisted PNC Synthesis

In this synthesis process, a template is used to create a PNC film as shown in Figure 4.6. This process is also very efficient to synthesize PNC films on a large

TABLE 4.4
PNC Prepared via Tape Casting Method.

PNC	Other Parameter	Conductivity	Ref
methyl ethyl ketone+ ethanol+ LSGM+ Polyethylene glycol+	speed of 20 mm/min		Nho et al. 2017
PEO+ $NaClO_4$+ DMMT+ PEG		4.4×10^{-6} S cm^{-1} (at 40°C)	Uchino, 2010
PEO_{25}-$NaClO_4$+ 5 wt % DMMT + x wt % PEG200		10^{-6} S cm^{-1}	Pradhan et al. 2008

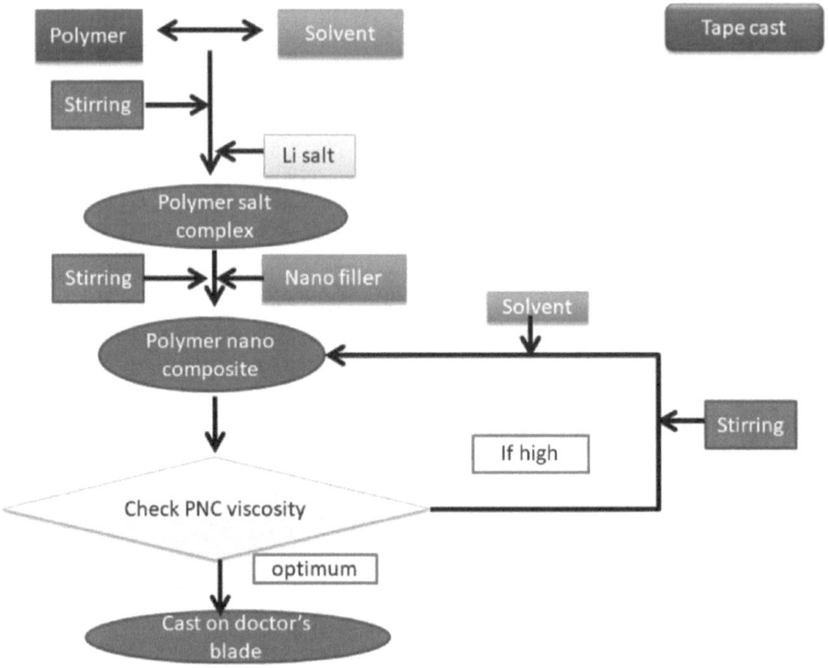

FIGURE 4.5 Tape casting method.

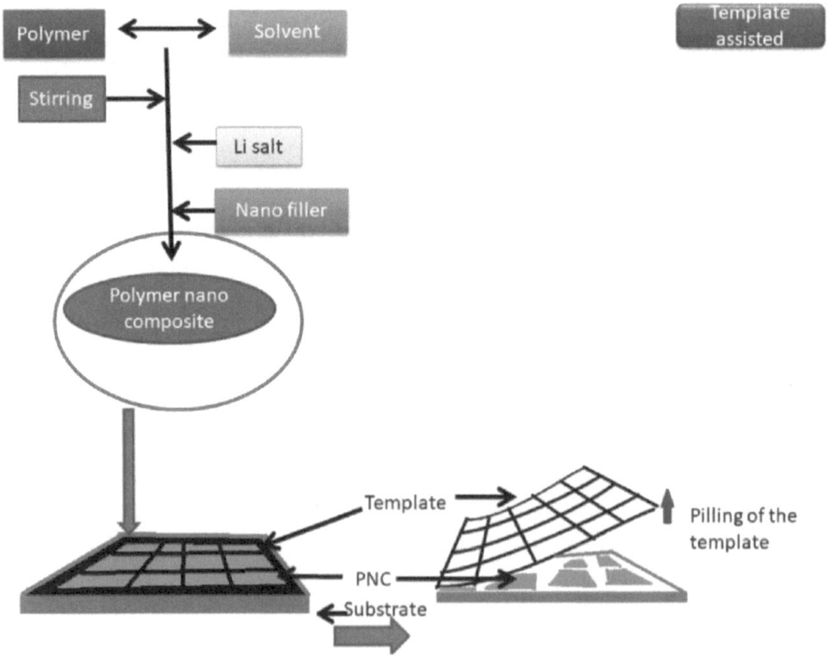

FIGURE 4.6 Template-assisted PNC synthesis method

scale. In this process, slurry of polymer, salt, and filler will be poured on a template. And after drying of the PNC film the template will be pilled off or dissolved in a suitable solution leaving a homogeneous thin PNC film for use.

4.2.6 Heat-Assisted Synthesis Method

4.2.6.1 Hot-press

In this method, initially polymer, salt, and filler components are dried under a vacuum chamber. Next, the dried components are thoroughly mixed by using a ball milling technique or motor pestle. Then they are placed inside a mold and will be kept inside a hot press as shown in Figure 4.7. The temperature of the hot press needs to be optimized based on polymer melting temperature. In this process, the PNC films are highly homogeneous and mechanically stable. A PEO-based polymer composite with different salt and filler has been prepared by using this technique. PEO-$LiCF_3SO_3$-SiO_2 based PNC sample has been prepared by Appetecchi et al. They have used 80–100 °C and aluminum mold for the hot press method (Appetecchi et al. 2003a, 2003b). PEO-$LiCLO_4$-nano Chitin has been prepared by Stephan et al. (Stephan et al. 2009).

4.2.6.2 Melt Intercalation

This process is another thermally assisted intercalation technique. These processes have been proved to be better than the solution intercalation process (following the solution cast technique), because of the absence of solvent in this technique. This lack of hazardous solvent makes it more environmentally friendly and establishes itself at the forefront of environmentally friendly battery technology. In this process, polymer and filler are mixed and heated close to polymer melting temperature. At this temperature region polymers reach a molten state so they easily intercalate inside the layered structure of the filler.

4.2.7 Electrospinning

Electrospinning has become very popular due to its ease of preparation method and scalability for 1D or 2D structure preparation. In this process, a polymer

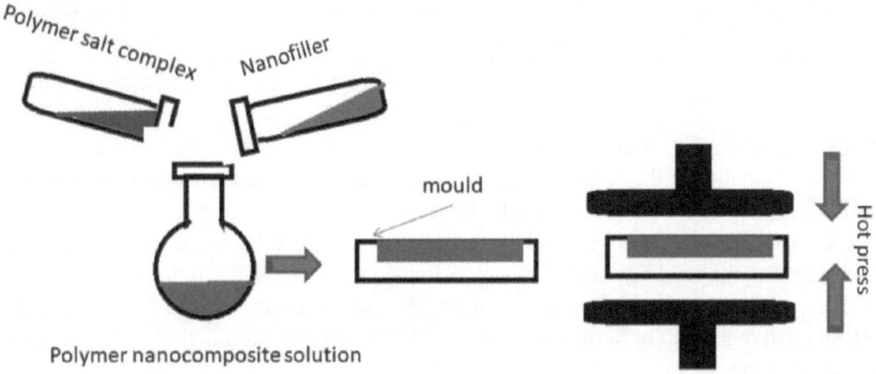

FIGURE 4.7 Hot press synthesis method.

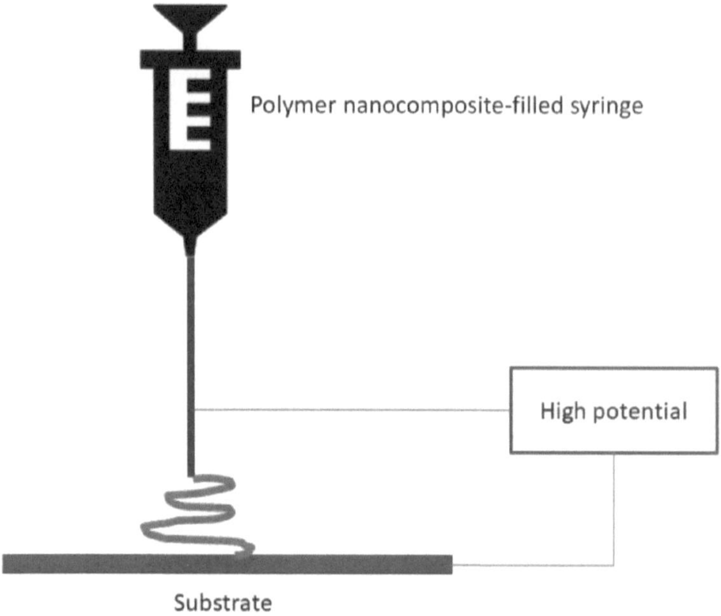

FIGURE 4.8 Electrospinning method.

TABLE 4.5
PNC Prepared via Electrospinning

PNC	Parameters	Reference
MFO@C+PAN	22 gauge needle, flow rate = 20 μL min^{-1}, voltage = 15 kV	Liu et al. 2016
PAN+PMMA+ $SnCl_2$	21 gauge needle, flow rate = 10 μL min^{-1}, voltage = 15 kV	Liu et al. 2015
$SnCl_2 \cdot 2H_2O$+PVP+DMF	5 mL plastic syringe, flow rate = 0.3 mL h^{-1} voltage = 18 kV	Xia et al. 2019
TiO_2/N-C NFs	10-mL syringe, flow rate = 0.36 mL h^{-1} voltage = 14 kV	Nie et al. 2018

nanocomposite solution is prepared and kept in a container connected with a syringe. Then that syringe will be kept over a substrate and a very high potential will be applied between syringe and substrate. It has been shown in Figure 4.8. The diameter of the nanofiber can be controlled by varying the potential between the syringe and substrate and the feed rate of the syringe. Hong et al. have obtained 50 μm thick fiber with a syringe to substrate potential of 15 kV and feed rate of 3 ml/h (Hong et al. 2018). Different process parameters for the PNC prepared via this method have been shown in Table 4.5.

4.3 CHARACTERIZATION TECHNIQUE

4.3.1 Structural

4.3.1.1 XRD

In the realm of material characterization, x-ray diffraction (XRD) is a very powerful and fundamental tool for both crystalline and non-crystalline phases. This method is particularly effective for determining the degree of crystallinity in amorphous, semi-crystalline, polymer, and polymer composite materials. Max von Laue created this technique in 1912, and it was later expanded to identify crystal structure by W. L. Bragg and W. H. Bragg in 1913.

Highly crystalline, semi-crystalline, and amorphous polymer/polymer composites have been developed. The amount and presence of crystallinity in these materials are determined by how they are formulated/processed. The following information can be extracted from a polymer nanocomposite using this technique.

1. Determine the composite formation and its nature
2. Alter interlayer spacing of the filler
3. Determine polymers' interchain spacing during the formation of the nanocomposite
4. Determine polymer crystallinity
5. Recognize the stress/strain caused by composite production
6. Analyze how ion dissociation changes as nanocomposites are produced

XRD is one of the strongest tools to establish the composite formation and identify the type of composite formation like intercalation, exfoliation, or dispersion. Sharma et al. have used host matrix (PAN) crystalline peak at 14.1° and clay (MMT) peak at 5.8° to establish composite formation and identify the type of composite formation. They have calculated the change of relative peak intensity of the host matrix, change in 'd' spacing and interchain separation (R) using the equations 4.1 and 4.2.

$$2d \sin\theta = n\lambda \tag{4.1}$$

$$R = \frac{5}{8}\left(n\lambda / \text{Sin}\,\theta \right) \tag{4.2}$$

d = interplaner spacing, R = interchain spacing, θ = Bragg's diffraction angle, n = order of maxima, and λ = wavelength of x-ray incident on sample.

Clay interlayer spacing calculated via equation 4.1 considering the clay crystalline peak indicates a significant increase in 'd' spacing from 15.1 Å (for pure clay), 19.5 Å (PNC with 10 wt % clay) and clay gallery width from 5.5 Å to 9.9 Å. This result establishes the insertion of a polymer chain inside the clay gallery (Sharma & Thakur, 2010b). Shukla & Thakur have shown changes in the n-YSZ, d_{001} peak profile and position, a significant change in the polymer interchain spacing and 'd' spacing upon dispersion of YSZ in the polymer salt system (Shukla & Thakur, 2010).

Das et al. have shown that polymer crystallinity can be estimated from XRD by using the following equation.

$$\chi_c = \left(\frac{S}{S_0}\right) \times 100\% \tag{4.3}$$

S = area under the crystalline peaks; S_0 = area under the crystalline peaks + amorphous hump.

They have shown that upon nanocomposite formation with MWCNT the polymer crystallinity drastically changes from 36.32% (polymer blend) to 12.82% (polymer nanocomposite) (Das & Thakur, 2017). Salt dissociation is a very important factor for polymer nanocomposite. An absence of salt peaks in XRD can establish the dissociation of salt upon polymer salt complex (PS) or polymer nanocomposite formation. Sharma et al. have shown in a PAN-$LiPF_6$-DMMT system that the disappearance of characteristic salt peaks indicates the complete dissociation of salt in the PS system (Sharma & Thakur, 2013). Das et al. have shown upon PNC formation the uncomplexed XRD peaks of $LiCF_3SO_3$ at 20.4° disappear. This indicates improved ion dissociation upon PNC formation (Das & Thakur, 2021). Polymer nanocomposite was created by Polu et al. utilizing Poly(vinyl alcohol) and magnesium nitrate. There is no XRD peak of magnesium nitrate in the PS sample, showing that the salt was completely dissociated during PS production (Polu & Kumar, 2013).

4.3.2 Microscopy

4.3.2.1 Transmission Electron Microscopy

Transmission electron microscopy (TEM) is a popular technique for studying surface/bulk properties as well as the internal structure of materials. To study the morphology of the samples under investigation, high-energy collimated electron beams (100–400 keV) are allowed to fall on the sample, and transmitted and diffracted signals are analyzed. The structural information obtained from x-ray diffraction (XRD) analysis is also confirmed and corroborated by this result. TEM analysis is a powerful tool to understand the nature of nanocomposite preparation for clay-based PNC.

Sharma et al. have observed homogeneously distributed clay platelets indicating exfoliation at low clay loading whereas at high clay loading the clay platelets are stacked in TEM (Sharma & Thakur, 2013). Mohapatra et al. have used TEM imaging to confirm the change in 'd' spacing obtained from XRD upon PNC formation. They have used PEO-$LiClO_4$-MMT for PNC preparation. In high clay concentration (>7.5 wt %) the presence of well-stacked clay platelets proves the occurrence of intercalation (Mohapatra et al. 2009).

4.3.2.2 Optical Microscopy

Optical microscopy (OM) is used to understand the bulk surface property. SEM, TEM images can only show a surface morphology in a limited area, whereas OM can identify it for a large surface area. OM is used to identify the following property in a PNC sample:

1. Homogeneous distribution of nanofiller over the entire PNC sample
2. Existence/absence of spherulites or other surface morphology

Xie et al. have prepared a polymer clay nanocomposite using two different kinds of polypropylene and closite clay. The OP shows no presence of agglomerated clay. This indicates a homogeneous distribution of filler in the polymer matrix (Xie et al. 2010). Sharma et al. have prepared a CNT-based polymer nanocomposite. OM shows a clear homogeneous dispersion of CNT in the polymer matrix (Sharma et al. 2009).

4.3.3 Spectroscopy

4.3.3.1 FTIR

Fourier Transform Infrared Spectroscopy (FTIR) is a sensitive analytical instrument for gaining a better understanding of the atomic/molecular vibrations present in the sample under study. It confirms the presence of numerous functional groups and backs up the structural results of the x-ray diffraction study. The theory of light-matter interaction and the ensuing molecular vibrations underpins this approach. The energy associated with molecular vibration is seen in the infrared (IR) spectrum of electromagnetic radiation. When infrared light is allowed to fall on the sample, the vibrating molecule absorbs it. It works on the concept of a change in the molecule's permanent dipole moment (μ) when it is excited by an infrared beam. FTIR can be used to

1. Determine ion pair and ion aggregate formation
2. Establish the polymer complex formation

Sharma et al. have used FTIR spectroscopy to investigate the cation-anion interaction, cation-cation coordination site interaction, and cation-cation coordinated site-filler interaction in a PAN-$LiCF_3SO_3$-DMMT-based PNC system. A significant change has been observed in the stretching and deformation vibrational mode of the methylene group upon PS formation and PNC formation. This shows evidence of cation coordination with the polymer host site. Also a clear change in anion ($CF_3SO_3^-$) characteristic peaks in terms of peak shift and asymmetry in the peak profile is attributed to the ion-pair interaction. Further, the ion dissociation effect upon PNC formation can be proved with FTIR spectra by using the following equation.

$$Fraction\ of\ free\ anion = \left(\frac{Area\ of\ free\ anion\ peak}{Total\ peak\ area} \right) \tag{4.4}$$

$$Fraction\ of\ ion\ pair = \left(\frac{Area\ of\ ion\ pair\ peak}{Total\ peak\ area} \right) \tag{4.5}$$

With PNC formation the increase of the fraction of free anions indicates better salt dissociation (Sharma & Thakur, 2010a).

Das et al. have used FTIR spectra for understanding the ion dissociation effect in a PEO-PDMS-$LiCF_3SO_3$-MWCNT-based PNC system (Das & Thakur, 2017).

4.3.3.2 RAMAN Spectroscopy

Raman spectroscopy is a very sensitive technique that can detect structural changes caused by lattice strain and local disorder caused by sample preparation and foreign element substitution. This approach is based on light scattering by vibrating molecules of the sample under investigation and is a supplement to IR spectroscopy. When a monochromatic source of light causes a change in atomic polarizability, a material is said to be Raman active. Instead of using a wide variety of IR sources in IR analysis, a monochromatic light source with a set wavelength in the Raman spectrum is typically employed to investigate the sample. In Raman spectroscopy, monochromatic laser beams such as an Ar laser ion (514 nm), He-Ne laser (632.8 nm), and solid-state diode laser (785 nm) are commonly utilized as excitation sources.

C. V. Raman devised this technique in 1928 and was awarded the Noble Prize for it in 1930.

The important applications of Raman spectroscopy are as follows:

1. Composite preparation conformation
2. Ion dissociation effect in PNC

Das et al. said a change in peak profile of CH_2 and C-O-C will indicate nanocomposite formation in a PEO-PDMS-$LiCF_3SO_3$-Hectorite PNC sample. Further, a change in peak profile CF_3 and SO_3 peak indicates an enhancement of ion dissociation effect (Das et al. 2017). Edman have used Raman characteristic peak TFSI at 740 cm^{-1} to understand the ion dissociation effect concerning salt concentration and temperature in a PEO-LITFSI system (Edman, 2000). Caruso et al. have used PEO-$LiCF_3SO_3$ as a PS system. They indicated that the stretching vibration of triflate ion at 1033 cm^{-1} and ion pair vibration at 1040–1045 cm^{-1} is very sensitive to the local environment. So a slight change in the local environment of the triflate ion can be identified as a change in its Raman spectroscopy (Caruso et al. 2002).

4.3.4 Electrical

The impedance spectroscopy technique is widely used to understand the electrical properties of different polymer or crystalline materials. It is typically based on measuring the ac response of the materials under investigation about an externally applied ac field and then estimating the impedance as a function of frequency. Generally, the following information can be extracted from the impedance experimental data:

1. AC/DC conductivity
2. Dielectric response

Equations used to analyze impedance data:

$$\text{DC conductivity: } \sigma_{dc} = \frac{l}{R \times A} \tag{4.6}$$

$$\text{AC conductivity: } \sigma_{ac} = \sigma_o \left\{ 1 + \left(\frac{w}{w_p} \right)^n \right\} \tag{4.7}$$

$$\text{Conductivity master curve: } \frac{\sigma_{ac}}{\sigma_{dc}} = \frac{w}{w_p} \tag{4.8}$$

$$\text{Carrier concentration: } k = \frac{\sigma_{dc \times T}}{w_p} \tag{4.9}$$

σ_{dc} = DC conductivity, l = thickness of the sample, A = area of the sample, R = resistivity of the samples (Das & Thakur, 2018; Das & Thakur, 2020).

Pritam et al. have used complex impedance spectroscopy to investigate the electrical property of a novel Na^+ conducting blended solid polymer electrolyte (Pritam et al. 2021). Klongan et al. have prepared a polymer salt complex by using PEO-$LiCF_3SO_3$. They have analyzed the role of nano alumina as filler, and PEG and DOP as a plasticizer in a PEO-$LiCF_3SO_3$ based PS system. The conductivity obtained from complex impedance spectra indicates improvement of two orders (~10^{-7} S cm^{-1} for PS to ~10^{-5} S cm^{-1}) for an Al_2O_3 based PNC; three orders (~10^{-4} S cm^{-1}) for a DOP-based system (Klongkan & Pumchusak, 2015). Ibrahim et al. have used CNT as filler in a PEO-$LiPF_6$–EC-based PS system. Using the impedance spectroscopy they have shown a 5 wt % filler content in PNC can show a dc conductivity as high as ~10^{-3} S cm^{-1} (Ibrahim et al. 2012). Das et al. have used complex impedance spectra to extract the dielectric relaxation parameter with varying temperature and salt concentration. Using the insight obtained from the relaxation parameter they have explained the near-constant loss phenomenon in solid polymer blend electrolytes (Das et al. 2014b; Das et al. 2015).

4.3.5 Electrochemical Property Analysis

Electrochemical property analysis is a very important tool for analyzing the electrochemical performance of a cell using PNC as an electrolyte. In general for electrochemical property analysis, the PNC electrolyte is sandwiched between symmetric or asymmetric electrode assemblies. Generally, cyclic voltammetry and charge-discharge analysis are the two most popular electrochemical characterization techniques for any electrochemical cell.

The capacitance value of the electrochemical cell can be calculated by using the following equation (Gao et al. 2018).

$$C = \frac{I \times dt}{m \times (V_c - V_a)} \tag{4.10}$$

m = mass loading of active material in gm, $V_c - V_a$ = potential window. Hashim et al. have used PVA-H_3PO_4 as an electrolyte for supercapacitor and has obtained ~ 90% efficiency and long cycle life (Hashim et al. 2012). Tamilarasan et al. have prepared an all solid state supercapacitor. They have used PAN/[BMIM][TFSI] electrolyte and HEG electrodes. The electrochemical characterization shows a capacitance of 98F/g @ 10A/g (Tamilarasan & Ramaprabhu 2013). Kumar et al. have used a PVdF-LiTf–EC-PC as electrolyte and MWCNT-AC-PVdF-based electrode to prepare the supercapacitor. The electrochemical characterization reveals that MWCNT-based electrodes show significantly lower specific capacitance (32 F/g) compared to AC-based electrodes (157 F/g) (Kumar et al. 2012).

4.3.6 Ion Transport Number

Ion transport no analysis technique is generally used to establish the ionic conduction nature of the electrolyte material. To calculate the ion transport no the PNC samples are sandwiched between two blocking electrodes and then a constant voltage (~ 50 mV) is applied to the system. The polarization current obtained from that system is then plotted concerning time and then the following equation is used.

$$t_{ion} = \frac{I_T - I_e}{I_T}, I_T = I_{ion} + I_e, t_{ion} + t_e = 1 \tag{4.11}$$

Using equation 4.11 Mohapatra et al. have calculated ~ 99% ionic transport for PS and PNC samples (PEO-$LICLO_4$-MMT) (Mohapatra et al. 2009). Shukla et al. have prepared PNC by using PMMA-$LiCLO_4$-CeO_2. The transport no analysis of the PNC shows a ~ 99% cationic conduction for PS as well as PNC samples (Shukla & Thakur, 2011).

To further understand the contribution by cation in transport no analysis, the following equation has been used by Sharma et al. (Sharma & Thakur, 2010b).

$$t_{Li} = \left[\frac{I_s(V - I_o R_o)}{I_o(V - I_s R_s)}\right] \tag{4.12}$$

I_o = current before polarization, I_s = current after polarization, R_o = initial resistance, R_s = steady state resistance.

4.3.7 Thermal Stability Analysis

This is another very important analysis tool to understand temperature-dependent phenomena inside a polymer nanocomposite. Ion migration inside a PS or PNC heavily depends on its polymer backbone mobility. Further, the chain mobility is predominantly temperature-dependent. So thermal stability analysis is a very essential tool to understand and correlate electrical results obtained via complex impedance analysis. In general two techniques are used for this purpose, as given below.

4.3.8 Differential Scanning Calorimetry

In this process change in entropy is calculated over a temperature range. DSC characterization technique is very important for PNC. It revels the change in thermal phase transition temperature over PNC formation. The following information can be extracted from this technique.

1. Glass transition, and melting temperature
2. Estimation of crystallization

Salehan et al. have shown that the introduction of Al_2O_3 filler in a corn starch-LiI electrolyte system significantly decreases the crystalline melting temperature. A significant decrease in crystalline melting temperature indicates improved interaction between polymer host and nanofiller (Salehan et al. 2021). This eventually improves the amorphous content in the system. Pandey et al. have prepared PEO: NH_4HSO_4 (80:20 w/w) + $x\%$ SiO_2. The DSC thermogram indicates a significant change in glass transition temperature and crystalline melting temperature concerning the PS system (Pandey et al. 2008). Chen & Chang have calculated the crystallization percentage of PEO-$LiCF_3SO_3$-Clay by using the following equation.

$$\textit{Crystallization percentage} = \frac{\begin{array}{c}\textit{Heat enthalpy of crystalline polymer}\\ \textit{phase under study}\end{array}}{\begin{array}{c}\textit{Heat enthalpy of crystalline polymer}\\ \textit{phase of pure polymer}\end{array}} \tag{4.13}$$

It shows the crystallinity significantly decreases as the clay concentration increases (Chen & Chang, 2001).

4.3.9 Thermogravimetric Analysis

TGA can be primarily be used in understanding the degradation temperature of PNC sample.

Li et al. have prepared a PNC sample using the PEO-PDMS-LiTFSI-OMMT system. They have shown the introduction of ionic liquid and OMMT has improved the thermal response of PNC. They have also concluded that the presence of OMMT has restricted the segmental motion of the polymer and improved its thermal stability (Li et al. 2013). Similar improved thermal stability has been observed in PEO-PMMA–$LiCLO_4$/LiTFSI-based PNC prepared by Liang et al. They have reported improved thermal stability (6.1% degradation) compared to their PS counterpart (7.3% degradation) (Liang et al. 2015). Sharma et al. have shown with a change in clay concentration in PAN-$LiCLO_4$-MMT-based PNC the thermal degradation temperature changes (Sharma et al. 2008).

4.3.10 NMR Analysis

The NMR technique is a very strong tool to investigate the proton transport property of any PNC samples. In a PNC sample cations propagate via different long-range and short-range relaxation assisted by polymer backbone in different atmospheric conditions. NMR is a very useful tool to separate the Li^+, or any other cation dynamics from polymer backbone dynamics. Wong et al. have used NMR spectroscopy in PEO+Li flurohectorite samples to understand the Li^+ migration and to calculate relaxation parameters of the cations (Wang et al. 1996). Saikia et al. have used NMR spectroscopy to understand the cation association with different coordination sites of polymers, plasticizers, and nanomaterials in a gel polymer electrolyte consisting of P(VdF-HFP) + (PC + DEC) + $LiClO_4$ + Silica aerogel. They have observed prominent changes in ^7Li peak profiles with a variation of silica aerogel (Saikia et al. 2009).

4.4 CONCLUDING REMARKS

A detailed review of polymer nanocomposite synthesis and its different characterization methods have been discussed here. Polymer nanocomposite properties and the possibility of their large-scale implication heavily depend upon preparation technique. So a detailed study of PNC thin film preparation by casting technique (solution casting, tape casting), coating technique (spin coating, cip coating), template-assisted method, heat-assisted method (hotpress, melt intercalation), and 1D/2D PNC nanofiber preparation via electrospinning method have been discussed, along with their advantages and disadvantages. Further, different PNC characterization techniques (structural, thermal, surface morphology, electrical, spectroscopy) have been discussed here. A different analytical expression for those characterizations techniques have been summarized here. Shortly PNC is likely to replace conventionally and presently used liquid electrolyte technology for energy storage devices. So a suitable preparation methodology is very much necessary which can successfully transfer technology from lab scale to large scale fabrication without compromising its properties.

ACKNOWLEDGMENT

Dr. Avirup Das is very much thankful to VIT Bhopal University for providing necessary help during the research. Dr. Atma Rai is thankful to the National Project Implementation Unit/AICTE (MoE Govt of India) for providing financial assistance through CRS project (Id:1–5722352205).

REFERENCES

Appetecchi, G.B.; Croce, F.; Hassoun, J.; Scrosati, B.; Salomon, M.; Cassel, F.; Bai, J. (2003a) Hot-pressed, dry, composite, PEO-based electrolyte membranes I. Ionic conductivity characterization. *J. Power Sources*, 114, 105–112.

Appetecchi, G.B.; Hassoun, J.; Scrosati, B.; Croce, F.; Cassel, F.; Salomon, M. (2003b) Hot-pressed, solvent-free, nanocomposite, PEO-based electrolyte membranes II. All solid-state Li/$LiFePO_4$ polymer batteries. *J. Power Sources*, 124(1), 246–253.

Arya, A.; Sharma, A.L. (2017) Polymer electrolytes for lithium ion batteries: A critical study. *Ionics*, 23, 497–540.

Brinker, C.J.; Hurd, A. (1994) Fundamentals of sol-gel dip-coating. *Journal de Physique III, EDP Sciences*, 4(7), 1231–1242.

Caruso, T.; Capoleoni, S.; Cazzanelli, E.; Agostino, R.G.; Villano, P.; Passerini, S. (2002) Characterization of PEO-Lithium Triflate Polymer Electrolytes: Conductivity, DSC and Raman Investigations—Lonics 8.

Chapi, S. (2020) Optical, electrical and electrochemical properties of PCL5/ITO transparent conductive films deposited by spin-coating: Materials for single-layer devices. *Journal of Science: Advanced Materials and Devices*, 5, 322–329.

Chapi, S.; Niranjana, M.; Devendrappa, H. (2016) Synthesis and characterization of nanocomposite polymer blend electrolyte thin films by spin-coating method. *AIP Conference Procedings*, 1731, 080010. doi: 10.1063/1.4947888.

Chen, H.W.; Chang, F.C. (2001) The novel polymer electrolyte nanocomposite composed of poly(ethylene oxide), lithium triflate and mineral clay. *Polymer*, 42(24), 9763–9769.

Das, A.; Thakur, A.K. (2017) Effect on ion dissociation in MWCNT-based polymer nanocomposite. *Ionics*, 23(10), 2845–2853.

Das, A.; Thakur, A.K. (2018) AC conductivity and dielectric spectra study of a blended solid polymer electrolyte. *Int. J. Materials Engineering Innovation*, 9(3), 208–217.

Das, A.; Thakur, A.K. (2020) AC Conductivity and dielectric property analysis of a MWCNT based polymer nano composite. AIP Conference Proceedings, 080041.

Das, A.; Thakur, A.K. (2021) Role of clay intercalation in the structural, and thermal property of a polymer blend electrolyte materials today proceedings. https://doi.org/10.1016/j.matpr.2021.05.339

Das, A.; Thakur, A.K.; Kumar, K. (2013) Exploring low temperature Li+ ion conducting plastic battery electrolyte. *Ionics*, 19, 1811–1823.

Das, A.; Thakur, A.K.; Kumar, K. (2014a) Conductivity scaling and near-constant loss behavior in ion conducting polymer blend. Solid State Ionics, 268, 185–190.

Das, A.; Thakur, A.K.; Kumar, K. (2014b) Evidence of low temperature relaxation and hopping in ion conducting polymer blend. *Solid State Ionics*, 262, 815–820.

Das, A.; Thakur, A.K.; Kumar, K. (2015) Origin of near constant loss (NCL) in ion conducting polymer blends. *J of Physics and Chemistry of Solids*, 80, 62–66.

Das, A.; Thakur, A.K.; Kumar, K. (2017) Raman spectroscopic study of ion dissociation effect in clay intercalated polymer blend nano composite electrolyte. *Vibrational Spectroscopy*, 92, 14–19.

Edman, L. (2000) ion association and ion solvation effects at the crystalline-amorphous phase transition in PEO-LiTFSI. *J. Phys. Chem. B*, 104, 7254–7258.

Gao, Y.; Ying, J.; Xu, X.; Cai, L. (2018) Nitrogen-enriched carbon nanofibers derived from polyaniline and their capacitive properties. *Applied Sciences*, 8, 1079. doi: 10.3390/app8071079

Hashim, M.A.; Sa'adu, L.; Dasuki, K.A. (2012) Supercapacitors based on activated carbon and polymer electrolyte. *International Journal of Sustainable Energy and Environment Research*, 1(1), 1–6.

Hong, K.; Yuk, J.; Kim, H.J.; Lee, J.Y.; Kim, S.; Leed, J.L.; Lee, K.H. (2018) Electrospun polymer electrolyte nanocomposites for solid-state energy storage. *Composites Part B*, 152, 275–281.

Ibrahim, S.; Mohd Yasin, S.M.; Nee, N.M.; Ahmad, R.; Mohd, J.R. (2012) Conductivity and dielectric behaviour of PEO-based solid nanocomposite polymer electrolytes. *Solid State Communications*, 152(5), 426–434. doi: 10.1016/j.ssc.2011.11.037

Klongkan, S.; Pumchusak, J. (2015) Effects of nano alumina and plasticizers on morphology, ionic conductivity, thermal and mechanical properties of PEO-LiCF3SO3 solid polymer electrolyte. *Electrochimica Acta*, 161, 171–176.

Krebs, F.C. (2009) Fabrication and processing of polymer solar cells: A review of printing and coating techniques. *Sol Energy Mater Sol Cells*, 93(4), 394–412.

Kumar, Y.; Pandey, G.P.; Hashmi, S.A. (2012) Gel Polymer electrolyte based electrical double layer capacitors: Comparative study with multiwalled carbon nanotubes and activated electrodes. *The J of Phy Chemistry C*, 116, 26118–26127.

Kwon, Y.; Han, Y. (2020) Fabrication of electrolyte-supported solid oxide fuel cells using a tape casting process. *Journal of the Ceramic Society of Japan*, 128(6), 310–316.

Li, Y.J.; Wu, F.; Chao, H.R.; Chen, S. (2013) A new composite polymer electrolyte based on poly (ethyleneoxide)/polysiloxane/BMImTFSI/organo montmorillionite. *Chinese Chemical Letters*, 24, 70–72.

Liang, B.; Tang, S.; Jiang, Q.; Chen, C.; Chen, X.; Li, S.; Yan, X. (2015) Preparation and characterization of PEO-PMMA polymer, composite electrolytes doped with nano-Al2O3. *Electrochimica Acta*, 169, 334–341.

Liu, Y.; Zhang, N.; Jiao, L.; Tin, J.C. (2015) Nanodots encapsulated in porous nitrogen-doped carbon nanofibers as free standing anode for advanced sodium-ion batteries. *Adv. Mater.*, 27, 6702–6707.

Liu, Y.; Zhang, N.; Yu, C.; Jiao, L.; Chen, J. (2016) $MnFe_2O_4$@C nanofibers as high-performance anode for sodium-ion batteries. *Nano Lett.*, 16(5), 3321–3328.

Mohapatra, S.R.; Thakur, A.K.; Choudhary, R.N.P. (2009) Effect of nanoscopic confinement on improvement in ion conduction and stability properties of an intercalated polymer nanocomposite electrolyte for energy storage applications. *Journal of Power Sources*, 191(2), 601–613.

Nguyen-Tri, P.; Nguyen, T.A.; Carriere, P.; Xuan, C.N. (2018) Nanocomposite coatings: Preparation, characterization, properties, and applications. *International Journal of Corrosion*, 19. https://doi.org/10.1155/2018/4749501

Nho, Y.C.; Sohn, J.Y.; et al. (2017) Preparation of nanocomposite γ-Al_2O_3/polyethylene separator crosslinked by electron beam irradiation for lithium secondary battery. *Radiation Physics and Chemistry*, 132, 65–70.

Nie, S.; Liu, L.; Liu, J.; Xie, J.; Zhang, Y.; Xia, J.; Yan, H.; Yuan, Y.; Wang, X. (2018) Nitrogen-doped TiO_2—C composite nanofibers with high capacity and long-cycle life as anode materials for sodium-ion batteries. *Nano-Micro Lett*, 10, 71.

Norrman, K.; Ghanbari-Siahkali, A.; Larsen, N. B. (2005) Studies of spin-coated polymer films: Annual reports section C". *Physical Chemistry*, 101, 174–201.

Padmaraj, O.; Venkateswarlu, M.; Satyanarayana, N. (2013) Effect of ZnO filler concentration on the conductivity, structure and morphology of PVdF-HFP nanocomposite solid polymer electrolyte for lithium battery application. *Ionics*, 19, 1835–1842.

Pandey, G.P.; Hashmi, S.A.; Agrawal, R.C. (2008) Hot-press synthesized polyethylene oxide based proton conducting nanocomposite polymer electrolyte dispersed with SiO_2 nanoparticles. *Solid State Ionics*, 179(15–16), 543–549.

Park, C.H.; Park, M.; Yoo, S.I.; Joo, S.K. (2006) A spin-coated solid polymer electrolyte for all-solid-state rechargeable thin-film lithium polymer batteries. *Journal of Power Sources*, 158, 1442–1446.

Polu, A.R.; Kumar, R. (2013) Preparation and characterization of pva based solid polymer electrolytes for electrochemical cell applications. *Chinese J of Polymer Science*, 31(4), 641–648.

Pradhan, D.K.; Choudhary, R.N.P.; Samantaray, B.K. (2008) Studies of dielectric relaxation and AC conductivity behavior of plasticized polymer nanocomposite electrolytes. *Int. J. Electrochem. Sci.*, 3, 597–608.

Pradhan, D.K.; Samantaray, B.K.; Choudhary, R.N.P.; Karanv, N.K.; Thomas, R.; Katiyar, R.S.; (2011) Effect of plasticizer on structural and electrical. *Ionics*, 17, 127–134.

Pritam, Arya A.; Sharma, A.L. (2021) *Conductivity and Dielectric Spectroscopy of Na^+ Ion Conducting Blended Solid Polymer Nanocomposites, Recent Research Trends in Energy Storage Devices*. Springer, Singapore. https://doi.org/10.1007/978-981-15-6394-2_14

Ratna, D.; Divekar, S.; Patchaiappan, S.; Samui, A.B.; Chakraborty, B.C. (2007) Poly(ethylene oxide)/clay nanocomposites for solid polymer electrolyte applications. *Polym. Int.*, 56(7), 900–904.

Saikia, D.; Chen-Yang, Y.W.; Chen, Y.T.; Li, Y.K.; Lin, S.I. (2009) 7Li NMR spectroscopy and ion conduction mechanism of composite gel polymer electrolyte: A comparative study with variation of salt and plasticizer with filler. *Electrochimica Acta*, 54, 1218–1227.

Salehan, S.S.; Nadirah, B.N.; Saheed, M.S.M.; Yahya, W.Z.N.; Shukur, M.F.; (2021) Conductivity, structural and thermal properties of corn starch-lithium iodide nanocomposite polymer electrolyte incorporated with Al_2O_3. *J Polym Res*, 28, 222.

Sharma, A.L.; Kumar, S.; Tripathi, Singh B.M.; Vijay, Y.K. (2009) Aligned CNT/polymer nanocomposite membranes for hydrogen separation. *International Journal of Hydrogen Energy*, 34(9), 3977–3982

Sharma, A.L.; Shukla, N.; Thakur, A.K. (2008) Studies on structure property relationship in a Polymer: Clay nanocomposite film based on $(PAN)_8LiClO_4$. *J of Polymer Sci Part B: Polymer Physics*, 46, 2577–2592.

Sharma, A.L.; Thakur, A.K. (2010a) Improvement in voltage, thermal, mechanical stability and ion transport properties in polymer-clay nanocomposites. *J. of Appl Poly Sci*, 118, 2743–2753.

Sharma, A.L.; Thakur, A.K. (2010b) Polymer-ion-clay interaction based model for ion conduction in intercalation-type polymer nanocomposite. *Ionics*, 16, 339–350.

Sharma, A.L.; Thakur, A.K. (2013) Plastic separators with improved properties for portable power device applications. *Ionics*, 19(5), 795–803.

Shukla, N.; Thakur, A.K. (2010) Nanocrystalline filler induced changes in electrical and stability properties of a polymer nanocomposite electrolyte based on amorphous matrix. *J.Mater Sci*, 45, 4236–4250.

Shukla, N.; Thakur, A.K. (2011) Enhancement in electrical and stability properties of amorphous polymer based nanocomposite electrolyte. *Journal of Non-Crystalline Solids*, 357(22–23), 3689–3701.

Stephan, A.M.; Kumar, T.P.; Kulandainathan, M.A.; Lakshmi, N.A. (2009) Chitin incorporated poly(ethylene oxide)-based nanocomposite electrolytes for lithium batteries. *J. Phys. Chem. B*, 113(7), 1963–1971.

Tamilarasan, P.; Ramaprabhu, S. (2013) Graphene based all-solid state supercapacitor with ioniq liquid incorporated polyacrylonitrile electrolyte. *Energy*, 51, 374–381.

Uchino, K. (2010) Multilayer technologies for piezo-ceramic materials. *Advanced Piezoelectric Materials*, 387–411.

Wang, S.; Vaiab, R.A.; Giannelis, E.P.; Zax, D.B. (1996) Dynamics in a poly(ethylene oxide)-based nanocomposite polymer electrolyte probed by solid state NMR. *Solid State Ionics*, 86–88, 547–557.

Wang, X.; Hu, Y.; Li, L.; Fang, H.; Fan, X.; Li, S. (2019) Preparation and performance of polypropylene separator modified by SiO_2/PVA layer for lithium batteries, *e-Polymers*, 19(1), 470–476. https://doi.org/10.1515/epoly-2019-0049

Wilson, M. S.; Gottesfeld, S. (1992) Thin-film catalyst layers for polymer electrolyte fuel cell electrodes. *J Appl Electrochem*, 22(1), 1–7.

Wua, J.; et al. (2019) Functional composite polymer electrolytes with imidazole modified SiO2 nanoparticles for high-voltage cathode lithium ion batteries. *Electrochimica Acta*, 320, 10, 134567.

Xia, J.; Liu, L.; Jamil, S.; Xie, J.; Yuan, Y.; Zhang, Y.; Nie, S.; Pan, J.; Wang, X.; Cao, G.; Yan, H. (2019) Free-standing SnS/C nanofiber anodes for ultralong cycle-life lithium-ion batteries and sodium-ion batteries. *Energy Storage Materials*, 17, 1–11.

Xie, S.; Jones, E.H.; Shen, Y.; Hornsby, P.; McAfee, M.; McNally, T.; Patel, R.; Benkreira, H.; Coates, P.; (2010) Quantitative characterization of clay dispersion in polypropylene-clay nanocomposites by combined transmission electron microscopy and optical microscopy. *Materials Letters*, 64(2), 185–188.

Part II

Application

5 Polymer Composites for Supercapacitors

Atma Rai, Shweta Tanwar, Avirup Das, and A. L. Sharma

CONTENTS

DOI: 10.1201/9781003208662-7

5.1 INTRODUCTION

The high demand of energy in the last few years attracted a flurry of research activity in the field of alternative sources of energy (renewable energy) (Tanwar et al., 2022b). As the population is growing rapidly, the optimisation between the production and demand of energies is facing challenges in the coming days (Tanwar et al., 2021). Fossil fuels such as coal, petrochemicals, natural gas and nuclear energy are supplying the majority of energy demand globally, and they are considered lifelines for various utilities including automobiles and other sectors (Wilberforce et al., 2019). Though they contribute ~70% of electricity production globally, their non-renewable nature and continuous depletion have created big issues (Tanwar et al., 2022a). Many studies on fossil fuels have indicated that these major contributors of energy on earth are exhaustible and depleting continuously. In addition to that, these sources of energy have shown a harmful impact on the environment via carbon emissions, which has led to the quest of alternative sources of energy generation for domestic and industrial operations (Singh et al., 2022). The world body's striving for the reduction of carbon emissions has also forced different countries to think about other sources of energy (Wilberforce et al., 2017). In view of these issues, renewable energy production and storage is urgently required to fulfil the energy needs of domestic and industrial evolution. Some storage devices such as electric capacitors, batteries and fuel cells have been developed. Typically, the electric capacitor is made of two metal plates separated by a dielectric medium. The storage capacities of capacitors are found to be very low. Though batteries are used to store a large amount of charges, the power density of the battery is found to be low as it takes sufficient time to charge and discharge. So, in order to fill the gap between capacitors and batteries, supercapacitors have received tremendous worldwide attention due to their potential applications. Supercapacitors are considered as a bridge between batteries and electric capacitors (Tanwar et al., 2021). The charge capacity and power density of supercapacitors are relatively high (i.e. they are capable of charging (taking energy) and discharging (releasing energy) very fast). The present chapter is fully devoted to polymer electrolytes for supercapacitor applications. Properties of supercapacitors are tailored by selecting appropriate electrodes and electrolytes (Arya et al., 2021). Wide research is being carried out globally in this field to identify/search for electrodes and electrolyte materials to optimise the property of supercapacitors to make them cost effective, eco-friendly.

5.2 BACKGROUND OF SUPERCAPACITORS

The concept of energy storage is not new but rather started back in the eighteenth century. The roadmap of development of research in the supercapacitor field is depicted in Figure 5.1a. The first attempt in this direction came in to existence by a German engineer Ewald George von Kliest in 1745. Further on the same year, a Dutch scientist Pieter van Musschenbroek invented the first capacitor on the same principle, and it was named the Leyden jar (Keithley, 1999). The Leyden jar was a simple device containing a glass jar filled with water and lined with a conducting foil inside and out. Daniel Gralath, a Poland physicist, attempted to improve charge

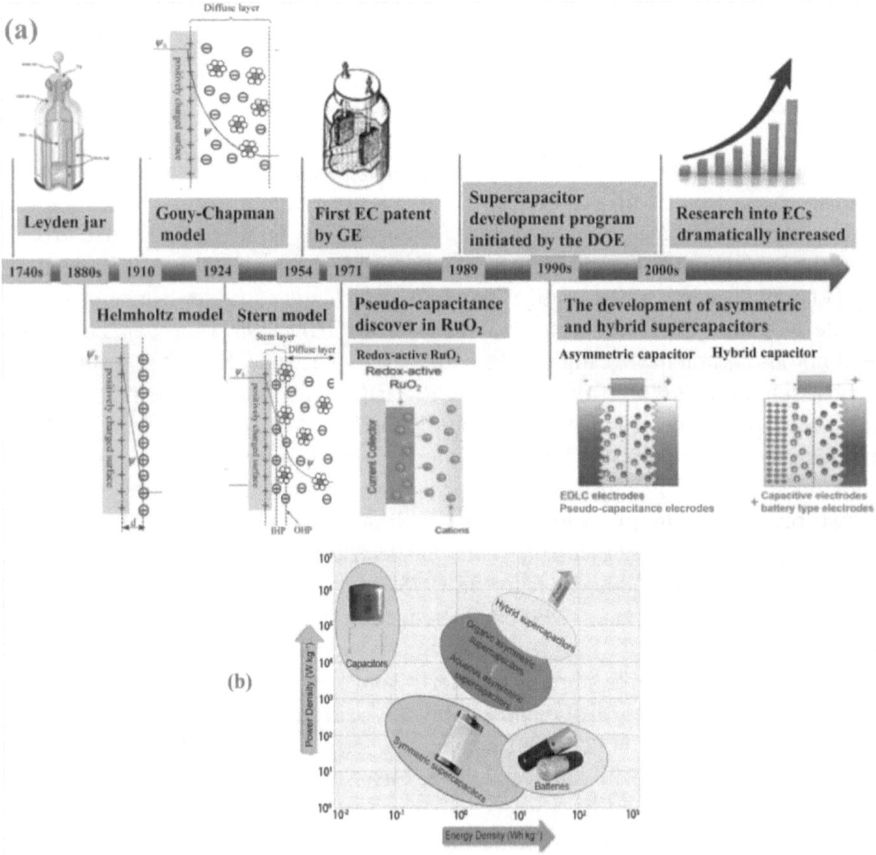

FIGURE 5.1 (a) Progress of the supercapacitor; (b) sketch of Ragone plot for different energy storage and conversion devices.

Source: (a) Reproduced with permission from Jingyuan Zhao & Burke (2021, Elsevier); (b) Forouzandeh et al. (2020).

storage capacity by demonstrating such jars in parallel combination. Benjamin Franklin studied extensively on the Leyden jar and concluded that the electric charges were stored on glass plates rather in the water as presumed earlier (von Kleist, n.d.). He also demonstrated the combination of several jars and named it a 'battery'. The term 'capacitor' was adopted in 1926 and previously these capacitors were termed as condensers and coined by Allesandro Volta in 1782 in reference to their ability to store more density of charges as compared to that of isolated conductors. Later, in the 1830s a renowned English scientist Michael Faraday observed that the materials used between the plates plays a key role in storing the charges. He made major contributions in the field of capacitors including the concept of a dielectric constant. The unit of capacitance 'Farad' was named to give honour to Michael Faraday. Later on several other attempts were made with various dielectric materials such as mica, porcelain and paper to improve the charge storage capacity of the capacitor. Charles Pollack invented the first electrolytic (alkaline electrolyte) capacitors. Bell Laboratories invented a new type of tantalum capacitor in the 1950s having solid electrolytes and capability of storing more electric charges in less space.

Several capacitors were developed using different dielectric materials to store electric charges but their potential applications are limited. These capacitors are mainly used as circuit elements and filter to stop dc signals. After the Second World War, the industrial revolution started globally and it was realised there was need of high energy production and storage. The first electrolytic capacitor (supercapacitor) was developed by H. Becker from General Electric (GE) in 1957 (Ho et al., 2010). He used porous carbon having high surface area as an electrode. Robert A. Rightmire, a chemist at Standard Oil Company Ohio (SOHIO), came up with the first electrochemical capacitor design in 1966 (Schindall, 2007). Later, Donald L. Boos patented the electrochemical capacitor, having activated carbon (high surface area) as an electrode in 1970. These electrochemical capacitors were made of electrodes consisting of two pieces of aluminium foil covered with activated carbon. SOHIO patented his findings to a Japanese company (Nippon Electric Company) which brought it in market with the name 'supercapacitor' in 1975.

B. E. Conway (Pell et al., 1999) conducted extensive research on supercapacitors with oxide materials (RuO_2) as electrodes, and further, he successfully described the basic mechanism of charge storage in batteries and electrochemical supercapacitors. He also observed that two different mechanisms—surface adsorption (non-Faradic), forming a double layer, and charge transfer between electrode-electrolyte interfaces (Faradic)—are involved in supercapacitors. His work on supercapacitor provided extensive understanding of the charge storage mechanism.

The supercapacitor market is increasing very fast. Researchers working on supercapacitors came up with different electrodes and electrolytes in order to improve their storage capacity. Supercapacitors can be compared with batteries in view of charge storage capacity, energy density and power density. The charge storage mechanism differs in these two energy storage devices. A typical battery consists of two electrodes, namely cathodes and anodes. The potential energy is released through redox processes which take place between electrodes and electrolytes. The anode oxidises when the current passes through the circuit whereas it reduces at the cathode. The chemical reaction that takes place between electrodes and electrolytes also affects its functionality after several cycles of charge discharge. Battery discharge is also found to be exothermic which in turn gives off a substantial amount of heat that affects its operation. The basic difference between capacitors, supercapacitors, batteries and fuel cells can be understood by the Ragone plot.

A Ragone plot (Figure 5.1b) indicating energy density and power density is very much helpful to compare various energy storage devices. In this plot the energy density (Wh kg^{-1}) is plotted to the power density (W kg^{-1}). Actually, this plot was initially designed for batteries. Now, it is applicable for all kinds of energy storage devices. In this plot, the horizontal axis clearly indicates the energy density of various devices such as capacitors, supercapacitors, batteries and fuel cells. Whereas, the vertical axis reveals how fast these devices deliver stored energies.

5.3 DIFFERENT CHARGE STORAGE MECHANISMS OF SUPERCAPACITORS

Typically, the quantum of charge storage capacity is found to be very high as compared to normal electric capacitors. The charges are stored via Faradic and/or non-Faradic method, depending on the type of materials used as electrodes and electrolytes in the supercapacitor. A supercapacitor is also divided into three categories on the basis of charge storage mechanism: electric double layer capacitor (EDLC), where the charges are being stored via non-Faradic mode; ultra capacitor (pseudocapacitor), where the charges are stored via Faradic mode; and hybrid supercapacitor that possesses both Faradic and non-Faradic processes to store charge.

The charge storage mechanism in EDLC-based supercapacitors takes place via non-Faradic mode where charges are adsorbed at surfaces similar to that found in normal capacitors.

A sketch of all the electrochemical capacitors is presented in Figure 5.2, which clearly shows the charge storage mechanism in all three types of supercapacitors. In the EDLC-type of electrochemical capacitor, charges/ions from electrolytes form a double layer similar to the polarisation effect seen in normal capacitors containing dielectric materials as electrolytes. On the other hand, the charge storage in transition metal oxide-based materials (RuO_2, MnO_2, Co_3O_4, ZrO_2, Fe_3O_4, etc.) takes place via Faradic process. In this process redox reaction occurs at electrode materials in accordance to the following equation:

$$Ox + Ze \rightleftarrows \mathrm{Re}\, d$$

Actually, an electric double layer-type electrochemical capacitor is complemented by the capacitors which are based on pseudo-capacitance that typically arises out in some electro-sorption processes or in redox reactions taking place at electrode surfaces. Such an ultracapacitor has shown excellent capacitive behaviour. When mixed valent oxide materials are being used as an electrode, the oxidation-reduction (Redox) peaks are clearly visualised in cyclic Voltammogram.

In addition to the EDLC and pseudocapacitor, there is another possibility where both Faradic and non-Faradic mechanisms are employed to store charges. Such type of energy storage device is called a hybrid supercapacitor. In hybrid supercapacitors, the electrode materials are preferred in such a way that it promotes both types of storage mechanism that in turn enhance the capacitive property of the device. Various composite material is prepared as an electrode including polymer composite with carbonaceous and transition metal-based oxide to achieve high energy density.

5.4 FEASIBILITY OF POLYMER COMPOSITE AS ELECTROLYTE IN SUPERCAPACITORS

The electrolyte is an important component of a supercapacitor for deciding the overall performances of the devices. Electrolytes behave as a platform for ions and hinder the electrode movement within. A polymer electrolyte is a composition of dissolved salts in a suitable polymer matrix with a property of high ionic conductivity.

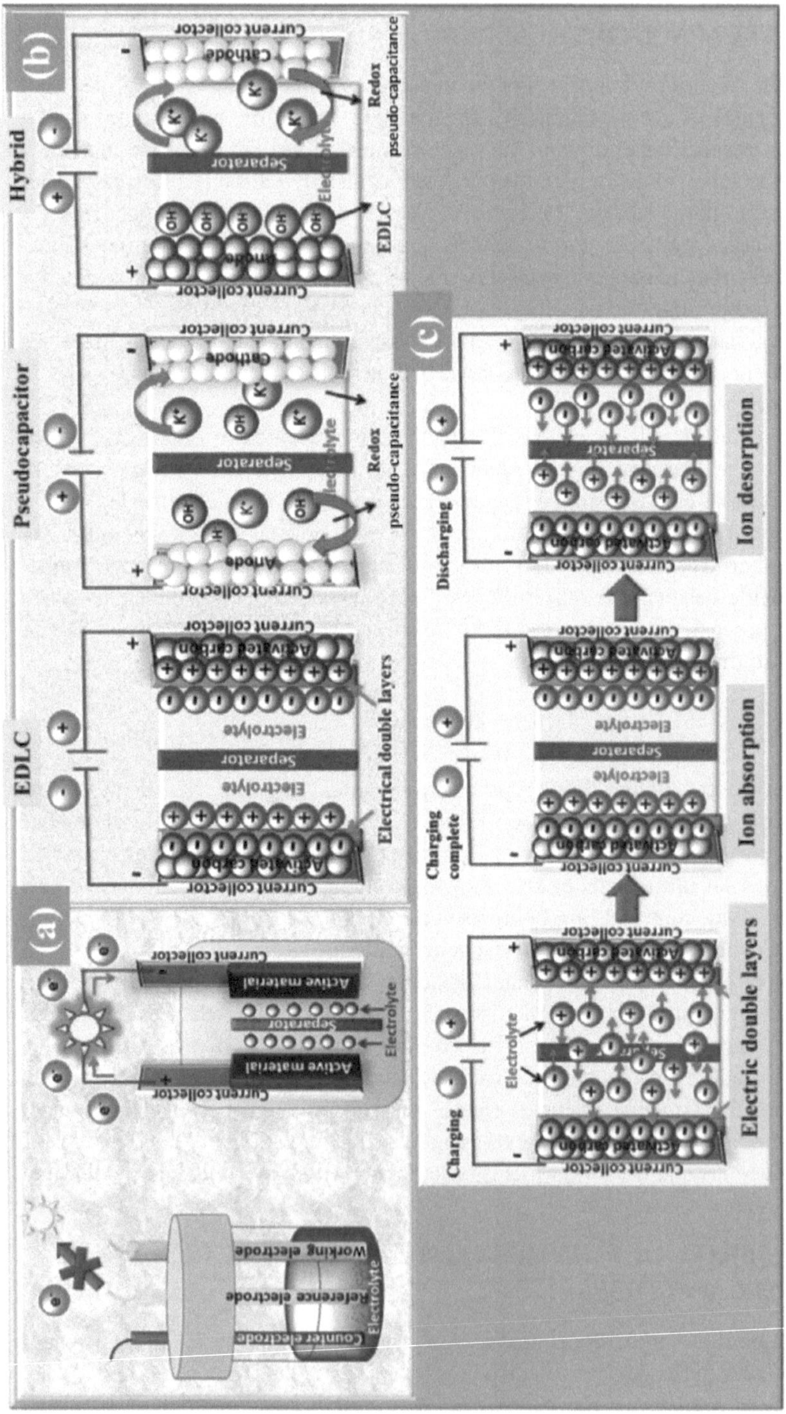

FIGURE 5.2 (a) Schematics of the three-electrode configuration and two-electrode configuration; (b) type of supercapacitors: EDLC, pseudocapacitor and hybrid; (c) charge storage mechanism in EDLCs, ion absorption and desorption on the surface.

Source: Reproduced with permission from Bhupender Pal et al. (2019, Royal Society of Chemistry).

5.4.1 Key Features of Desirable Polymer Electrolyte

The electrolyte to be compatible for device fabrication needs to have certain specific characteristics like conductivity (including both electronic and ionic), low cost, lightweight, stability (chemical, thermal and mechanical) and so on.

5.4.1.1 Ionic Conductivity

The polymer matrix ion dynamics can be easily studied via ionic conductivity equation, $\sigma = \sum ne\mu$, where n represents free charge carries number, e is charge of the ion, μ reveals the ion mobility. The ionic conductivity of the polymer-based electrolyte is high, based on the ion mobility and number of free charge carriers' mobilisation into the polymer matrix. The value of ionic conductivity is determined via complex impedance spectroscopy (CIS) tool along an ac signal of range 10–100 mV applied across the cell (electrode||polymer electrolyte||electrode) formed under study. The ionic conductivity also was calculated using a Nyquist plot (graph of real and imaginary parts of impedance) using relation; $\sigma = \frac{t}{A * R_b}$ (t refers to thickness of the polymer electrolyte, A is area of the electrode across the electrolyte and R_b represents the bulk resistance, its value extracted from the intercept of the Nyquist plot on the real axis).

The ionic conductivity is a temperature dependent parameter. The lowering of activation energy along the activation of the charge carriers takes place as the temperature is increased. Based on the temperature taken, the variation of ionic conductivity with the temperature exhibits three behaviours: (i) Vogel-Tamman-Fulcher (VTF) behaviour, (ii) Williams-Landel-Ferry (WLF) behaviour (MacCallum & Vincent, 1989; Adam & Gibbs, 1965; Williams et al., 1955; Watanabe & Ogata, 1988).

5.4.1.1.1 Vogel-Tamman-Fulcher (VTF) Behaviour

The VTF depicts a plot of ionic conductivity versus reciprocal of temperature which is non-linear in nature. The ion migration happens using segmental motion of the polymer-chain-cum-hopping. The ion motion mainly depends on the free volume offered by polymer chains in the polymer matrix. The VTF relation is represented as $\sigma = AT^{-1/2}e^{\left(-\frac{B}{T-T_o}\right)}$, where σ refers to ionic conductivity, B is a constant, A represents the pre-exponential factor, T_0 is the temperature near to T_g of the material under study when entropy is close to zero.

5.4.1.1.2 Arrhenius Behaviour

The Arrhenius behaviour supports the idea of high ion dynamics with the help of the hopping mechanism. It is more valid at the temperature lower than the transition temperature of the material. The upliftment in the temperature leads to activation of charge carries and increase in flexibility to fasten the ion mobilisation through coordinating sites. The Arrhenius expressions is $\sigma = \sigma_o e^{\left(-\frac{E_a}{kT}\right)}$; here σ_o is the pre-exponential factor, k refers to the Boltzmann constant. The lower value of activation energy favours the high ion migration and thus high ionic conductivity.

5.4.1.2 Cation/Ion Transference Number

The main contribution is of ion migration in a polymer electrolyte (PE). Thereby, to determine the accurate contribution of cations and ions, the transference number of the cation (t_c) and (t_i) are evaluated via cell assembly (Electrode|PE|Electrode). The cation transference number is calculated using AC impedance while ion transference number is determined via Wagner DC polarization technique (Sen et al., 2016; Mindemark et al., 2018). The equation to calculate cation transference number is $t_c = \frac{I_s\left(V - I_i I_i\right)}{I_i\left(V - I_s I_s\right)}$; here V is the potential applied across the cell, I_s and I_i are the steady-state and initial currents, R_i and R_s represent the resistance at interfaces before and after polarisation. The ion transference number is obtained using equation $t_i = \frac{(I_t - I_e)}{I_t} \times 100$; here I_t and I_e represent the total and residual current and the relation relating them is $I_t = I_i + I_e$.

5.4.1.3 Electrochemical Stability Window (ESW)

The potential window is a crucial parameter to decide the suitability of the polymer electrolyte for supercapacitor application. The main focus of the research community is to enhance the energy density of the supercapacitor for practical application which directly depends in the potential window of the electrolyte and complete cell. The ESW of the polymer electrolyte under study is determined via linear sweep voltammetry (LSV) technique, which is a graph between current and voltage.

5.4.1.4 Thermal Stability

The thermal stability of the electrolyte is a major aspect for supercapacitor application. The role of the electrolyte is avoiding the short-circuiting of the cell via maintaining the gap between the electrodes which are placed across the polymer electrolyte. So, it becomes essential for a polymer electrolyte to work for wider ranges of temperature (–20 to 80 °C). The popularly used liquid electrolyte in supercapacitors gets decomposed and releases excess heat that accelerates temperature and pressure inside the cell during its operation. The thermal stability of polymer electrolyte is much higher for normally popular liquid electrolytes. The thermal stability and voltage window range for a polymer electrolyte can be improved using appropriate nanofillers into the polymer matrix.

5.4.2 Different Types of Polymer Electrolytes

The polymer electrolyte used in various energy storage applications are of two types based on the source criterion as shown in Figure 5.3. They are (i) natural and (ii) synthetic PE. The natural obtained PE are from chitosan (Kadir et al., 2011), corn starch (Liew & Ramesh, 2013) and rice starch (Khanmirzaei & Ramesh, 2014; Ramesh, Shanti, et al., 2012). There are four kinds of PE based on the physical state and chemical composition: (i) solid polymer electrolyte (SPE), (ii) composite polymer electrolyte (CPE), (iii) plasticised polymer electrolyte, (iv) gel polymer electrolyte (GPE).

5.4.2.1 Solid Polymer Electrolytes (SPEs)

Wright et al. was first to report the research work on SPE, which was 30 years ago (Wright, 1975). The application of SPEs in electrochemical devices was launched by Armand et al. long back in 1979 (Vashishta et al., 1979). The first reported SPE was prepared by taking PEO as polymer host and was a solvent-free film (Fenton, 1973). But, the ionic conductivity showed by this PEO-based SPE was less at room temperature. The SPE plays a dual role as electrolyte plus separator for electrochemical devices. It can also be used as a binder for electrical contact between electrodes (Armand & Van Tendeloo, 1992). The major features of SPEs are high thermal, mechanical, electrochemical electrical stabilities. Other important features are leak proof, solvent-free, low volatility, light weight, and high flexibility (Sharma et al., 2018, 2022). The merits of SPEs are their ability to reduce the chance of release of hazardous gas or leakage of solvent liquids, and to operate at a wide temperature range (Raghavan et al., 1998). But SPEs have certain major disadvantages like high interface resistance, low ionic conductivity as compared to liquid electrolytes (Ngai et al., 2016).

5.4.2.2 Composite Polymer Electrolytes (CPEs)

The efforts to overcome the demerits of SPEs encouraged the discovery of composite polymer electrolytes. The various methods utilised to prepare CPEs are binary salt systems, doping of nanomaterials, cross-linking polymer matrices (Wen et al., 2003), polymer blending (J. Hu et al., 2011), addition of plasticisers (Ramesh & Bing, 2012) and incorporation of ionic liquids (Hao et al., 2015). The merits of using CPE are their high ionic conductivity, high thermal and mechanical stability, high flexibility and better interfacial contact (Kam et al., 2014). The ionic and electronic conductivities' value for CPEs highly rely on the parameters such as size of the particle, concentration, surface area, pore size and the level of interaction among incorporated particles and the polymer matrices (Stevens & Wieczorek, 1996).

5.4.2.3 Gel Polymer Electrolytes (GPEs)

Gel polymer electrolytes are popularly also known as plasticised polymer electrolytes proposed by Feuillade and Perche firstly in 1975 (Feuillade & Perche, 1975). The procedure to synthesise GPEs is very easy, with a blend mixture of polymer like poly(ethylene oxide) (PEO), an alkali metal salt of lithium or sodium, and a suitable solvent. The clear and viscous mixture is casted into a petri dish and dried properly to form a thin film. The GPE has merits of both liquid and solid electrolytes (Stephan, 2006). The peculiar features of GPE are better chemical, mechanical, electrochemical, photochemical, structural stability; high ionic conductivity, low reactivity, safety in use and low volatility (Liew et al., 2011). The use of GPEs in supercapacitor can reduce the possibilities of leakage and internal short-circuiting and hence uplifts the device life (Kim et al., 2003). The major drawback of GPE is its poor mechanical stability. This demerit can be overcome by incorporating fillers or nanomaterials into the polymer matrices (Zhang et al., 2011).

The polymer electrolyte is prepared using a minimum of one polymer as host for incorporation of other materials for the enhancement of electrochemical properties, especially ionic conductivity. Table 5.1 showcases the various polymer hosts.

TABLE 5.1
List of Polymer Hosts with Their Respective Ionic Conductivity at Room Temperature.

Polymer Host	Repeat Unit	Ionic Conductivity (S cm^{-1})	Reference
Poly(ethylene host)	$-(CH_2CH_2O)_n-$	2.6×10^{-4}	(Itoh et al., 2003)
Poly(vinylidene fluoride)	$-(CH_2CH_2O)_n-$	4.5×10^{-3}	(Song et al., 2004)
Poly(vinylidene fluoride-hexafluoro propylene)	$-[(CH_2-CF_2)-(CF_2-CF-(CF_3)]_n-$	4.3×10^{-3}	(J.-K. Kim et al., 2008)
Poly(methyl methacrylate)	$-(CH_2C(-CH_3)(-COOCH_3))_n-$	7.3×10^{-6}	(Rajendran et al., 2002)
Poly(vinyl chloride)	$-(CH_2-CHCl)_n-$	2.8×10^{-6}	(Ramesh, Yin et al., 2011)
Poly(vinyl alcohol)	$-(CH_2-CH(-OH))_n$	4.8×10^{-2}	(C.-C. Yang & Lin, 2002)
Poly(acrylic acid)	$-CH_2-CH(-COOH))_n-$	2.2×10^{-4}	(Kam et al., 2014)
Poly(acrylonitrile)	$-(CH_2-CH(-CN))_n-$	5.7×10^{-4}	(Chen-Yang et al., 2002)
Poly(ethyl methacrylate)	$-(CH_2-C(-CH_3)(-COOCH_2CH_3))_n-$	1.2×10^{-4}	(Ramesh et al., 2014)
Poly(ε-caprolactone)	$-(O-(CH_2)_5-CO)_n-$	3.8×10^{-5}	(Woo et al., 2014)
Chitosan	$-(CH_2-(C_5H_2NO_3)_5-OH)_n-$	1.6×10^{-3}	(Kadir et al., 2010)

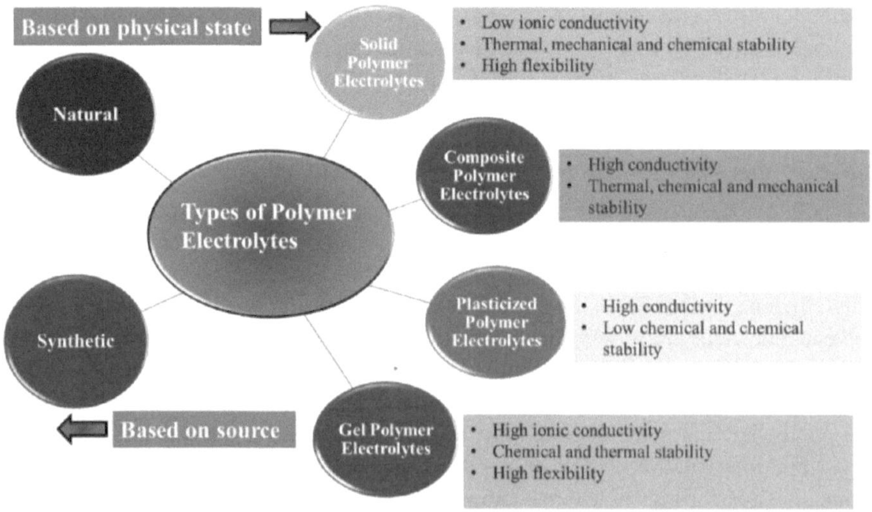

FIGURE 5.3 Different types of polymer electrolytes used in supercapacitor applications.

The insertion of the small particles as fillers into the matrix of a host polymer improves the overall properties of the polymer electrolyte. The additional properties that are improvised after adding filler to PE are higher ionic conductivity, better mechanical and thermal stability, lower value of glass transition temperature and reduced crystallinity and so on (Rajendran et al., 2002). It further decreases the crystallinity of the PE and uplifts the amorphous phase with its better stability (Croce et al., 1992). The mostly used high surface area fillers are titanium oxide (TiO_2), aluminium oxide (Al_2O_3), hydrophobic-fumed silica and zirconium oxide (ZrO_2) (Raghavan et al., 1998).

5.5 WAYS TO IMPROVE ELECTROCHEMICAL PERFORMANCES OF POLYMER ELECTROLYTES

Various techniques can be utilised to enhance the properties and performance of the polymer electrolytes. The popular methods used are incorporation of inorganic fullers, polymer blending, addition of plasticisers and dopants and use of comb-branched copolymers (Hashmi et al., 2000).

5.5.1 Inorganic Fillers

PEs are also synthesised by adding nanomaterial as fillers into the matrices of the host polymer. The incorporation of nanofillers improves the mechanical stability of PE along its ionic conductivity (Zhang et al., 2011).

5.5.2 Blending

The polymer blending is the most easily practiced technique. It includes the step of mixing a minimum of two polymers with/without chemical bonding between them. It is a way to mix two polymers/copolymers physically. The blending of polymers improves the mechanical stability of the PE over the entire temperature range. The merits of using a blend-based PE system are its high ionic conductivity, physical and electrical properties (Ahmad et al., 2007). The blending method helps to lower the crystallinity of the prepared PE along better increments of the amorphous fractions of the polymer matrices.

5.5.3 Plasticisers

The use of plasticisers in the PE helps to improve the plasticity and ionic mobility along with enhancing ionic conductivity at room temperature (Ramesh & Bing, 2012). The popular liquid plasticisers that have the benefit of a low molecular weight used for PE synthesis are diocthyl adipate (DOA), dimethyl formamide (DMF), DBP, dimethyl carbonate (DMC), diethyl phthalate (DEP), PC, g-butyrolactone (BL), glycol sulphite (GS), methylethyl carbonate (MEC) and EC (Shukur et al., 2013). The other benefits of using plasticisers are their ability to provide a maximum charge carrier by providing an easy path for ion mobilisation, to uplift PE ionic conductivity, to improve thermal and mechanical stabilities (Ramesh, Yin, et al., 2011). The demerits of plasticisers are their low flash point, narrow region of potential window, narrow working voltage and high vapour pressure (Raghavan et al., 2010).

5.5.4 Copolymers

Copolymers are defined as the mixture of two different kinds of monomers, which are mixed via cross-linking method (Wen et al., 2003). The most common copolymer used is PVdF-HFP which is the resultant of the co-polymerisation of PVdF and HFP respectively. The features of copolymers are more enhanced in many aspects for use in the polymer electrolyte than the standard monomers.

5.5.5 Cross-Linked Polymers

The better and higher ionic conductivity of a PE highly depends on the amorphous nature of the polymer host used. The cross-linked polymers-based PEs showcase high and improved ionic conductivity at room temperature as they exhibit greater amorphous features (Nishimoto et al., 1999). They also show low elasticity, brittleness and processability. The research on high molecular weight cross-linked polymers indicates that the PEs based on them exhibit high ionic conductivity. The ionic conductivity and flexibility of PEs highly rely on the monomer content used for cross-linked polymers.

5.5.6 Doping Salts

The PEs prepared using doping salts show high conductivity and electrical stability. The interface among the polymer host and dopant salt in the synthesised PE play a crucial role in deciding the ionic conductivity, chemical stability and mechanical stability of the PE (Ramesh, Lu, et al., 2012). The main reason of high conductivity of doped-salt-based PE is its low lattice energy and its high dielectric constant.

5.5.7 Room Temperature Ionic Liquids (RTILS)

RTILS are comprised of molten salts with organic cations and inorganic anions (Ramesh, Lu, et al., 2012). They are classified as aprotic ionic liquids and protic ionic liquids (Luo et al., 2012). The important properties of RTILS are low viscosity; recyclability; less flammability; minimum vapour pressure; low melting point; wider potential window; high ionic conductivity; mechanical, chemical and thermal stability (Liew et al., 2012). The research interest in RTILS increased due to their ability to make PEs suitable for various applications like catalytic cracking and radical polymerisation of the polymers. The RTILS-based PEs are environmentally friendly and reduce the harmful gases released into the atmosphere (Jain et al., 2005). The merit of RTILS-based PEs are their extremely improved ionic conductivity. Whereas the drawbacks related to RTILS-based PEs are their not-so-good mechanical stability (Ramesh & Liew, 2012).

5.6 PREPARATION METHODS

The preparation of the desired PE for supercapacitor application is synthesised utilising various methods such as spin coating, solution cast technique, dip coating, melt intercalation and hot press technique. *Spin coating*—the step involved in this method is the dropping of a small quantity of solution on a suitable substrate and placing the substrate on the spin coater to rotate at constant speed. The uniform layer of the solution is formed on the substrate. This method is more popular for less viscous mixture. *Solution cast technique*—this is the most widely practiced method for PE synthesis. The main reasons are its easy, fast and wide range thick (50–300 μm) PE fabrication. *Dip coating*—this is the lowest-cost technique used to produce high uniform film coated on either side of a substrate. It is a three-step process, namely immersion, deposition and drainages respectively. *Melt intercalation*—this is a useful method due to its environmental friendly and low-cost nature without requiring any solvent. *Hot press technique*—the hot press technique is a cost-effective, easy method and a solvent free technique.

5.6.1 Characteristics of the Techniques

Various characterisation methods are utilised for analysing the suitability of the prepared PE for supercapacitor application. The essential techniques used for PE characterisations are mentioned as follows: *Field Emission Scanning Electron Microscopy (FESEM)*—this is an essential microstructural and morphology analysis technique for the polymer electrolyte. *X-Ray Diffraction (XRD)*—the structural structure of the prepared PE studied via XRD method. The parameters which can be determined using a XRD plot are d-spacing, inter chain separation, crystallite size and so on. *Fourier Transform Infrared spectroscopy (FTIR)*—this is a useful technique to study the various interactions among different constituents present in the prepared PE. It also helps to determine the free ions' contribution to the total ionic conductivity of the PE sample. The various transport parameters are also estimated via FTIR spectra. *Differential Scanning Calorimetry (DSC)*—the different transition in the polymer electrolyte with the change in the temperature is determined via DSC technique. *Nuclear Magnetic Resonance imaging (NMR)*—the ion dynamics and ionic conductivity via line-width of the NMR spectra are determined for the prepared polymer electrolyte respectively. *Linear Sweep Voltammetry (LSV)*—the decomposition voltage window of the prepared PE is evaluated via LSV technique. *Thermogravimetric Analysis (TGA)*—it is important to check the thermal stability of the polymer electrolyte for the supercapacitor application. *I-t characteristics (Transference Number)*—the i-t characteristics techniques for the PE is used to analysis the ionic and electronic transference number of the polymer electrolyte.

5.6.1.1 The Parameters to Evaluate Supercapacitor Applications

The use of prepared polymer electrolyte for supercapacitor applications could be confirmed via calculating the following mentioned parameters.

The specific capacitance measured in F g^{-1} of the fabricated cell determined from Nyquist plot of impedance spectroscopy:

$$C_{EIS} = \frac{1}{2*3.14*f*m*Z''} \tag{5.1}$$

Here, m in mg is the active mass of the electrode pasted on the current collector, f in Hz represents the frequency range applied using noting readings and Z'' refers to the complex impedances measured in ohms.

The specific capacitance calculated from cyclic voltammetry profile is given as:

$$C_{CV} = \frac{\int I * dV}{2 * m * S * V} \tag{5.2}$$

Here $\int I * dV$ refers to the area enclosed by CV curve, m is the mass loaded, S is the scan rate applied and V represents potential applied cell.

The capacitance value is evaluated using galvanostatic charging/discharging curves using the equation:

$$C_{GCD} = \frac{I * t}{m * V} \tag{5.3}$$

Here I, t, m and V represent the current, discharging time, mass of the electrode and potential window of the cell under study. The specific capacitance of the single electrode is evaluated to be four times of the fabricated cell.

The equivalent series resistance (ESR) of the fabricated cell could be determined from the GCD profile via equation indicated as follows:

$$ESR = \frac{V_{drop}}{2 \times I} \tag{5.4}$$

Here, V_{drop} refers to voltage drop and I is the applied discharge current.

The energy, power density and coulombic retention of the cell is determined as

$$E = \frac{CV^2}{7.2} \text{ (Wh kg}^{-1}\text{)} \tag{5.5}$$

$$P = \frac{E * 3600}{t_{dis}} \text{ (kW kg}^{-1}\text{)} \tag{5.6}$$

$$\eta = \frac{t_{dis}}{t_{ch}} \times 100\% \tag{5.7}$$

Here C, V, E, t_{dis} and t_{ch} refer to the specific capacitance, potential window, energy density, discharge time and charging time, respectively.

5.7 LATEST DEVELOPMENTS

The target of safe and suitable supercapacitors is fulfilled via preparing appropriate polymer electrolyte. The selection of the perfect polymer electrolyte for a supercapacitor depends on its properties such as low crystallinity, high ionic conductivity, and improved energy-cum-power-density. Next we discuss the recent findings of versatile polymer electrolytes for energy storage devices, specially supercapacitors.

Zhao et al. (Zhao et al., 2021b) prepared PVA-based cotton gel polymer electrolyte (Figure 5.4) which exhibited 1.6 V potential window along ionic conductivity

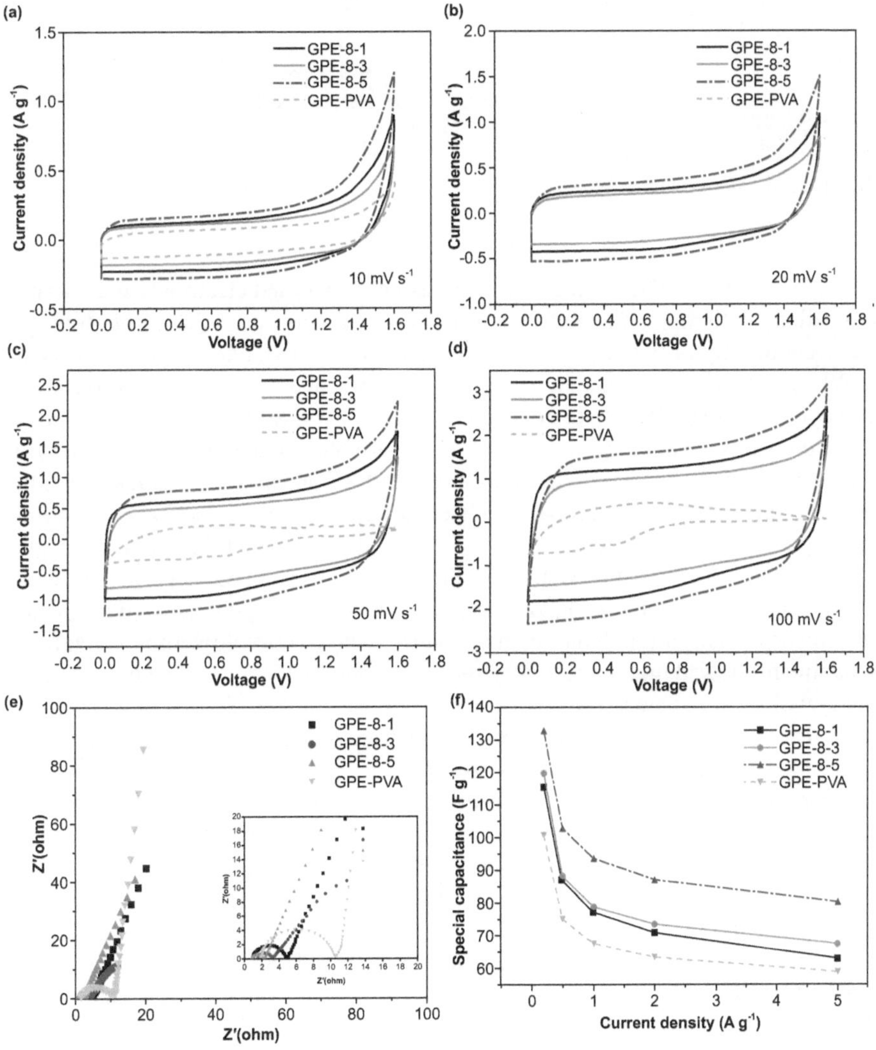

FIGURE 5.4 (a–d) Comparative cyclic voltammetry curves at different scan rates, (e) Nyquist plot of all samples, (f) variation of specific capacitance with current densities, (g) capacitance retention of prepared samples.

Source: Reproduced with permission from Zhao et al. (2021b, Springer Nature); Z.

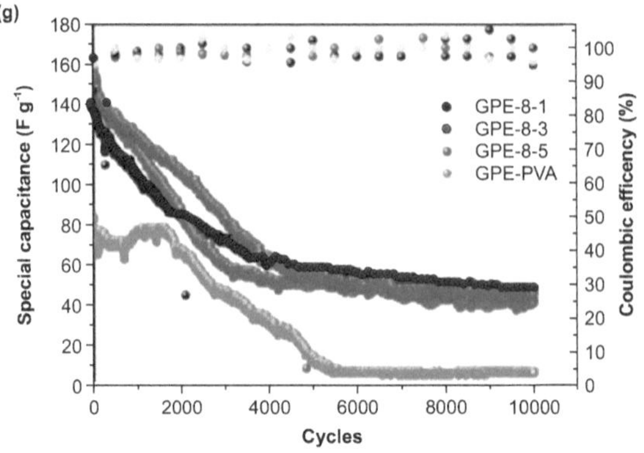

FIGURE 5.4 (Continued)

of 28×10^{-3} S cm^{-1}. The device fabricated using PVA-based electrolyte with a ratio of 8:5 along with dipping in KOH solution showed specific capacitance of 160 F g^{-1} at current density of 0.5 A g^{-1}. The supercapacitor also exhibited energy density of 12 Wh kg^{-1}. Yanfang Xu, et al. (Xu et al., 2021) fabricated a supercapacitor device by employing polymer electrolyte of poly(vinylidene) along lithium triflate, and epoxy. The dip-coating technique was utilised to prepare the suitable polymer electrolyte. The device displayed specific capacitance of 12 mF g^{-1} and high mechanical strength. The mechanical properties of the fabricated device were tested using cyclic loading and a quasi-static test respectively. The devices also presented the improved cyclic stability around 96% via a galvanostatic charging/discharging (GCD) test. Devi et al. (Devi et al., 2021) reported a polymer electrolyte based on a TiO_2 nanowire as nanofiller incorporated in a polymer as host matrix. The electrolyte with 0.5 weight % of nanowire showed high ionic conductivity of order 10^{-5} S cm^{-1} along the potential window of approximately 5 V. The ion transference number of the film was near to unity. The optimised polymer electrolyte was used to prepare a prototype supercapacitor device which demonstrated a specific capacitance of 58 F g^{-1} at 10 mV s^{-1}. The corresponding emerging and power density exhibited were 1.38 Wh kg^{-1} and 0.709 kW kg^{-1}. The Coulombic efficiency of the device was maintained to 98% after 500 GCD cycles respectively. Kumar et al. (Kumar et al., 2020) investigated $BaTiO_3$ (nanofiller) incorporated polymer electrolyte. Increased electrical conductivity via addition of nanofiller was reported. The stable potential window was about 5 V for the optimised sample (PPS5). The optimised polymer electrolyte was used to make an EDLC device, which exhibited specific capacitance of 134 F g^{-1} at a scan rate of 10 mV s^{-1}. The reported energy and power density of the device was 5.99 Wh Kg^{-1}, and 27.03 kW kg^{-1} at current density 0.02 A g^{-1}, respectively.

Jianghe Liu et al. (Liu et al., 2021) synthesised a gel-polymer-based electrolyte of PVdF-HFP/EMITf/Al$(Tf)_3$ materials (Figure 5.5). The film prepared provided better ionic conductivity of value 1×10^{-3} S cm^{-1} along the stable potential window of 4.2 V at ambient temperature. The supercapacitor device fabricated from the prepared electrolyte exhibited specific capacitance of 323.9 F g^{-1} and showed cyclic stability for 5000 GCD cycles respectively at room temperature. The device fabricated was flexible in nature and showed tolerance at low temperature and good performance under various harsh conditions. Kamboj et al. (Kamboj et al., 2021) reported the polymer electrolyte with ZrO_2 as nanofiller in the PEO matrix via solution cast method. The confirmation of the polymer electrolyte was done via X-ray diffraction, Fourier transform infrared spectra and so on. The ionic conductivity of the reported electrolyte was around 3×10^{-4} S cm^{-1}, which increased with the temperature. The stable potential window for the electrolyte was around 4 V and the ion transference number was near to unity. The optimised film showed the highest dielectric constant and lowest relaxation time respectively. Further, the optimised electrolyte was utilised to prepare a solid flexible ultracapacitor which exhibited specific capacitance of value 42 F g^{-1} at scan rate 10 mV s^{-1}. The energy and power density of the fabricated EDLC device were 13 Wh kg^{-1} and 2.8 kW kg^{-1} respectively.

Shujahadeen B. Aziz et al. (Aziz, et al., 2021) investigated biodegradable polymer-based electrolyte comprised of chitosan incorporated with ammonium thiocyanate

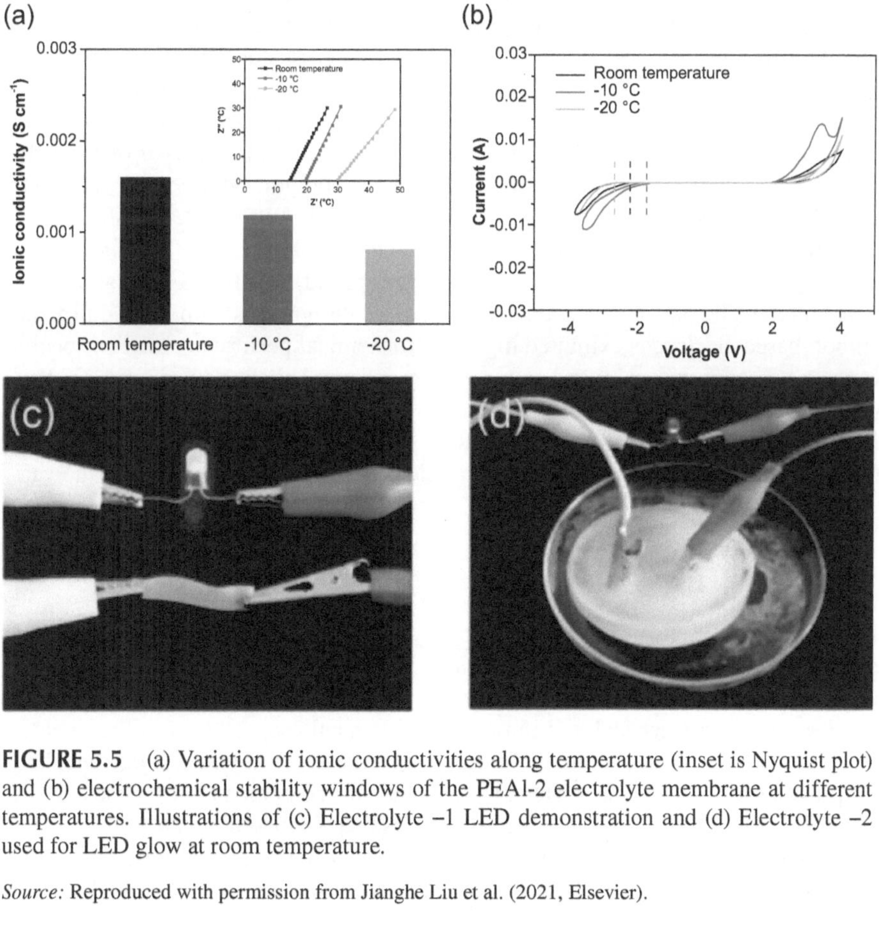

FIGURE 5.5 (a) Variation of ionic conductivities along temperature (inset is Nyquist plot) and (b) electrochemical stability windows of the PEA1-2 electrolyte membrane at different temperatures. Illustrations of (c) Electrolyte –1 LED demonstration and (d) Electrolyte –2 used for LED glow at room temperature.

Source: Reproduced with permission from Jianghe Liu et al. (2021, Elsevier).

using the solution cast technique. The ionic conductivity of the electrolyte was found to be 9×10^{-4} S cm^{-1} with stable potential window 2 V. The evaluated ionic transference number was equal to 0.95 and could be considered as a high ionic conductive system. The optimised film was used to fabricate the EDLC device, which depicted specific capacitance of 41 F g^{-1} at 50 mV s^{-1}. The specific energy and power density was determined to be 15 Wh kg^{-1} and 750 W kg^{-1}. The equivalent series resistance of the device was found to be in the range of 75–330 Ω. Mohamad A. Brza et al. (Brza et al., 2021) prepared poly(vinyl alcohol) (PVA)-based polymer electrolyte with nanofiller ammonium thiocyanate dispersed uniformly. The reported film displayed the ionic conductivity around 2×10^{-5} S cm^{-1} along the stable potential window 2 V. The parameters such as ions number density, mobility and diffusion coefficient were incremented with more glycerol content. The synthesised polymer electrolyte was utilised to design an EDLC-type of supercapacitor. The parameters like specific capacitance, energy density and power density were evaluated to be 18 F g^{-1}, 2 Wh kg^{-1} and 318 W kg^{-1} respectively. The cycling stability was tested for 450 GCD cycles. Shujahadeen B. Aziz et al. (Aziz, Dannoun, et al., 2021) reported the glycerol-based polymer electrolyte which showed improved properties such as high ionic conductivity of 1×10^{-3} S cm^{-1} along potential window of 2 V. The supercapacitor build-up via using the prepared electrolyte-cum-separator exhibited specific capacitance of 50 F g^{-1}. The stability was cross checked using 500 GCD cycles. The calculated energy density of the EDLC device was about 6 Wh kg^{-1}. The equivalent series resistance and power density of the device was found to be 63 Ω and 1642 W kg^{-1} respectively. Maryam Hina et al. (Hina et al., 2021) investigated a hydrogel-polymer-based composite electrolyte of poly(acrylamide)/poly(3,4-ethylenedioxythiophene):poly(styrene sulfonate) (PEDOT: PSS) system with sodium montmorillonite clay doped via the free radical polymerisation technique. The electrolyte showed improved ionic conductivity around 14×10^{-1} S cm^{-1} at ambient temperature. The prototype supercapacitor device designed using prepared film and carbon-based electrodes exhibited high electrochemical performances. The specific capacitance, energy density and power density delivered by the device were 327 F g^{-1} (at 3 mV s^{-1}), 54 Wh kg^{-1} and 100 W kg^{-1} respectively. The fabricated device was further utilised to light up the light emitting diode (LED) respectively.

Table 5.2 represents the comparison of electrochemical performance of the various polymer-based electrolytes.

5.8 CONCLUDING REMARKS

Energy storage systems like supercapacitors are needed to fulfil society's demand for solving the energy crisis. The polymer-based supercapacitor is one of the best options to replace the current aqueous electrolyte-based supercapacitors. The lightweight, flexibleness, high electrical and improved mechanical properties of polymer-based electrolytes make them in demand for supercapacitor applications. The current chapter provides an overview of different supercapacitor components. It focuses on different polymer nanocomposite-based electrodes (carbon-based, and transition metal oxide-based) materials and summarises their performance parameters. A detailed discussion on different charge storage mechanisms (Faradic, non-Faradic) has been

TABLE.5.2
Various Polymer Electrolytes and Their Supercapacitor Performance.

Polymer Electrolyte	Ionic Conductivity ($S\ cm^{-1}$)	Specific Capacitance ($F\ g^{-1}$)	Energy Density ($Wh\ kg^{-1}$)	Power Density ($W\ kg^{-1}$)	Capacitance retention (%)	Reference
Cotton/PVA	28×10^{-3}	60 @ 0.5 A g^{-1}	11.8	-	37.5	(Z. Zhao et al., 2021)
PVdF/LiTf/epoxy	-	162	-	-	96%	(Xu et al., 2021)
PEO/$NaPF_6$/$BaTiO_3$	2.61×10^{-6}	135.16 @ 0.02 A g^{-1}	5.99	27.03	-	(Kumar et al., 2020)
PVdF-HFP/EMITf/$Al(Tf)_3$	1.6×10^{-3}	323	-	-	67% @ 50,000 cycles	(Liu et al., 2021)
PEO-$NaPF_6$-ZrO_2	2.9×10^{-4}	42 @10 mV/s	13	2.8	-	(Kamboj et al., 2021)
PNC-TiO_2	10^{-5}	57.5 @ 10 mV/s	1.38	0.71	97.7% @ 500 cycles	(Devi et al., 2021)
Chitosan (CS) Chitosan (CS)-NH_4SCN	8.5×10^{-4}	41 @ 50 mV/s	15	750	-	(Aziz, Nofal, et al., 2021)
PVA–NH_4SCN	1.82×10^{-5}	18.30	2.06	318	95.39% @ 200 cycles	(Brza et al., 2021)
Chitosan—methylcellulose—NH_4NO_3–glycerol	1.31×10^{-4}	50	6.97	1941	90% @ 500 cycles	(Aziz, Dannoun, et al., 2021)
PVdF-HFP	7.8×10^{-2}	115	27	-	-	(Redda et al., 2021)
PEDOT: PSS	13.7×10^{-3}	385.4	53	100	-	(Hina et al., 2021)
Poly(arylene ether ketone) (PAEK)/ poly(ethylene glycol)-grafted	2×10^{-2}	134.38	-	-	54% after 5000 cycles	(F. Hu et al., 2021)
CNT fibers/polymer electrolyte	3.5×10^{5}	88	37.5	30	96% after 5000 cycles	(Senokos et al., 2018)
Carboxylated chitosan hydrogel	8.69×10^{-2}	45.9	5.2	226.6	-	(H. Yang et al., 2019)
PVA—$LiClO_4$	-	114	1.9	150	94% after 5000 cycles	(Le et al., 2019)
Poly(vinyl alcohol) (PVA)-based gel polymer electrolyte	61.1×10^{-3}	392.1	43.1	-	87.9% after 10000 cycles	(Fan et al., 2020)
Poly(ethylene glycol) diglycidyl ether and hydrophobic bisphenol	2.41×10^{-3}	19.8	-	-	100% for 10,000 cycles	(Lim et al., 2019)
Boron-containing GPE	5.1×10^{-3}	-	27.62	6.91	91.2% after 5000 cycles	(Jin et al., 2018)
CNF membrane	-	625.6	21.7	1100	75.4% after 5000 cycles	(Hou et al., 2019)

discussed. In brief, the different electrolytes used for supercapacitors are discussed in brief. Hence, polymer-based electrolytes have great potential for supercapacitor applications.

ACKNOWLEDGEMENT

Dr. Atma Rai is very much thankful to the National Project Implementation Unit/ AICTE (MoE Govt of India) for providing financial assistance through CRS project (Id:1–5722352205). Dr. Avirup Das is very much thankful to VIT Bhopal University for providing necessary help during research. Shweta Tanwar is thankful to UGC for providing the JRF fellowship.

REFERENCES

Adam, G., & Gibbs, J. H. (1965). On the temperature dependence of cooperative relaxation properties in glass-forming liquids. *The Journal of Chemical Physics*, *43*(1), 139–146.

Ahmad, Z., Al-Awadi, N. A., & Al-Sagheer, F. (2007). Morphology, thermal stability and visco-elastic properties of polystyrene: Poly (vinyl chloride) blends. *Polymer Degradation and Stability*, *92*(6), 1025–1033.

Armand, M., & Van Tendeloo, G. (1992). *Solid polymer electrolytes-fundamentals and technological applications: By Fiona M. Gray, VCH, Weinheim 1991, X, 245 pp., DM 128, hardcover, ISBN 3-527-27925-3*. Wiley Online Library, USA.

Arya, A., Gaur, A., Kumar, V., Tanwar, S., & Sharma, A. L. (2021). Nanostructured metal oxides for hybrid supercapacitors. In *Advanced ceramics for energy and environmental applications* (pp. 82–103). CRC Press, UK.

Aziz, S. B., Dannoun, E., Hamsan, M. H., Ghareeb, H. O., Nofal, M. M., Karim, W. O., Asnawi, A. S. F. M., Hadi, J. M., & Kadir, M. F. Z. A. (2021). A polymer blend electrolyte based on CS with enhanced Ion transport and electrochemical properties for electrical double layer capacitor applications. *Polymers*, *13*(6), 930.

Aziz, S. B., Nofal, M. M., Abdulwahid, R. T., Kadir, M. F. Z., Hadi, J. M., Hessien, M. M., Kareem, W. O., Dannoun, E. M. A., & Saeed, S. R. (2021). Impedance, FTIR and transport properties of plasticized proton conducting biopolymer electrolyte based on chitosan for electrochemical device application. *Results in Physics*, *29*, 104770.

Brza, M. A., Aziz, S. B., Anuar, H., Alshehri, S. M., Ali, F., Ahamad, T., & Hadi, J. M. (2021). Characteristics of a plasticized PVA-based polymer electrolyte membrane and H+ conductor for an electrical double-layer capacitor: Structural, morphological, and ion transport properties. *Membranes*, *11*(4), 296.

Chen-Yang, Y. W., Chen, H. C., Lin, F. J., & Chen, C. C. (2002). Polyacrylonitrile electrolytes: 1. A novel high-conductivity composite polymer electrolyte based on PAN, LiClO4 and α-Al2O3. *Solid State Ionics*, *150*(3–4), 327–335.

Croce, F., Scrosati, B., & Mariotto, G. (1992). Electrochemical and spectroscopic study of the transport properties of composite polymer electrolytes. *Chemistry of Materials*, *4*(6), 1134–1136.

Devi, C., Swaroop, R., Arya, A., Tanwar, S., Sharma, A. L., & Kumar, S. (2021). Fabrication of energy storage EDLC device based on self-synthesized TiO2 nanowire dispersed polymer nanocomposite films. *Polymer Bulletin*, 1–19.

Fan, L.-Q., Tu, Q.-M., Geng, C.-L., Huang, J.-L., Gu, Y., Lin, J.-M., Huang, Y.-F., & Wu, J.-H. (2020). High energy density and low self-discharge of a quasi-solid-state supercapacitor with carbon nanotubes incorporated redox-active ionic liquid-based gel polymer electrolyte. *Electrochimica Acta*, *331*, 135425.

Fenton, D. E. (1973). Complexes of alkali metal ions with poly (ethylene oxide). *Polymer, 14*, 589.

Feuillade, G., & Perche, P. (1975). Ion-conductive macromolecular gels and membranes for solid lithium cells. *Journal of Applied Electrochemistry, 5*(1), 63–69.

Forouzandeh, P., Kumaravel, V., & Pillai, S. C. (2020). Electrode materials for supercapacitors: A review of recent advances. *Catalysts, 10*(9), 969.

Hao, J., Li, X., Yu, S., Jiang, Y., Luo, J., Shao, Z., & Yi, B. (2015). Development of proton-conducting membrane based on incorporating a proton conductor 1, 2, 4-triazolium methanesulfonate into the Nafion membrane. *Journal of Energy Chemistry, 24*(2), 199–206.

Hashmi, S. A., Upadhyaya, H. M., Thakur, A. K., & Verma, A. L. (2000). Experimental investigations on poly (ethylene oxide) based sodium ion conducting composite polymer electrolytes dispersed with SnO 2. *Ionics, 6*(3), 248–259.

Hina, M., Bashir, S., Kamran, K., Iqbal, J., Ramesh, S., & Ramesh, K. (2021). Fabrication of aqueous solid-state symmetric supercapacitors based on self-healable poly (acrylamide)/ PEDOT: PSS composite hydrogel electrolytes. *Materials Chemistry and Physics, 273*, 125125.

Ho, J., Jow, T. R., & Boggs, S. (2010). Historical introduction to capacitor technology. *IEEE Electrical Insulation Magazine, 26*(1), 20–25.

Hou, M., Xu, M., Hu, Y., & Li, B. (2019). Nanocellulose incorporated graphene/polypyrrole film with a sandwich-like architecture for preparing flexible supercapacitor electrodes. *Electrochimica Acta, 313*, 245–254.

Hu, F., Liu, Y., Shao, W., Zhang, T., Liu, S., Liu, D., Zhang, S., & Jian, X. (2021). Novel poly (arylene ether ketone)/poly (ethylene glycol)-grafted poly (arylene ether ketone) composite microporous polymer electrolyte for electrical double-layer capacitors with efficient ionic transport. *RSC Advances, 11*(24), 14814–14823.

Hu, J., Luo, J., Wagner, P., Agert, C., & Conrad, O. (2011). Thermal behaviours and single cell performance of PBI-OO/PFSA blend membranes composited with Lewis acid nanoparticles for intermediate temperature DMFC application. *Fuel Cells, 11*(6), 756–763.

Itoh, T., Ichikawa, Y., Uno, T., Kubo, M., & Yamamoto, O. (2003). Composite polymer electrolytes based on poly (ethylene oxide), hyperbranched polymer, $BaTiO_3$ and LiN $(CF_3SO_2)_2$. *Solid State Ionics, 156*(3–4), 393–399.

Jain, N., Kumar, A., & Chauhan, S. M. S. (2005). Metalloporphyrin and heteropoly acid catalyzed oxidation of CNOH bonds in an ionic liquid: Biomimetic models of nitric oxide synthase. *Tetrahedron Letters, 46*(15), 2599–2602.

Jin, M., Zhang, Y., Yan, C., Fu, Y., Guo, Y., & Ma, X. (2018). High-performance ionic liquid-based gel polymer electrolyte incorporating anion-trapping boron sites for all-solid-state supercapacitor application. *ACS Applied Materials & Interfaces, 10*(46), 39570–39580.

Kadir, M. F. Z., Aspanut, Z., Majid, S. R., & Arof, A. K. (2011). FTIR studies of plasticized poly (vinyl alcohol): Chitosan blend doped with NH_4NO_3 polymer electrolyte membrane. *Spectrochimica Acta Part A: Molecular and Biomolecular Spectroscopy, 78*(3), 1068–1074.

Kadir, M. F. Z., Majid, S. R., & Arof, A. K. (2010). Plasticized chitosan: PVA blend polymer electrolyte based proton battery. *Electrochimica Acta, 55*(4), 1475–1482.

Kam, W., Liew, C.-W., Lim, J. Y., & Ramesh, S. (2014). Electrical, structural, and thermal studies of antimony trioxide-doped poly (acrylic acid)-based composite polymer electrolytes. *Ionics, 20*(5), 665–674.

Kamboj, V., Arya, A., Tanwar, S., Kumar, V., & Sharma, A. L. (2021). Nanofiller-assisted Na^+-conducting polymer nanocomposite for ultracapacitor: structural, dielectric and electrochemical properties. *Journal of Materials Science, 56*(10), 6167–6187.

Keithley, J. F. (1999). *The story of electrical and magnetic measurements: From 500 BC to the 1940s*. John Wiley & Sons, USA.

Khanmirzaei, M. H., & Ramesh, S. (2014). Nanocomposite polymer electrolyte based on rice starch/ionic liquid/TiO2 nanoparticles for solar cell application. *Measurement*, *58*, 68–72.

Kim, H.-S., Kum, K.-S., Cho, W.-I., Cho, B.-W., & Rhee, H.-W. (2003). Electrochemical and physical properties of composite polymer electrolyte of poly (methyl methacrylate) and poly (ethylene glycol diacrylate). *Journal of Power Sources*, *124*(1), 221–224.

Kim, J.-K., Cheruvally, G., Li, X., Ahn, J.-H., Kim, K.-W., & Ahn, H.-J. (2008). Preparation and electrochemical characterization of electrospun, microporous membrane-based composite polymer electrolytes for lithium batteries. *Journal of Power Sources*, *178*(2), 815–820.

Kumar, A., Madaan, M., Arya, A., Tanwar, S., & Sharma, A. L. (2020). Ion transport, dielectric, and electrochemical properties of sodium ion-conducting polymer nanocomposite: Application in EDLC. *Journal of Materials Science: Materials in Electronics*, *31*, 10873–10888.

Le, P.-A., Nguyen, V.-T., Yen, P.-J., Tseng, T.-Y., & Wei, K.-H. (2019). A new redox phloroglucinol additive incorporated gel polymer electrolyte for flexible symmetrical solid-state supercapacitors. *Sustainable Energy & Fuels*, *3*(6), 1536–1544.

Liew, C.-W., & Ramesh, S. (2013). Studies on ionic liquid-based corn starch biopolymer electrolytes coupling with high ionic transport number. *Cellulose*, *20*(6), 3227–3237.

Liew, C.-W., Ramesh, S., & Durairaj, R. (2012). Impact of low viscosity ionic liquid on PMMA-PVC-LiTFSI polymer electrolytes based on AC-impedance, dielectric behavior, and HATR-FTIR characteristics. *Journal of Materials Research*, *27*(23), 2996–3004.

Lim, J. Y., Kang, D. A., Kim, N. U., Lee, J. M., & Kim, J. H. (2019). Bicontinuously crosslinked polymer electrolyte membranes with high ion conductivity and mechanical strength. *Journal of Membrane Science*, *589*, 117250.

Liu, J., Khanam, Z., Ahmed, S., Wang, H., Wang, T., & Song, S. (2021). A study of low-temperature solid-state supercapacitors based on Al-ion conducting polymer electrolyte and graphene electrodes. *Journal of Power Sources*, *488*, 229461.

Luo, J., Van Tan, T., Conrad, O., & Vankelecom, I. F. J. (2012). 1H-1, 2, 4-triazole as solvent for imidazolium methanesulfonate. *Physical Chemistry Chemical Physics*, *14*(32), 11441–11447.

MacCallum, J. R., & Vincent, C. A. (1989). *Polymer electrolyte reviews* (Vol. 2). Springer Science & Business Media, London.

Mindemark, J., Lacey, M. J., Bowden, T., & Brandell, D. (2018). Beyond PEO: Alternative host materials for Li^+-conducting solid polymer electrolytes. *Progress in Polymer Science*, *81*, 114–143.

Ngai, K. S., Ramesh, S., Ramesh, K., & Juan, J. C. (2016). A review of polymer electrolytes: Fundamental, approaches and applications. *Ionics*, *22*(8), 1259–1279.

Nishimoto, A., Agehara, K., Furuya, N., Watanabe, T., & Watanabe, M. (1999). High ionic conductivity of polyether-based network polymer electrolytes with hyperbranched side chains. *Macromolecules*, *32*(5), 1541–1548.

Pal, B., Yang, S., Ramesh, S., Thangadurai, V., & Jose, R. (2019). Electrolyte selection for supercapacitive devices: A critical review. *Nanoscale Advances*, *1*(10), 3807–3835.

Pell, W. G., Conway, B. E., Adams, W. A., & De Oliveira, J. (1999). Electrochemical efficiency in multiple discharge/recharge cycling of supercapacitors in hybrid EV applications. *Journal of Power Sources*, *80*(1–2), 134–141.

Raghavan, P., Zhao, X., Manuel, J., Chauhan, G. S., Ahn, J.-H., Ryu, H.-S., Ahn, H.-J., Kim, K.-W., & Nah, C. (2010). Electrochemical performance of electrospun poly (vinylidene fluoride-co-hexafluoropropylene)-based nanocomposite polymer electrolytes incorporating ceramic fillers and room temperature ionic liquid. *Electrochimica Acta*, *55*(4), 1347–1354.

Raghavan, S. R., Riley, M. W., Fedkiw, P. S., & Khan, S. A. (1998). Composite polymer electrolytes based on poly (ethylene glycol) and hydrophobic fumed silica: dynamic rheology and microstructure. *Chemistry of Materials*, *10*(1), 244–251.

Rajendran, S., Mahendran, O., & Kannan, R. (2002). Ionic conductivity studies in composite solid polymer electrolytes based on methylmethacrylate. *Journal of Physics and Chemistry of Solids*, *63*(2), 303–307.

Ramesh, S., & Bing, K. N. (2012). Conductivity, mechanical and thermal studies on poly (methyl methacrylate)-based polymer electrolytes complexed with lithium tetraborate and propylene carbonate. *Journal of Materials Engineering and Performance*, *21*(1), 89–94.

Ramesh, S., & Liew, C.-W. (2012). Exploration on nano-composite fumed silica-based composite polymer electrolytes with doping of ionic liquid. *Journal of Non-Crystalline Solids*, *358*(5), 931–940.

Ramesh, S., Liew, C.-W., & Ramesh, K. (2011). Evaluation and investigation on the effect of ionic liquid onto PMMA-PVC gel polymer blend electrolytes. *Journal of Non-Crystalline Solids*, *357*(10), 2132–2138.

Ramesh, S., Lu, S.-C., & Morris, E. (2012). Towards magnesium ion conducting poly (vinylidenefluoride-hexafluoropropylene)-based solid polymer electrolytes with great prospects: Ionic conductivity and dielectric behaviours. *Journal of the Taiwan Institute of Chemical Engineers*, *43*(5), 806–812.

Ramesh, S., Shanti, R., & Morris, E. (2012). Studies on the thermal behavior of CS: LiTFSI:[Amim] Cl polymer electrolytes exerted by different [Amim] Cl content. *Solid State Sciences*, *14*(1), 182–186.

Ramesh, S., Uma, O., Shanti, R., Yi, L. J., & Ramesh, K. (2014). Preparation and characterization of poly (ethyl methacrylate) based polymer electrolytes doped with 1-butyl-3-methylimidazolium trifluoromethanesulfonate. *Measurement*, *48*, 263–273.

Ramesh, S., Yin, T. S., & Liew, C.-W. (2011). Effect of dibutyl phthalate as plasticizer on high-molecular weight poly (vinyl chloride): Lithium tetraborate-based solid polymer electrolytes. *Ionics*, *17*(8), 705–713.

Redda, H. G., Nikodimos, Y., Su, W.-N., Chen, R.-S., Jiang, S.-K., Abrha, L. H., Hagos, T. M., Bezabh, H. K., Weldeyohannes, H. H., & Hwang, B. J. (2021). Enhancing the electrochemical performance of a flexible solid-state supercapacitor using a gel polymer electrolyte. *Materials Today Communications*, *26*, 102102.

Schindall, J. (2007). The charge of the ultracapacitors. *IEEE Spectrum*, *44*(11), 42–46.

Sen, S., Jayappa, R. B., Zhu, H., Forsyth, M., & Bhattacharyya, A. J. (2016). A single cation or anion dendrimer-based liquid electrolyte. *Chemical Science*, *7*(5), 3390–3398.

Senokos, E., Ou, Y., Torres, J. J., Sket, F., González, C., Marcilla, R., & Vilatela, J. J. (2018). Energy storage in structural composites by introducing CNT fiber/polymer electrolyte interleaves. *Scientific Reports*, *8*(1), 1–10.

Sharma, S. K., Sharma, G., Gaur, A., Aray, A., Mirsafi, F. S., Abolhassani, R., . . . & Mishra, Y. K. (2022). Progress in Electrode and Electrolyte Materials: Path to All-solid-state Li-ion Batteries (ASSLIB). *Energy Advances*. e378.

Sharma, B., Malik, P., & Jain, P. (2018). Biopolymer reinforced nanocomposites: A comprehensive review. *Materials Today Communications*, *16*, 353–363.

Shukur, M. F., Ithnin, R., Illias, H. A., & Kadir, M. F. Z. (2013). Proton conducting polymer electrolyte based on plasticized chitosan: PEO blend and application in electrochemical devices. *Optical Materials*, *35*(10), 1834–1841.

Singh, N., Tanwar, S., Yadav, B. C., & Sharma, A. L. (2022). High efficient activated carbon-based asymmetric electrode for energy storage devices. *Materials Today: Proceedings, 57:5-10.*

Song, M.-K., Kim, Y.-T., Cho, J.-Y., Cho, B. W., Popov, B. N., & Rhee, H.-W. (2004). Composite polymer electrolytes reinforced by non-woven fabrics. *Journal of Power Sources*, *125*(1), 10–16.

Stephan, A. M. (2006). Review on gel polymer electrolytes for lithium batteries. *European Polymer Journal*, *42*(1), 21–42.

Stevens, J. R., & Wieczorek, W. (1996). Ionically conducting polyether composites. *Canadian Journal of Chemistry*, *74*(11), 2106–2113.

Tanwar, S., Arya, A., Gaur, A., & Sharma, A. L. (2021). Transition metal dichalcogenide (TMDs) electrodes for supercapacitors: a comprehensive review. *Journal of Physics: Condensed Matter*, *33*(30), 303002. https://doi.org/10.1088/1361-648x/abfb3c

Tanwar, S., Arya, A., Singh, N., Yadav, B. C., Kumar, V., Rai, A., & Sharma, A. L. (2021). High efficient carbon coated TiO_2 electrode for ultra-capacitor applications. *Journal of Physics D: Applied Physics,* 55:055501.

Tanwar, S., Singh, N., & Sharma, A. L. (2022a). Aging impact of Se powder on the electrochemical properties of Molybdenum selenide: Supercapacitor application. *Materials Today: Proceeding, 57:94-99.*

Tanwar, S., Singh, N., & Sharma, A. L. (2022b). Structural and electrochemical performance of carbon coated molybdenum selenide nanocomposite for supercapacitor applications. *Journal of Energy Storage*, *45*, 103797.

Vashishta, P., Mundy, J. N., & Shenoy, G. (1979). *Fast ion transport in solids: Electrodes and Electrolytes,* Elsevier North Holland, Inc; New York.

Kleist, E. G." Britannica.com, https://www.britannica.com/biography/E-Georg-von-Kleist; and "Pieter van Musschenbroek," Britannica.com,

Watanabe, M., & Ogata, N. (1988). Ionic conductivity of polymer electrolytes and future applications. *British Polymer Journal*, *20*(3), 181–192.

Wen, Z., Itoh, T., Uno, T., Kubo, M., & Yamamoto, O. (2003). Thermal, electrical, and mechanical properties of composite polymer electrolytes based on cross-linked poly (ethylene oxide-co-propylene oxide) and ceramic filler. *Solid State Ionics*, *160*(1–2), 141–148.

Wilberforce, T., Baroutaji, A., Soudan, B., Al-Alami, A. H., & Olabi, A. G. (2019). Outlook of carbon capture technology and challenges. *Science of the Total Environment*, *657*, 56–72.

Wilberforce, T., El-Hassan, Z., Khatib, F. N., Al Makky, A., Baroutaji, A., Carton, J. G., & Olabi, A. G. (2017). Developments of electric cars and fuel cell hydrogen electric cars. *International Journal of Hydrogen Energy*, *42*(40), 25695–25734.

Williams, M. L., Landel, R. F., & Ferry, J. D. (1955). The temperature dependence of relaxation mechanisms in amorphous polymers and other glass-forming liquids. *Journal of the American Chemical Society*, *77*(14), 3701–3707.

Woo, H. J., Liew, C.-W., Majid, S. R., & Arof, A. K. (2014). Poly (ε-caprolactone)-based polymer electrolyte for electrical double-layer capacitors. *High Performance Polymers*, *26*(6), 637–640.

Wright, P. V. (1975). Electrical conductivity in ionic complexes of poly (ethylene oxide). *British Polymer Journal*, *7*(5), 319–327.

Xu, Y., Pei, S., Yan, Y., Wang, L., Xu, G., Yarlagadda, S., & Chou, T.-W. (2021). High-performance structural supercapacitors based on aligned discontinuous carbon fiber electrodes and solid polymer electrolytes. *ACS Applied Materials & Interfaces*, *13*(10), 11774–11782.

Yang, C.-C., & Lin, S.-J. (2002). Alkaline composite PEO-PVA-glass-fibre-mat polymer electrolyte for Zn-air battery. *Journal of Power Sources*, *112*(2), 497–503.

Yang, H., Liu, Y., Kong, L., Kang, L., & Ran, F. (2019). Biopolymer-based carboxylated chitosan hydrogel film crosslinked by HCl as gel polymer electrolyte for all-solid-sate supercapacitors. *Journal of Power Sources*, *426*, 47–54.

Zhang, P., Yang, L. C., Li, L. L., Ding, M. L., Wu, Y. P., & Holze, R. (2011). Enhanced electrochemical and mechanical properties of P (VDF-HFP)-based composite polymer electrolytes with SiO2 nanowires. *Journal of Membrane Science*, *379*(1–2), 80–85.

Zhao, J., & Burke, A. F. (2021). Review on supercapacitors: Technologies and performance evaluation. *Journal of Energy Chemistry*, *59*, 276–291.

Zhao, Z., Huang, Y., Qiu, F., Ren, W., Zou, C., Li, X., Wang, M., & Lin, Y. (2021). A new environmentally friendly gel polymer electrolyte based on cotton-PVA composited membrane for alkaline supercapacitors with increased operating voltage. *Journal of Materials Science*, *56*(18), 11027–11043.

6 Polymer Composites for Lithium-Ion Batteries

Ravi Vikash Pateriya, Shweta Tanwar, Anil Arya, and A. L. Sharma

CONTENTS

6.1 INTRODUCTION

In the 21st century, energy shortages and environmental issues are major problems that are being faced worldwide and forces the researchers to eliminator these issues by adopting latest technology. Efforts have been made by researchers to replace non-renewable sources of energy with renewable sources (Tanwar et al. 2021). The finiteness of fossil fuels and major dependency on fossil fuels is also a key factor for environmental problems. This leads us towards the search for sustainable, clean, and eco-friendly sources of energy. A renewable energy source is a possible replacement for fossil fuels. But the problem with renewable sources of energy is that they may not be there when needed, like solar energy on a cloudy day. Hence energy

DOI: 10.1201/9781003208662-8

storage devices like batteries, supercapacitors, superconducting magnetic energy storage, etc. play a crucial role in fulfilling our energy demands. In the current scenario, if structured and profitable energy storage devices are put into practice, they can be very beneficial to the power world (Arya and Sharma 2020). Among various energy storage options available, chemical energy storage devices like batteries, supercapacitors, fuel cells, etc. are very important due to their enhanced properties. As compared to electrostatic capacitors which come up with high power density for a short period, and fuel cells which come up with high-energy density for a long time, batteries are perfect candidates for filling the gap between the two, providing balanced energy and power density (Arya and Sharma 2019). Batteries are of two types: primary and secondary. Secondary batteries are preferred over primary batteries as after getting discharged they can be recharged again and again and hence can be used for a long period, while primary batteries can only be used once. Over many secondary batteries available, Li-ion batteries (LIBs) are of prominent use as they have high power density, high storage capability, better leakage current, constant voltage, cost-effectiveness, etc., making them the most favorable contender for energy storage. Nowadays, lithium-ion batteries have various uses in different fields such as plug-in hybrid electric vehicles, clean personal transportation, power tools, smartphones, medical field, etc. (Nayak et al. 2022). Figure 6.1 represents the use of lithium-ion batteries in different fields. Continuous efforts from worldwide researchers and scientists have encouraged the development in Li-ion batteries. In recent years the different components of batteries such as electrodes, separators, and electrolytes have seen rapid progress. Out of various components of Li-ion batteries, the present chapter will mainly throw light on the role of polymer electrolytes (D. Zhou et al. 2019). Additionally, the history of Li-ion battery components and

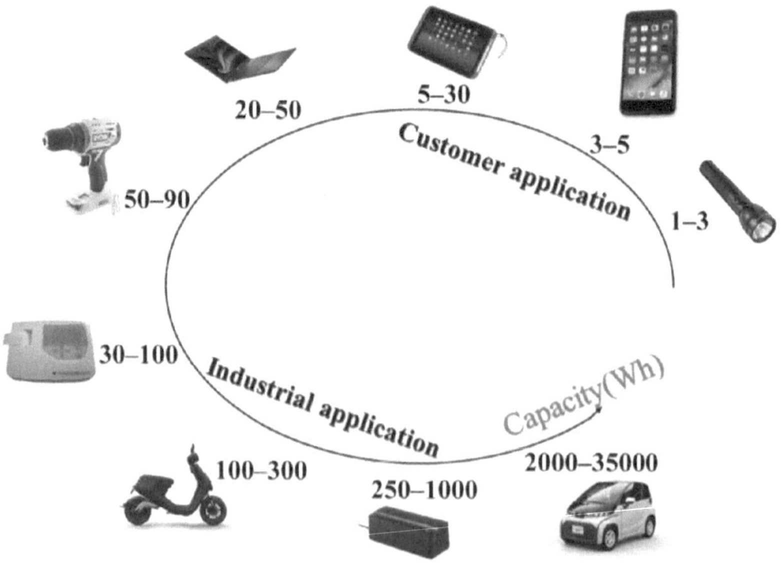

FIGURE 6.1 Application of Li-ion batteries in different fields.

working mechanisms is discussed briefly. Performance parameters are key factors for polymer electrolytes in Li-ion batteries. So, key performance parameters related to polymer electrolytes with their importance are discussed in the present chapter. Overall, this chapter explores the major advancements in the area of polymer composites for Li-ion batteries. The polymer electrolyte (PE) plays an important role in ion relocation between electrodes in a Li-ion battery. Therefore, polymer electrolytes are an interesting prospect that fulfils needs because of their superior electrochemical performances and mechanical properties. Based on the recent developments, we have also shone light on strategies adopted to alter the properties of polymer composites for future development.

6.1.1 A Brief History of Li-Ion Batteries

Electricity is an omnipresent product, and life, as we are fully aware, is very hard to imagine without a convenient power supply (Yabuuchi et al. 2011). In rechargeable batteries, Li-ion battery is a kind of battery in which ions move back and forth across an electrolyte from anode to cathode and cathode to anode during discharging and charging respectively (Zhang et al. 2015). The different lithium compounds are used in Li-ion batteries as cathode materials, and as an anode material generally, graphite is used. Originally, in 1985 Akira Yoshino evolved Li-ion batteries on the basis of research performed in the 1970–1980s by John Goodenough, M. Stanley Whittingham, Koichi Mizushima, and Rachid Yazami (Institute of Electrical and Electronics Engineers 2016; Masaki Yoshio 2009). In 1991, a spark ignited in the field of Li-ion batteries with the introduction of a commercial Li-ion battery by a team of Sony and Asahi Kasei assisted by Yoshio Nishi. Development in the Li-ion battery is shown in Figure 6.2. In today's scenario portable devices are widely used in day-to-day life. For that, there is a need for cheaper, smaller, and lighter portable devices that can

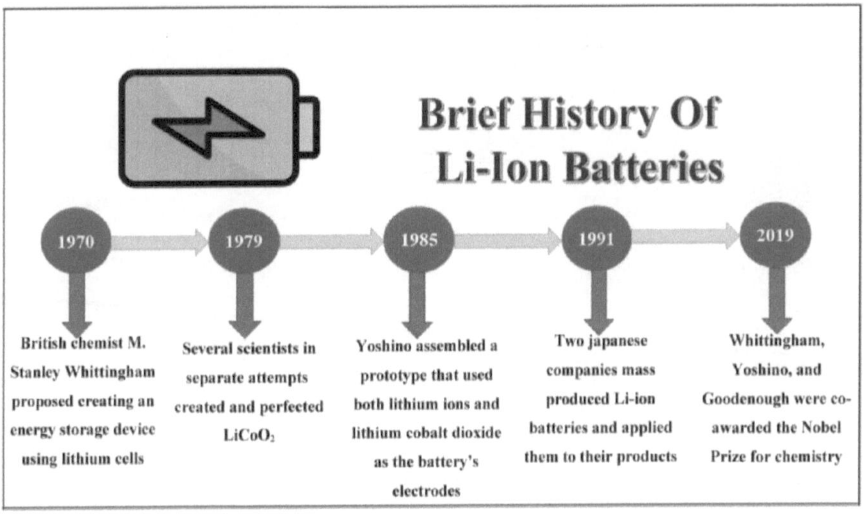

FIGURE 6.2 Development in Li-ion batteries.

be used with confidence. Lithium-ion batteries are very good candidates for fulfilling the requirements due to their high working voltage, good cyclability, and higher energy density (Arya and Sharma 2017). Intensive research continues by researchers for the betterment of Li-ion batteries. So, there is much expectation for the improvement in properties of Li-ion batteries. The Nobel Prize in Chemistry 2019 was awarded to John B. Goodenough, M. Stanley Whittingham, and Akira Yoshino for the development of the lightweight, rechargeable and powerful Li-ion battery. This invention was also recognized as foundation stone for the traditional fuels free community.

6.1.2 Components and Working Mechanism of Li-Ion Batteries

Lithium-ion batteries have provided breakthrough electrochemical cell technology for industry as well as for consumption, where size, weight, and safety are the navigating factors. A battery is made up of two electrodes (positive and negative), an electrolyte, and a separator. In a battery, the electrodes are dipped in an electrolyte which supplies a way for ion migration and a separator. The major components of a Li-ion battery are mentioned here:

1. Anode (negative electrode)
2. Cathode (positive electrode)
3. Electrolyte
4. Separator

The cathode is normally an electron acceptor with high electronegativity. Figure 6.3 shows the charging and discharging of a battery. During discharging, a negative

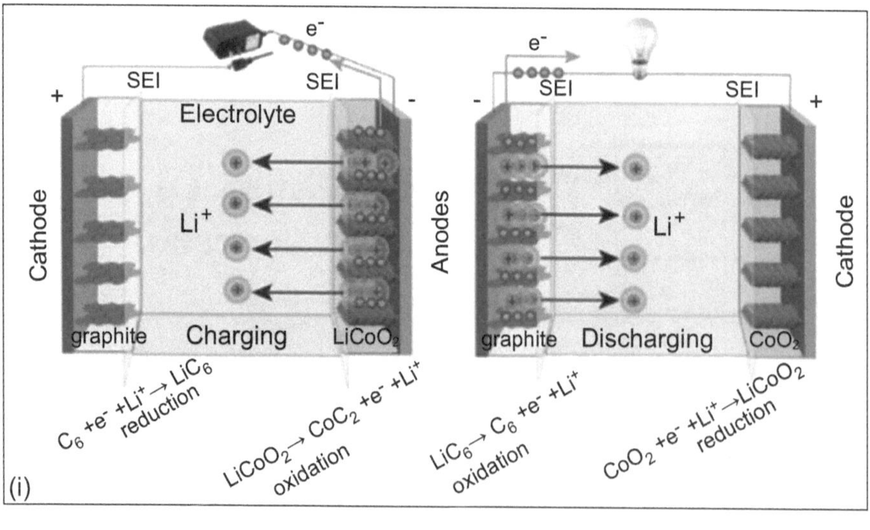

FIGURE 6.3 Charging and discharging mechanisms of Li-ion batteries.

Source: Reproduced with permission from Gulzar et al. (2016, Copyright Royal Society of Chemistry 2016).

electrode electrochemically oxidizes and releases electrons which reach the anode through the external circuit. During charging, the movement of Li ions is from cathode to anode through electrolyte which causes oxidation of the metal present in the cathode material. The role of electrolytes is to conduct ions in a battery. The separator separates the electrodes, stopping them from connecting and causing short-circuiting. There is a type of electrolyte known as polymer electrolyte which has been developed to enhance the properties of batteries, as they themselves work as a separator resulting in decreasing the size and weight of the battery. In a lithium-ion battery, polymer electrolyte performs two things: the first is to conduct ions effectively and the other is electrodes separation to get rid of short-circuiting. Material can be defined as a polymer electrolyte if the system is free from liquid. In it, the conduction is done by dissolving salts in a heavy molecular weight matrix of polymer which is in polar form.

High-performance electrolytes have been a huge reason in the success of lithium-ion batteries due to their excellent property of mediating ion movement between cathode and anode. Replacement of traditional liquid electrolytes with polymer electrolytes in Li-ion batteries has been a very encouraging method for safety, performance, etc. Various properties of polymer electrolytes make them beneficial candidates over conventional liquid electrolytes. Batteries having polymer electrolytes are easy to fabricate, have a long life span, low cost, and are non-toxic. There is no internal shorting in such devices. Dendrite growth is also suppressed and there is no leakage problem in Li-ion batteries which use polymer electrolytes for the movement of ions. Polymer electrolytes have many benefits over conventional electrolytes like high-energy density, solvent-free environment, high ionic conductivity, low volatility, electrochemically stability, and light-weight.

6.2 KEY PERFORMANCE PARAMETERS OF POLYMER ELECTROLYTES

Li-ion batteries are so far commanding the market, from electrical equipment to automobiles (Niu et al. 2021). Hence the electrolyte must be a good insulator for the proper movement of the electron through the outer circuit. Mutilation of the electrolyte may hamper the safety and thus make it hard to manufacture. Therefore, the surroundings for operation, thermal stability, consistency between battery components, and overcharging need particular observation as they determine safety and performance of battery (Tomaszewska et al. 2019). Hence structural stability of polymer electrolytes comes out to be the main element in Li-ion batteries because it plays an important part in the movement of ions.

6.2.1 Ionic Conductivity

A crucial parameter on which performance of a battery is heavily dependent is its ionic conductivity. In polymer electrolytes generally, cations are movable with respect to anions. In upgrading the ionic conductivity of lithium-ion batteries, the role of polymer electrolyte is very critical because it provides the medium for the motion of ions in between electrodes (Zhao et al. 2019). The choice of a perfect polymer electrolyte can provide enhanced ionic conductivity to a Li-ion battery, succeeded by a high Li-ion transference number, guiding to a decrease in concentration gradients, and obstructing growth of dendrite (Dias et al. 2000). Impedance spectroscopy technique is preferred for the estimation of ionic conductivity. The IS technique is performed in a range of frequency of 1 Hz to 1 MHz. In this technique the polymer electrolyte is kept between stainless-steel (SS) electrodes and a ac signal (10-20 mV) is passed across the cell. The ionic conductivity is calculated by this equation:

$$\sigma = \frac{l}{R_b A} \tag{6.1}$$

where l is thickness of polymer film, A stands for area of SS electrode, and R_b represents bulk resistance.

6.2.2 Chemical Compatibility

Chemical compatibility is another parameter that has an effect on the Li-ion battery performance. The electrodes and electrolyte should be chemically stable. It can be explained as the equilibrium between chemical, physical and electrochemical properties of electrodes and electrolyte. For chemical compatibility, the material must have no disadvantageous consequence on the paired part or shape of the battery.

6.2.3 Coulombic Efficiency

In the capacity calibration experiment, due to consumption for discharge process, the capacity released by battery is not as much as its charging capacity. The Coulombic efficiency (CE) is defined to narrate the capacity let out by the battery. Coulombic efficiency is the ratio of capacity of discharge after getting fully charged to the capacity of charging of the same cycle. Its value is never to reach one. The reason behind this is that some electric discharge energy is used by internal resistance of the battery and hence total electric energy liberated by the battery is lower than total charged electric energy. The discharge efficiency of the battery can be damaged because of decomposition of electrolytes, atmospheric temperature, unlike rates of charge-discharge current. The Coulombic efficiency is calculated by the following formula:

$$\text{Coulombic efficiency (\%)} = \frac{C_{charging}}{C_{discharging}} \times 100 \tag{6.2}$$

6.2.4 Capacity Retention

Capacity retention has been utilized like a medium of long-term cycle life in a lithium-ion battery. Capacity retention is always explained in connection to initial discharge capacity. In defining long life of real devices, retention capacity is very helpful. Basically, it represents how long the device will work efficiently. It is the ratio of discharge capacity after 'n' cycles to the initial discharge capacity (Dias et al. 2000).

6.2.5 Electrochemical Potential Window

A potential window is a potential range in which the material and electrolyte are stable. The potential window depends on the material and electrolyte. The potential window material is gained by slow cyclic voltammetry (CV) and linear sweep voltammetry (LSV) measurement. The range between hydrogen and oxygen evolution is the potential window of your material. Cyclic voltammetry (CV) is a very common and well-known method used to look over the redox process occurring in the material. The purpose is to find out the degradation potential upon which the material would oxidize or reduce (Méry et al. 2021). Linear sweep voltammetry (LSV) is a simple electrochemical technique used to obtain electrochemical stability window (ESW). In CV the cycle depends linearly over potential range in both directions, while LSV includes only single direction.

6.2.6 Cation Transference Number

The differentiation of current carrier species is very crucial in ion-conducting solids because electrons and ions both participate in conduction in a conductor. With the help of ion transference number technique, the involvement of ions and electrons from the overall charge can be distinguished. The cation transference number (t_+) is obtained to extract details regarding the transportation of ions and relates it with ionic conductivity. Cation transference number t_+ is +1 for an ideal polymer electrolyte. Mainly t_+ is determined by two techniques: a.c. impedance and d.c. polarization. The Bruce-Vincent equation is used to express cation transference number, and is defined as:

$$t_{Li^+} = \frac{I_s\left(V - I_i R_i\right)}{I_i\left(V - I_s R_s\right)} \tag{6.3}$$

where I_s and I_i represent steady state and initial currents respectively, V stands for applied voltage along cell configuration, and R_s and R_i indicate interfacial resistance after and before polarization respectively. Figure 6.4 represents the main characteristics of a good polymer electrolyte.

6.3 PROGRESSION OF ELECTROLYTES IN LITHIUM-ION BATTERIES

Nowadays liquid organic electrolytes are found in commercially used lithium-ion batteries, which display remarkable superiority of high conductivity and magnificent coordination with surfaces of the electrode. There are many drawbacks faced by liquid electrolytes such as electrochemical uncertainty, incapability of ion refinement, etc., which are overcome by polymer electrolytes. In the advancement process, gel polymer electrolytes (GPEs) can provide a physical barrier that prevents cathodes and anodes from coming in contact and also restricts thermal explosion; therefore GPEs are safer and heat resistive in comparison to liquid electrolytes. The problems with liquid electrolytes were not completely rectified by GPE; hence solid polymer electrolytes were introduced with advance features.

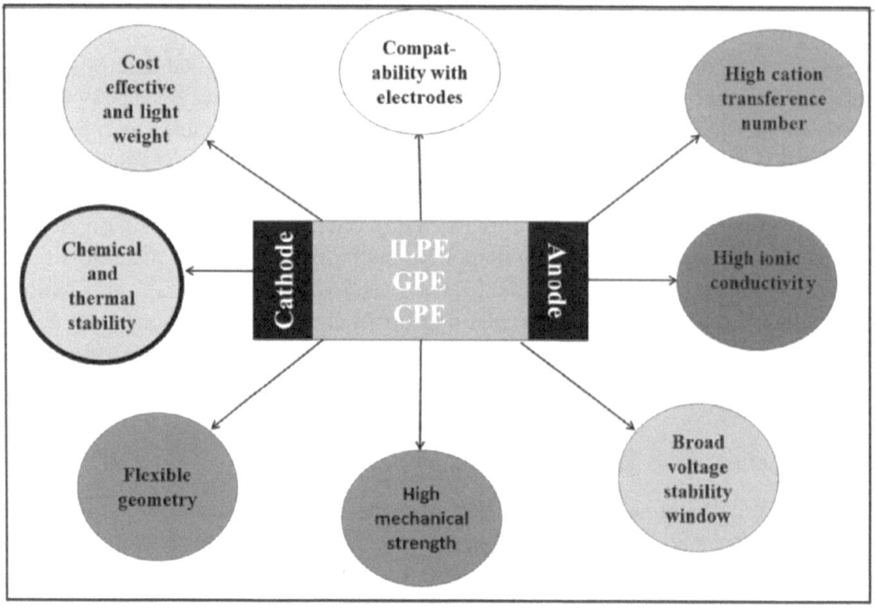

FIGURE 6.4 Desired properties of polymer electrolyte for high-energy Li-ion batteries.

Solid polymer electrolytes can constrain the dendrite formation from happening in GPE, making it possible to use lithium metal anodes. Regardless of their remarkable upper hand, there is room for improvement in some specific areas like low ionic conductivity and poor interface exposure. A large number of studies are being organized to overcome the shortcomings and advance new generations of solid-state LIBs. For commercialization purpose, solid electrolytes need good ionic conductivity, promising mechanical properties, and excellent accommodation firmness with electrodes as basic requirements (Wang et al. 2017). In lithium-ion batteries, liquid polymer electrolytes, gel polymer electrolytes, and solid polymer electrolytes are broadly studied.

6.3.1 Liquid Polymer Electrolytes (LPEs)

Liquid electrolytes are made up of at least one Li salt dissolved in some liquid non-aqueous polar solvent along with a separator to restrict short-circuiting. Liquid electrolytes acquire appealing characteristics like high ionic conductivity, relatively high electrochemical windows, etc. (Cheon et al. 2003; Fergus 2010).

6.3.2 Gel Polymer Electrolytes (GPEs)

GPEs are electrolytes formed by incorporation of organic solvents with a solid polymer matrix and dissolved Li salt (Cheon et al. 2003). The gel polymer electrolytes were introduced to overcome the electrolyte leakage problems in liquid polymer electrolytes. The GPEs exhibit various improved properties like good ionic conductivity, wide electrochemical stability windows and are compatible with electrodes because of increased ionic mobility. GPEs carry good mechanical strength and diffusive properties of solids (Kim et al. 2007).

6.3.3 Solid Polymer Electrolytes (SPEs)

SPEs have many benefits over LPEs and GPEs due to properties like high longevity, long life span, high-energy density, light weight, low reactivity towards electrodes. It gets rid of numerous difficulties generated by liquid electrolytes such as leakage problems and gets rid of the dendrite formation in gel polymer electrolyte. SPEs consist of dissolution of salt in a polymer matrix with an electron donor group and have advantages like design flexibility, improved safety, etc., and have advanced conductivity and stability properties under an atmospheric environment (Zhang et al. 2007). Sometimes required parameters are not obtained based on design, so different plans of action are advanced towards, including the unification of inorganic filler into solid polymer electrolyte matrix—called composite SPE (Arya et al. 2016). Further, solid polymer electrolytes also prevent the problem of dendrite formation. Concerning liquid electrolyte, solid electrolyte deals better with safety issues in the lithium-ion battery (Laghari et al. 2020). Hence, SPEs are proposed for a new generation of safe and less weighted energy storage devices (Fergus 2010; Laghari et al. 2020). SPEs are a favorite nowadays due to their ability to replace aqueous electrolytes.

6.4 PROGRESS IN POLYMER COMPOSITES FOR LITHIUM-ION BATTERIES

The development of polymer electrolytes has achieved rapid progress in recent years, because the performance of Li-ion batteries heavily depends upon polymer electrolytes. Various attempts have been done by researchers to modify the key performance parameters of polymer electrolytes. So, this section discusses the recent progress in different kinds of polymer electrolytes, on the basis of electrochemical performances like IS results, CV results, GCD results, coulombic efficiency, retention capacity, ion transference number, etc. The recent progress in different polymer electrolytes is discussed in a sequence of liquid polymer electrolyte, gel polymer electrolyte, and then solid polymer electrolyte.

6.4.1 EIS Results

In recent years electrochemical impedance spectroscopy (EIS) have seen an increase in popularity. By this characterization method, we can find ionic conductivity of the polymer electrolytes. The recent advancement in ionic conductivity of the various kinds of polymer electrolytes are discussed next:

In 2019 Tsao et al. (Tsao et al. 2019) reported the synthesis of liquid incorporated hybrid membranes which behave as a polymer and separator at the same time via sol-gel process. This ionic liquid functionalized mixed PE provides good ionic conductivity. In the synthesis of PE, first polyetheramine, IL-silane, and Glycidoxypropyl-trimethoxysilane were mixed and stirred vigorously for about 24 h. After that in sol-gel reaction different weight ratios 1:1:9 of acetic acid, methanol, DI water solution were added in silane mixture. Then, polymer membrane was fabricated by the solution cast technique. The prepared samples were mixed with 1.0M $LiPF_6$ in EC/DEC (1/1 v/v) liquid electrolyte to prepare PE. The EIS measurement of prepared PE was done on a CH instrument to calculate ionic conductivity by assembling it between SS electrodes. For EIS calculation the measurements were taken in the frequency range of 0.1 Hz to 1 MHz at a voltage range of 10 mV. The ionic conductivities of 9.4×10^{-3} S cm^{-1} and 6×10^{-3} S cm^{-1} were recorded at 80 °C and room temperature respectively by the prepared polymer electrolyte. The higher conductivity indicates that the ionized group plays a crucial part in the betterment of the ion transport properties.

In 2022, Wang et al. (Wang et al. 2021) synthesized trimethylolpropane trimethylacrylate (TMPTMA)-based polymer electrolyte with help of LiTFSI, DMC, and ionic liquid $N_{1,4,4,4}$TFSI by in situ polymerization method. The $N_{1,4,4,4}$TFSI was used due to its excellent capability to increase ion transportation by dissociating Li ions and their solid properties, therefore decreasing the ohmic resistance. The high dielectric constant of $N_{1,4,4,4}$TFSI was mainly responsible for increased ionic conductivity. Polymer electrolyte with 10% of ionic liquid showed the highest ionic conductivity of 6.15×10^{-3} S cm^{-1}. Approximate non-linear behavior of graph was plotted between log σ vs. T^{-1}, putting forward that GPE followed the Vogel-Tamman-Fulcher (VTF) empirical equation

$$\sigma = \sigma_o\, T^{-1/2} \exp\left(E_a/R\,(T - T_0)\right) \tag{6.4}$$

Here σ_o stands for pre-exponential factor, T for temperature, T_0 glass transition temperature, E_a activation energy, and R gas constant.

In 2019, Tan et al. (Tan et al. 2019) synthesized the GPEs made up of polyacrylonitrile/thermoplastic polyurethane/polystyrene (PAN/TPU/PS). To prepare the solution, different quantities of TPU, PS, PAN (5:1:5) were dissolved in DMF. For mixing, the solution was stirred at 60 °C for 12 h, and then it was carried into electro-spun with a syringe with a 0.5 mLh^{-1} rate of flow. A potential difference of 24 kV was maintained across the nozzle tip. The dried films were saturated in 1 mol L^{-1} $LiPF_6$-EC/DMC electrolyte to make it GPE in a glove box in an argon atmosphere. Afterward, these were used to assemble button cells. On performing EIS of prepared samples of PAN/TPU/PS GPE had the bulk resistance (R_b) of only 1.76 Ω, and ionic conductivity of 4.24×10^{-3} was obtained as shown in Figure 6.5b. Due to higher electrolyte uptake of electrolyte and porosity of film, the conductivity heavily depends on more amount of Li-ion concentration. 82% of porosity was calculated and uptake behavior 31.5%. High porosity and high uptake behavior was the reason for high conductivity of prepared GPE. In 2021, Jiang et al. (Jiang et al. 2021) prepared thermoplastic polyurethane (TPU) polymer electrolyte based on hydrophilic-lipophilic TiO_2. The TPU electrolyte incorporated with TiO_2 nanoparticles was fabricated by phase separation technology. As discussed earlier sufficient electrolyte uptake is a very important factor for increasing ionic conductivity. The ATP-TPU membrane could soak up more electrolyte than the PP membrane due to its enlarged surface area. The ATO-TPU membrane has an electrolyte absorbance of 259% which is much higher than the PP separator (72%). Due to these properties of ATO-TPU GPE, the ionic conductivity was recorded to be 1.59×10^{-3}, as shown in Figure 6.5d. Low crystallization is favorable for enhancing the ionic conductivity of polymer electrolytes (Solarajan et al. 2016).

In 2021, Zhang et al. (Zhang et al. 2021) prepared porous PVdF-HFP-PEO-SiO_2 gel polymer electrolyte through the universal immersion precipitation method. A very useful EIS technique was used for the calculation of ionic conductivity. GPE based on PVdF-HFP showed better ionic conductivity than the Celgard 2325 separator which has the property of absorbing liquid electrolyte. The ionic conductivities of 2.46×10^{-4} S cm^{-1}, 1.12×10^{-3} S cm^{-1} and 4.33×10^{-5} S cm^{-1} were measured for pure PVdF-HFP, PVdF-HFP-PEO-SiO_2 GPE and Celgard 2325 separator. There were the following reasons for the improvement of ionic conductivity. First, the high electronegativity of the fluorine atom of PVdF-HFP makes it easier for Li ions to move in the amorphous region. Another reason is, in comparison to empty orbitals of the lithium atom electron, the PEO has a segment of a function group having oxygen with excess electrons. Thus, it becomes easy for lithium ions to coordinate with ether oxygen functional groups (Cui et al. 2014; Fang et al. 2021). Third, an increase in electrolyte absorbance ratio was due to the high porous nature and good pore interconnectivity of PVdF-HFP-PEO-SiO_2.

In 2021 Whba et al. (Whba et al. 2021) reported the synthesis of acrylonitrile grafted epoxidized natural rubber (CAN-g-ENR) with LITFSI salt via UV curing technique. The maximum conductivity of CAN-g-ENR was reported to be 1.1×10^{-6} S cm^{-1} with 40% wt of LiTFSI and thereafter decreases as shown in Figure 6.5a. The increase in conductivity in electrolytes with salt content indicates an increasing number of charge carriers (Lee et al. 2019). In 2019, Borzutzki et al. (Borzutzki et al. 2019) synthesized single ion conducting gel polymer electrolyte based on polysulfonamide along with PVdF-HFP in a 3:1 ratio. At different temperatures 20 °C and 60 °C, the ionic conductivity of 0.5×10^{-3} S cm^{-1} and 1.08×10^{-3} S cm^{-1} respectively were achieved for polymer electrolyte membrane (1:3), which are amongst the highest values achieved by PVdF-HFP and polysulfonamide-based single Li-ion PEs (Zhang et al. 2014). Li et al. (2022) prepared an SPE by constructing a dual-range ionic conduction path at ambient temperature. They blended polymer with EO and PO block structure (B-PEG@DMC) with PEO to improve the conductivity of SPE. The ionic conductivity was recorded at 1.1×10^{-5} and 2.3×10^{-5} S cm^{-1} before and after blending respectively.

In 2020 Li et al. (Li et al. 2020) synthesized SPE in the presence of photoinitiator under UV irradiation comprised of different molecular weights of PEGDA and boric easter bonds by photopolymerization. Improved conductivity was shown by boronic ester bonds containing covalently cross-linked polymer-stabilized electrolyte. EIS technique was used for calculating ionic conductivity of polymer electrolytes. With 5:5 and 5:4 mass ratio of diGMPA to BA, the polymer electrolyte P(n-BA)-diGMPA delivered very less ionic conductivity of 9.5×10^{-6} S cm^{-1} and 5.7×10^{-6} S cm^{-1}

respectively which are a lot less for the practical application. Again, the ionic conductivity of P2K-diGMPA electrolyte with different components was plotted at 60 °C. The highest calculated ionic conductivity was 2.4×10^{-4} S cm^{-1} at 60 °C when PEGDA and diGMPA were combined with 3:5 mass ratio. In 2021 Cao et al. (Cao et al. 2021) fabricated an imine bond based on a new type of self-healing polymer electrolyte, solid polymer electrolyte (ShSPE). The ShSPE was synthesized by changing amounts of polyoxyethylene bis (amine) and terephthaladehyde by an easy Schiff base reaction. At 60 °C, the ShSPE with maximum amount of NH_2-PEG-NH_2 delivered highest ionic conductivity of 1.67×10^{-4} S cm^{-1}. It is clear that the conductivity highly depends on PEG concentration for salt dissociation capacity of the ethylene group. Meantime, the ionic conductivity is also affected by the lithium ion per ethylene oxide units present in the polymer matrix.

In 2021 Chen et al. (Chen et al. 2021) prepared semi-IPN SPE by thiol-tosylate polycondensation of pentaerythritol terakis (3-mercaptopropionate) (PETMP) with non-self-polymerized precursors, DTsPEG and TsPEGME, and then blended with PEO. Among all fabricated membrane SPEs, self-standing $SIS_{4k,5k}$ prepared from the $DTsPEG_{46=k}$ and $TsPEGME_{5k}$ possessed the highest ionic conductivity of value $\sigma = 1.3 \times 10^{-4}$ S cm^{-1}. It is observed that ionic conductivity is increasing with an increase in temperature thereby following VTF relation.

In 2021 Didwal et al. (Didwal et al. 2021) demonstrated the synthesis of solid polymer electrolyte (SPE) having mesoporous silica nanoparticles (MSN) as the filler and incorporated with poly(propylene carbonate) (PPC) as the host matrix. Different samples of CSPE were prepared with different amounts of MSN filler and their ionic conductivity was calculated. 4 wt% of MSN filler demonstrated the highest ionic conductivity of 8.48×10^{-4} S cm^{-1}.

In 2021 Hu et al. (Hu et al. 2021) demonstrated the fabrication of SPE with the function of single ion conduction by multi-nozzle electrospinning process and thereafter hot pressed treatment was given to it. The ionic conductivity of prepared SPE was increased to 5.4×10^{-5} S cm^{-1} by the introduction of hydrophilic and hydrophobic silica (SiO_2) which encourages dissociation of Li ions by decreasing the crystal behavior of PEO.

In 2021 Kalybekkyzy et al. (Kalybekkyzy et al. 2021) prepared a series of UV-photocross-linked flexible SPEs comprising PEGDA, ETPTA, and LITFSI salt, with the addition of polydimethylsiloxane with acryl-PDMS to reduce the crystallinity of the PEG chain. The conductivity of cross-linked SPE with a high salt amount ($[EO]/[Li^+] = 6$) was 1.75×10^{-6} S cm^{-1} at room temperature and 1.07×10^{-4} S cm^{-1} at 80 °C. In 2021 Karpagavel et al. (Karpagavel et al. 2021) prepared SPE consisting LiBr salt dispersed with PVdF-HFP/PVP blend polymer electrolyte by solution casting technique. Figure 6.5c shows the frequency-dependent conductivity spectra of SPE at various temperatures. The highest conductivity of 1.13×10^{-6} S cm^{-1} was recorded for 4 wt% of lithium bromide salt.

In 2021, Li et al. (Li et al. 2021) synthesized SPE with the help of PVdF and cellulose acetate as matrix and filler in the form of montmorillonite was used. The ionic conductivity of solid polymer electrolyte film was 3.40×10^{-4} S cm^{-1}. It was seen that with the increase in the concentration of MMT the ionic conductivity first increased and then decreased.

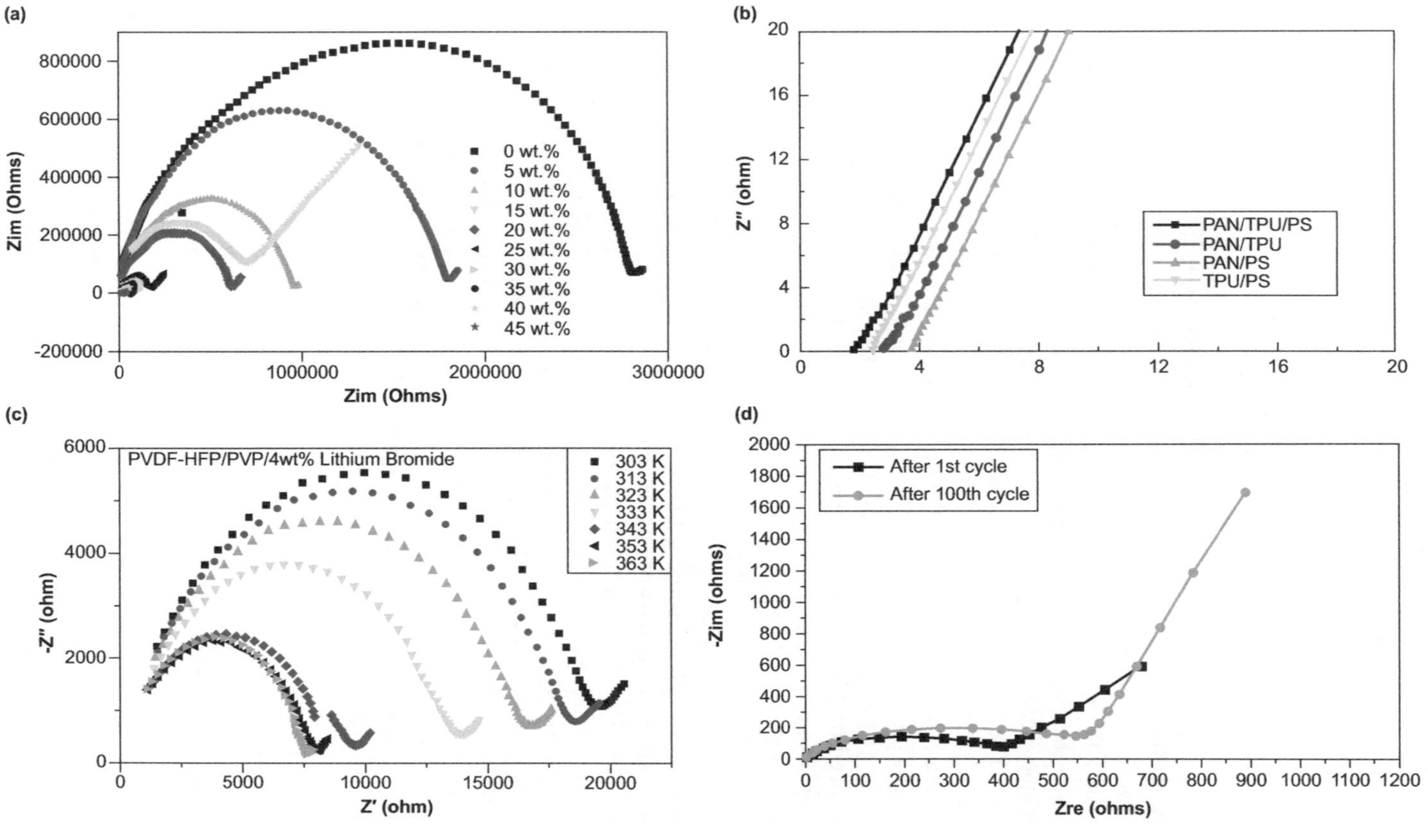

FIGURE 6.5 EIS results of (a) CAN-g-ENR PE; (b) PAN/TPU/PS GPE; (c) LiBr dopped PVdF-HFP/PVP blend PE; (d) ATO-TPU GPE.

Source: (a) Reproduced with permission from Whba et al. (2021, Copyright Elsevier 2021. (b) Reproduced with permission from Tan et al. (2019, Copyright Springer Nature 2019). (c) Reproduced with permission from Karpagavel et al. (2021, Copyright Springer Nature 2021). (d) Reproduced with permission from Jiang et al. (2021, Copyright Springer Nature 2020).

6.4.2 Electrochemical Potential Window

Electrochemical stability/potential window is a very important factor which measures the safe limit for battery operation. To check the decomposition behavior of polymer electrolyte, the LSV technique is performed. Some recent ESW measurements of polymer electrolytes are discussed next.

In 2018 Gupta et al. (Gupta et al. 2018) prepared liquid polymer electrolyte with PEO+ 20 wt% LITFSI + 20 wt% ionic liquid by solution casting technique. A cathode material $LiNi_{0.6}MN_{0.2}Co_{0.2}O_2$ was also synthesized along with this polymer electrolyte. LSV of the prepared sample was taken at a scan rate of 0.05 mVs^{-1}. From the LSV curve it is concluded that the electrochemical stable window of prepared electrolyte was 4.3V vs. Li^+/Li as shown in Figure 6.7d. The electrochemical reaction involved during the charging and discharging process was observed with the help of CV technique. The CV was carried out in voltage range 2.4–4.3V with scan rate of 0.05 mV s^{-1}. The redox peaks of prepared PE were observed at 3.3V and 3.9V as shown in Figure 6.6d.

In 2021 Zhou et al. (Zhou et al. 2021) examined polymer electrolyte film consisting of PVdF as a polymer, SPI and MMT inorganic filler as additive group. The Li/SPE/LFP half-cell was fabricated and the CV was taken in 2.6–4.5 voltage range at scanning speed of $0.5mVs^{-1}$. Figure 6.6a suggests that the reduction peak of prepared cell was obtained at 2.91V vs. Li/Li^+ during the discharging process and oxidation peak is observed at 3.94V vs. Li/Li^+ during the charging process. Figure 6.7b represents the LSV curve suggesting the electrochemical potential window of 5.1 V at 25 °C suggesting very steady electrochemical properties. In 2020 Srivastava et al. (Srivastava et al. 2020) prepared gel polymer electrolyte with the help of PVdF-HFP+PMMA with varying mass of EMIMTFSI and 20 wt% of LITFSI by solution casting technique. The electrochemical performance was performed using lithium rich nickel manganese cobalt oxide cathode and graphite anode. The CV curve of the cell indicates that at room temperature the reduction took place at −0.40V and oxidation at 0.40V vs. Li/Li^+ as shown in Figure 6.6b. In 2021 Wang et al. (Wang et al. 2021) reported the synthesis of prepared a solid polymer electrolyte with a property of healing by itself with the help of poly(HFBM-co-SBMA) network, ionic liquid based on imidazole (EMI-TFSI), and LiTFSI salt. The self-healing SPE exhibited a satisfying electrochemical potential window of 4.9V vs. Li/Li^+ which is quite appealing for a safer Li-ion battery. In 2021 Sadiq et al. (Sadiq et al. 2021) examined blend PE film based on PVA-PEG with dopant $NaNO_3$ salt by solution casting technique. The cyclic voltammetry (CV) of the samples was examined for three cycles at a scan rate of $0.01mVs^{-1}$ in –3V to +3V voltage range as shown in Figure 6.6c. Also, the linear sweep voltammetry (LSV) measurements were taken at a scan rate of 0.01 mVs^{-1} in a voltage range between –3V to +3V to measure the electrochemical stability window of prepared PE films. The prepared PB30 film showed an ESW of 3.9V.

In 2022 Guo et al. (Guo et al. 2022) used the solvent voltilization method for synthesis of PE of PVdF-HFP/PMMA/CMC. The LSV technique was used for confirming the electrochemical stability window of GPE at a scan rate of $5mVs^{-1}$ in a range of 0 to 6 V vs. Li/Li^+. The electrochemical stability of 4.8, 4.5, 4.6 V were observed for PVdF-HFP/PMMA/CMC, PP, PVdF-HFP/PMMA respectively. The

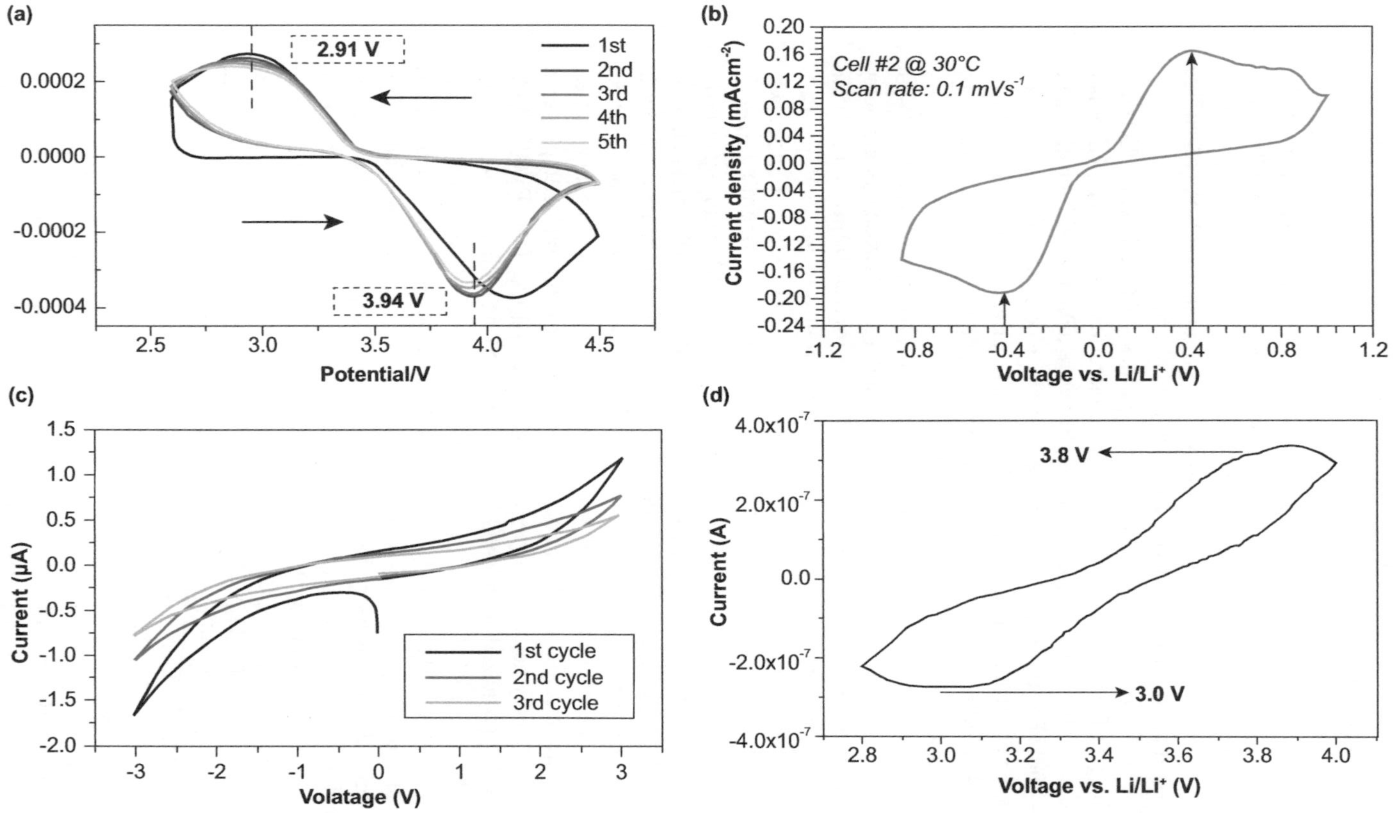

FIGURE 6.6 CV results of (a) Li/SPE/LPF half-cell; (b) PVdF-HFP+PMMA EMIMTFSI PE; (c) PB30film PE; (d) PEO and LiTFSI polymer electrolyte.

Source: (a) Reproduced with permission from Zhou et al. (2021, Copyright Springer Nature 2020). (b) Reproduced with permission from Srivastava et al. (2020, Copyright Elsevier 2020). (c) Reproduced with permission from Sadiq et al. (2021, Copyright Springer Nature 2021). (d) Reproduced with permission from Gupta et al. (2018, Copyright Elsevier 2018).

results show that PVdF-HFP/PMMA/CMC is very compatible with carbonate electrolyte. High electrochemical stability was attained by the addition of CMC. The optimized ESW concludes PE for commercial application for high-energy LIBs. In 2022 Gao et al. (Gao et al. 2022) reported the synthesis of polyurethane/cellulose PE by facile immersion approach. The prepared samples were marked as C-PU18, C-PU20, C-PU22, C-PU26, and C-Pu29. The electrochemical stability window of C-PU electrolyte was measured by LSV technique using Li/SS cells. The C-PU18 and C-PU29 exhibited electrochemical stability of approximately 4.8 V, concluding these PE can be applied for high-energy and -voltage batteries.

In 2021 Guo et al. (Guo et al. 2021) reported the synthesis of GPE having PEG electrolyte including TEP and nano-TiO_2 with dual acceleration conduction mechanism for LIBs. Better results of electrochemical stability were achieved with involvement of nano-TiO_2 and TEP. For the LSV scan the polymer electrolyte was sandwiched between working electrodes of stainless steel. For PEG-TEP, PEG-TiO_2 a satisfying electrochemical stability window of 5.18 V, 4.55 V respectively was achieved. The electrochemical potential window of PEG-TEP-TiO_2 GPE was 5.32 V vs. Li^+/Li.

In 2020 Jie et al. (Jie et al. 2020) fabricated NPH-GPE based on PVdF-HFP with a cautious electrolyte. Cyclic voltammetry (CV) and linear sweep voltammetry (LSV) techniques for assembled SS/GPE/Li cell were used at $0.1 mVs^{-1}$ to obtain electrochemical stability window. CV was performed from −0.2 to 2.5 V and LSV from 2.5–6 V. At room temperature electrochemical stability window of 5.15 V was observed for NPH-GPE and at 70 °C 5.2 V ESW was observed. In 2019 Tan et al. (Tan et al. 2019) fabricated and synthesized high-performance GPE made up of PAN/TPU/PS. The LSV technique was performed to study the electrochemical stability window (ESW). The electrochemical stability window of PAN/TPU/PS came out to be 5.8 V. The higher electrochemical stability makes polymer membrane a perfect candidate for use in higher voltage cathode materials and high-performance batteries. In 2019 Prasanna et al. (Sai Prasanna and Austin Suthanthiraraj 2019) prepared nanocomposite gel polymer electrolyte with nanofiller Al_2O_3, TiO_2 PVC/PEMA-based blend GPE plasticized with $Zn(OTf_2)$ salt and EMIMTFSI ionic liquid. The LSV studies of prepared GPE were carried out in a three-electrode system to find out electrochemical stability windows and to find their capacity to conduct zinc ions. The LSV of GPE distributed with 3 wt% Al_2O_3 and TiO_2 nanofillers was taken and electrochemical stability was found to be 4.27 and 4.09 V respectively. The received ESW is enough to use this GPE in advanced energy storage devices (Li et al. 2022).

In 2017 Li et al. (Li et al. 2021) prepared a gel polymer electrolyte membrane with a liquid electrolyte and PEO as host polymer. LSV of synthesized GPE was examined at a scan rate of $0.5 mVs^{-1}$ in range of 0 to 6 V vs. Li^+/Li for measurement of electrochemical stability window. The electrochemical stability of prepared GPE was found to be 5.2 V which is very satisfying for its practical application. In 2019 Zhou et al. (Zhou et al. 2019) demonstrated the fabrication of a self-healing PE with supermolecular network reinforcement via consolidation of SiO_2-UPy in the polymer matrix of PEG and LiTFSI (EO/Li^+ molar ratio 16:1). Different samples were prepared with SiO_2-UPy in different weights of the copolymer. The LSV technique was used at $1 mVs^{-1}$ scan rate to examine elctrochemical stability window of SHCPE-10. The current (oxidation) increases remarkably up to 5.1 V vs. Li^+/Li, referring a high electrochemical stability window of PE, making it applicable to common 5 V cathode materials and in high-voltage batteries. The LSV curve is shown in Figure 6.7a.

In 2021 Yu et al. (Yu et al. 2021) prepared P(PEGMA-co-MMA)-based PE having different [EO]:[Li] ratios through in situ polymerizations of electrolytes. The polymer electrolyte exhibited superb coordination with electrode material of prepared Li/$LiFePO_4$ cells. LSV curves were observed to find the electrochemical potential window of the prepared electrolyte. Among all prepared samples the PE based on P(PEGMA-co-MMA) had a better electrochemical stability of 5.20 V vs. Li/Li^+, indicating it is appropriate for higher-voltage battery applications. The LSV curve is shown in Figure 6.7c. CV and LSV curves of some of the previously discussed polymer electrolytes are presented next.

6.4.3 Galvanostatic Charge-Discharge Result

GCD is a mode of electrochemical measurement which identifies the mechanism and kinetics of reactions at electrodes. To an electrode, a constant current pulse is applied and by the galvanostatic method, the change in resulting current is measured. The reaction rate is measured by using GCD. In GCD a graph is plotted between potential and time at constant current. In 2021 Swiderska-Mocek et al. (Swiderska-Mocek and Kubis 2021) prepared PE comprising LiODFB and LiBOB salts with liquid $MePrPyrNTf_2$ or $MePrPyrNFT_2$ in polymer matrix TMS and PVdF by solution casting technique for lithium-ion batteries. Using $LiFePO_4$ as an electrode with different PE, the GCD outline of the prepared LPF|PE|Li cell was measured with current densities ranging from C/10 to 1C. The initial reversibility of 185 $mAhg^{-1}$ was observed at rate of C/10. With the increase in current density to C/2, the discharge capacity dropped to 132 $mAhg^{-1}$ with coulombic efficiency of 99%. The reversible discharge capacity of 164 $mAhg^{-1}$ was observed after 50 cycles at a C/10 rate as shown in Figure 6.8b.

In 2022 Long et al. (Long et al. 2022) reported the synthesis of thermotolerant and fireproof GPE consisting of diethyl vinyl phosphonate in cross-linked PEG diacrylate network and $LiPF_6$ salt. The battery was prepared with $LiFePO_4$ electrodes and prepared polymer electrolytes for examining charging/discharging results. The battery conveyed a capacity of 142.2 $mAhg^{-1}$ at a scan rate of 0.2C with capacity retention near 100% for 300 cycles at atmosphere temperature. The battery can attain balanced charge-discharge for 100 cycles with 95.5% of coulombic efficiency at a scan rate of 1C at 80 °C.

In 2021 Fang et al. (Fang et al. 2021) prepared PE by a ring-opening reaction of PEGDE with the ethoxy group and polyether amine (PEA) with amino groups, fabricated inter-connected polyethylene glycol-based resin (c-PEGR). The prepared PE reflected an excellent galvanostatic charge/discharge profile. The LCO||Li cell prepared with c-PEGR gel at a scan rate of 0.2C exhibited an opening capacity of 159.1 $mAhg^{-1}$. The capacity was retained to 146.3 $mAhg^{-1}$ after 100 cycles with a capacity retention of 91.95% and coulombic efficiency of 99.92%.

In 2021 Luo et al. (Luo et al. 2021) reported the preparation of PVdF-HFP-based GPE, having UV operable polymer matrix through UV-irradiation technique.

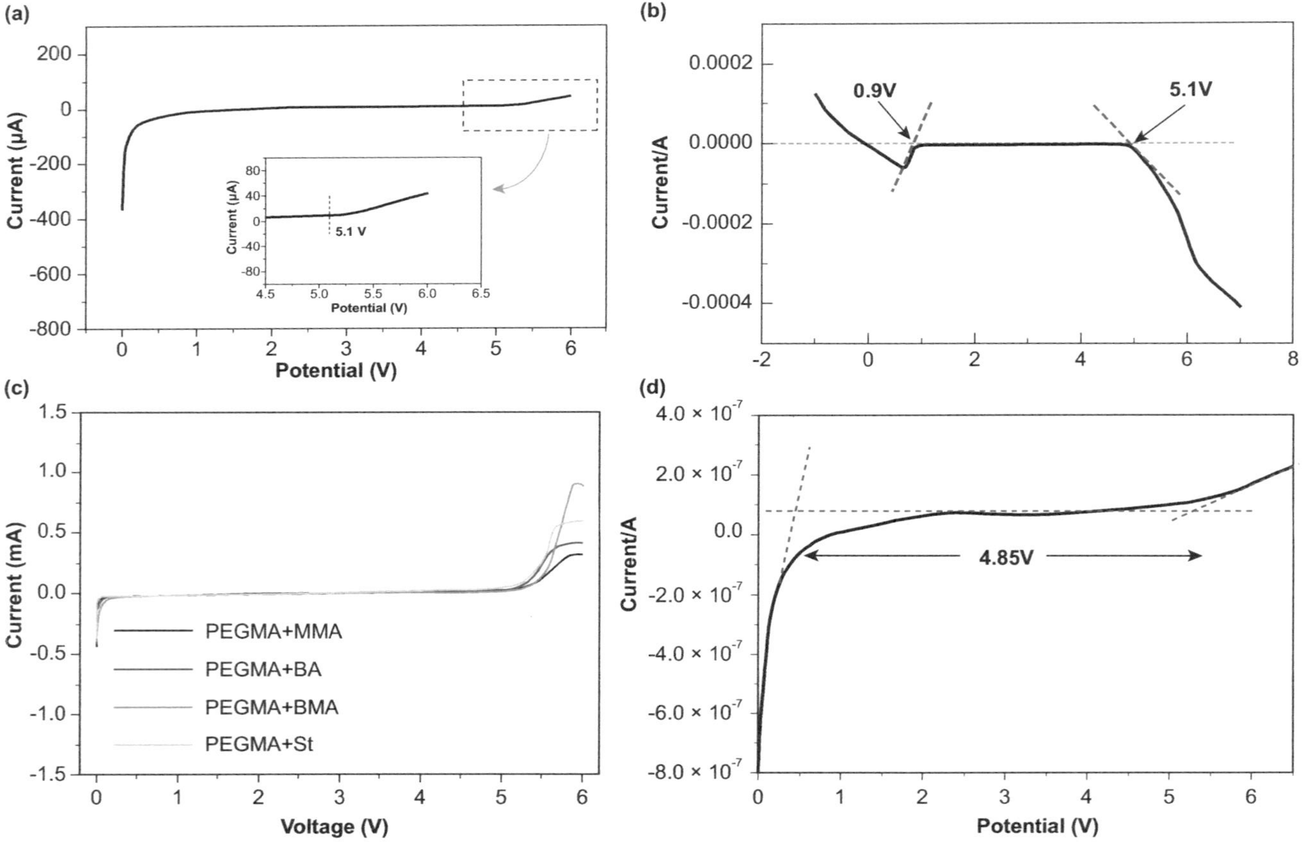

FIGURE 6.7 LSV results of (a) SHCPE-10 polymer electrolyte; (b) polymer electrolyte PVdF/SPI/MMT/LiTFSI; (c) P(PEGMA-co-MMA) PE; (d) PEO and LiTFSI polymer electrolyte.

Source: (a) Reproduced with permission from Deng et al. (2022, Copyright Elsevier 2022). (b) Reproduced with permission from Zhou et al. (2021, Copyright Springer Nature 2020). (c) Reproduced with permission from Yu et al. (2021, Copyright 2021 American Chemical Society). (d) Reproduced with permission from Gupta et al. (2018), Copyright Elsevier 2018).

$LiFePO_4$|PVdF-HFP/PETEA-GPE|Li battery was placed together to reveal the electrochemical performance of GPE. At a rate of 0.1C, the battery revealed a discharge specific capacity of 161mAhg^{-1}. The discharge specific capacity of the battery was reported to be 159, 151, 144 mAhg^{-1} at 0.2, 0.5, 1C rate respectively. Showing with increasing current density, discharge capacity decreases. The battery exhibited capacity retention of 148 mAh/g^{-1} after cycle and average high coulombic efficiency of 99%.

In 2021 Jiang et al. (Jiang et al. 2021) synthesized TPU gel polymer electrolyte filled with ultrafine hydrophilic-lipophilic TiO_2 nanoparticles (ATO). The initial specific discharge capacity of ATO-TPU GPE was recorded to be 145 mAhg^{-1} and after 50 cycles it dropped to 139 mAhg^{-1} having capacity retention of 95%. In 2019 Tan et al. (Tan et al. 2019) prepared GPE comprising PAN/TPU/PS. The first discharge and charge capacities of 161.44 mAhg^{-1} and 161.70 mAhg^{-1} were observed for the prepared Li/GPE/$LiFePO_4$ cell at a 0.1C rate as shown in Figure 6.8d. After 50 cycles, the capacity retention was 93.4%. The possible reason for the improved result is a huge number of pores in the polymer membrane which supply a compulsory way for ion transportation and ease fast ionic transport.

In 2021 Das et al. (Das et al. 2021) prepared PVdF-HFP-based polymer electrolyte with PVdF-HFP, 20 wt% LAGP, 20 wt% LITFSI, and 35 wt% EMITFSI. In GCD measurement of fabricated cell-II was calculated at higher and lower voltage range of 4.0 and 2.7V respectively at a constant 0.05C rate of current. The discharge and charge voltage levels of 3.49 and 3.33V respectively were observed. A lower value of resistance between electrode and electrolyte may be its predictable reason (Maia et al. 2022). The cell prepared was tested for 50 cycles. The maximum achieved capacity was 151 mAhg^{-1} at 0.05C. The capacity retention was 100% of the initial charge/discharge capacity as shown in Figure 6.8c. Table 6.1 shows recent advancement in polymer electrolytes for Li-ion batteries.

In 2022 Deng et al. (Deng et al. 2022) examined the electrochemical performance of synthesized self-healing polymer electrolyte (PBPE) membrane having NH_2-PEG-NH_2 with LIFSI, LIDFOB, $LiPF_6$ salts in EC/FEC/EPC. Excellent discharge capacity of 118.2 mAhg^{-1} was exhibited by LPF cell with PBPE at 5C rate and capacity retention of 97.8% over 125 cycles and coulombic efficiency close to 100%. In 2022 Deng et al. (Zhou et al. 2018) reported the preparation of adjustable, self-alleviating, and highly elastic PE for LIB. This polymer electrolyte exhibited a special property of self-healing after getting damaged. The polymers were prepared with a variable amount of UPyMA and PEGMA with the help of reversible addition-fragmentation chain transfer polymerization technique. PE was prepared by solution casting technique with UPy units and LiTFSI salt. The novel shPE was fabricated to Li metal batteries using Li metal anode and $LiPO_4$ cathode. A lithium metal battery was assembled using $LiPO_4$ as cathode, Li metal as anode and a prepared self-healing PE. The LPF/shPE/Li cell conveyed an earlier discharge capacity of 157 mAhg^{-1} with coulombic efficiency of 92.4% at a 0.1C rate as shown in Figure 6.8a. With the increase in current density rate to 0.2C and 0.5C, the battery maintained a high capacity of 152 mAhg^{-1} and 132 mAhg^{-1} respectively. After 100 cycles, capacity retained to 143 mAhg^{-1} and 97.9% of coulombic efficiency. In 2021 Ding et al. (Ding et al. 2021) prepared PEO-based PE in the composition of PEO, LITFSI, and

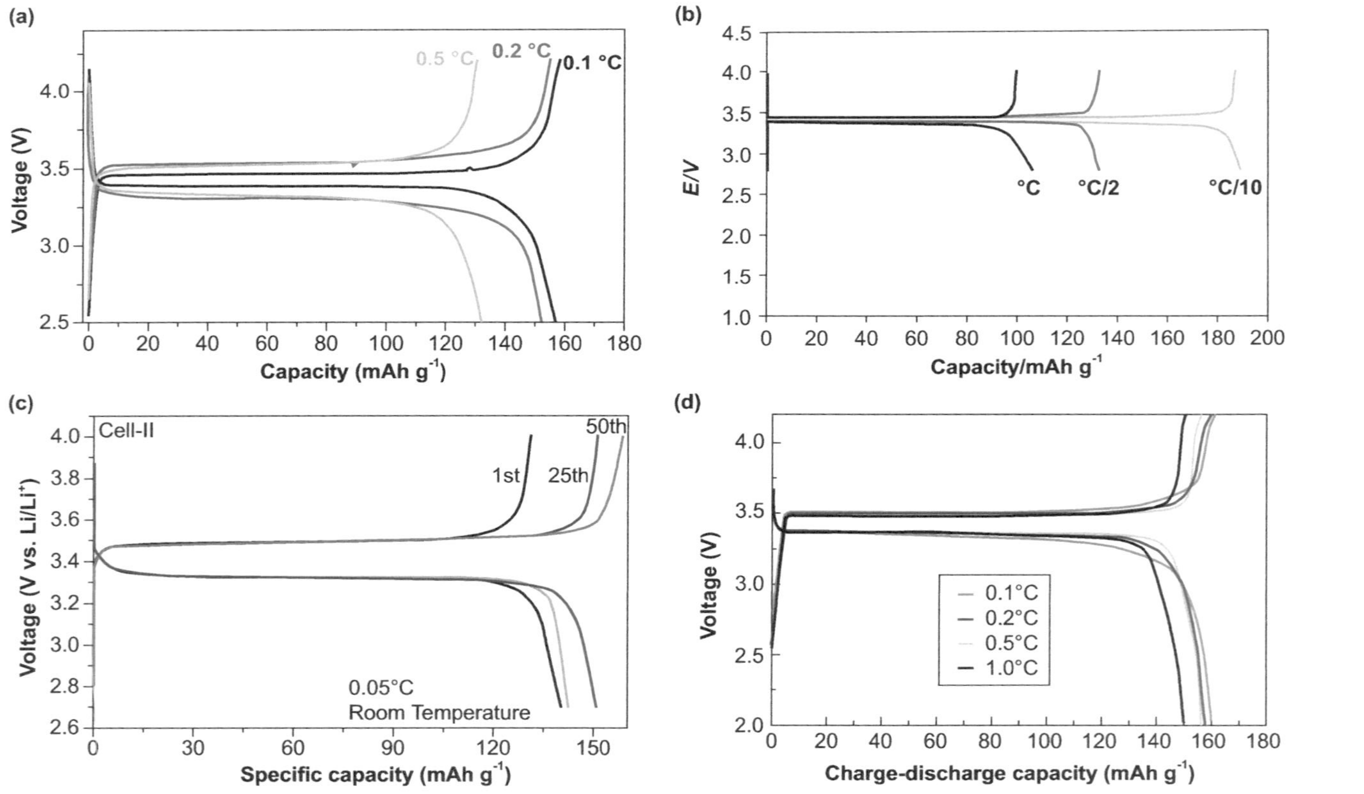

FIGURE 6.8 GCD results of (a) LPF/shPE/Li cell; (b) $LiFePO_4$|PE 2|Li cell; (c) LAGP CSE; (d) PAN/TPU/PS.

Source: (a) Reproduced with permission from Zhou et al. (2018, Copyright the Royal Society of Chemistry 2018). (b) Reproduced with permission from Swiderska-Mocek and Kubis (2021, Copyright Elsevier 2021). (c) Reproduced with permission from Das et al. (2021 Copyright Elsevier 2021). (d) Reproduced with permission from Tan et al. (2019, Copyright Springer Nature 2019).

TABLE 6.1
Recent Advancement in Polymer Electrolytes for Li-Ion Batteries.

Sr no.	Material Used	Ionic Conductivity ($\times 10^{-3}$ S cm^{-1})	Cation Transference Number t^+	ESW (V)	Capacity (mAhg^{-1})	Coulombic Efficiency (%)	Retention Capacity (%)	Year	Reference
1	IPDI + DEG + DBTDL + DMF + FPC3018	2.4	0.42	4.8	134	99	96	2022	(Gao et al. 2022)
2	PVdF-HFP + PMMA + CMC	4.43	-	4.2	162	-	93.5	2022	(Guo et al. 2022)
3	PEO + B-PEG@DMC	0.023	-	-	109	100	73	2022	(Li et al. 2022)
4	PEGDA + DEVP	0.6	0.43	5	142	100	96.6	2022	(Long et al. 2022)
5	PVA + PEG + NaNO3	0.0153	0.97	3.9	-	-	-	2021	(Sadiq et al. 2021)
6	(PVdF + PHFP) + LiTFSI + EMITFSI + LAGP	4.49		1.8	151	-	100	2021	(Das et al. 2021)
7	NH_2-PEG-NH_2 + $LiPF_6$ + LiFSI + LiDFOB	4.79	0.38	4.5	118	100	97	2021	(Deng et al. 2022)
9	GMA + PEGDA + BP	0.24	-	5.3	-	-	-	2021	(Zhang et al. 2021)
10	TPA + DGEBA + NH2-PEG-NH2	0.167	0.39	4.5	141	-	97.7	2021	(Cao et al. 2021)
11	UPyMA + PETEA + LITFSI + NMA + FEC	1.79	0.79	4.5	117	-	86.1	2021	(Jaumaux et al. 2020)
12	PEG + PEGME + PETMP + TEA + LITFSI	0.13	0.24	4.3	143	97.1	93.6	2021	(Chen et al. 2021)
13	PPC + LITFSI + MSN	0.85	0.86	4.8	171	100	86	2021	(Didwal et al. 2021)
14	PEO + LITFSI	0.28		3.5	155	-	95	2021	(Deng et al. 2022)
15	PEGDE + PEA	0.7	0.47	4.36	146	99.92	91.	2021	(Fang et al. 2021)
16	PVdF + soybean protein isolate + montmorillonite (MMT)	0.256	0.77	5.1	100	-	-	2021	(Zhou et al. 2021)
17	PVdF-HFP-PEO-SiO_2(gel)	1.12	0.48	4.7	147	94	100	2021	(Zhang et al. 2021)
18	PEGDA + LITFSI + AIBN + DME + TEP + TiO_2	1	-	5.32	125	95.05	-	2021	(Guo et al. 2021)
19	Li_2CO_3 + $CuSO_4$ + SiO_2 + PVdF-HFP	0.054	0.96	3.8	145	-	88.1	2021	(Hu et al. 2021)
20	ATO + TPU + $LiPF_6$ + DMC + TiO_2	1.59		4.3	139	100	95	2021	(Jiang et al. 2021)

(Continued)

TABLE 6.1 *(Continued)*
Recent Advancement in Polymer Electrolytes for Li-Ion Batteries.

Sr no.	Material Used	Ionic Conductivity ($\times 10^{-3}$ S cm^{-1})	Cation Transference Number t^+	ESW (V)	Capacity (mAhg^{-1})	Coulombic Efficiency (%)	Retention Capacity (%)	Year	Reference
21	ETPTA + HMPP + $LiPF_6$ + DEC + PVdF-HFP	0.91		5.1	151	-	-	2021	(Luo et al. 2021)
22	LiTFSI/OMMT/CA/PVdF	0.34	0.3	4.2	112	100	99	2021	(Li et al. 2021)
23	LiTFSI-N 1,4,4,4 TFSI-DMC(70:20:10)	6.15	0.59	5.3	183	97	80	2021	(Wang et al. 2021)
24	PVdF + LiODFB + $EtMeImNTf_2$ + TMS	3.21	0.03	3.4	168	99.4		2021	(Swiderska-Mocek and Kubis 2021)
25	PVdF-HFP + PETEA + HMPP	0.5	-	4.8	151	99	99	2021	(Luo et al. 2021)
26	PVdF-HFP + NMP	0.724	0.57	5.2	150	-	98.65	2020	(Jie et al. 2020)
27	PVdF HPF + SIPE	0.5	0.9	4.0	163	92.2	-	2019	(Borzutzki et al. 2019)
28	PEO + 1-methyl-3-trimethoxysilane	5.9	0.57	5	156	-	-	2019	(Tsao et al. 2019)
29	PAN + TPU + PS	3.9	-	5.8	161	-	93.4	2019	(Tan et al. 2019)

low viscous solvent. The electrochemical performance of electrolytes was examined using $LiFePO_4$/Li cell. The fabricated electrolyte exhibited discharge of 155.4 and 157.8 $mAhg^{-1}$ at 1C for designed and carbonate electrolyte. After 100 cycles the cyclic performance of designed and carbonate electrolytes was high with capacity retention of 98.5% and 75.5% respectively.

In 2021 Kim et al. (Kim et al. 2021) optimized synthesis of the ion-conductive track in UV-healed SPE via a semi-permeate polymer matrix having minimum liquid content. The USPE had ETPTA as backbone film with ion solvated GPE (Li^+/PVdF-HFP) for increasing Li^+ conduction. The charging-discharging capacities of the cell were examined by GCD profile. With various C-rates 0.2, 0.5, 1.0, 2, 3, 5, and 10 C, a charge/discharge curve was obtained. USPEs having 3, 6, 9, and 12 wt% of HFP content in PVdF-HFP showed extra steady C-rate ability and a high-rise discharge capacity.

6.5 CONCLUSIONS

In conclusion, we reviewed the role and recent work by researchers/scientists in polymer electrolytes for lithium-ion batteries (LIBs). This chapter mainly discusses the important role of polymer electrolytes in Li-ion batteries. Li-ion batteries were developed from 1970s, and the introduction of commercial Li-ion batteries in 1991 changed the dynamics of the energy storage world. The different components of Li-ion batteries are discussed in detail. The electrolyte has a very important role in a battery, and a polymer electrolyte can rectify the problems with conventional electrolytes. Different polymer electrolytes have their specific advantages and disadvantages. And hence intense research is conducted by researchers. Here recent developments in electrochemical properties of polymer electrolytes have been discussed. There is the enhancement of many properties of polymer electrolytes like better conductivity, self-healing properties after damage, etc.

Lithium-ion batteries have attracted researchers because of the growing need for energy. Polymer electrolytes have a very crucial responsibility in enhancing electrochemical performance of LIBs. The important factors that are required to be communicated are ionic conductivity, electrochemical potential window, and charge-discharge capacity, although there is scope for modification of a few parameters like ionic conductivity, thermal stability, coulombic efficiency, and cyclic stability. So, distinct schemes have to be implemented for improving energy density and efficiency for their practical use in commercial LIBs.

ACKNOWLEDGMENT

Ravi Vikash Pateriya and Shweta Tanwar are very thankful to UGC for providing the JRF fellowship.

REFERENCES

Arya, Anil, and A. L. Sharma. 2017. "Polymer Electrolytes for Lithium Ion Batteries: A Critical Study." *Ionics* 23:497–540.

Arya, Anil, and A. L. Sharma. 2019. "Electrolyte for Energy Storage/Conversion (Li^+, Na^+, Mg 2^+) Devices Based on PVC and Their Associated Polymer: A Comprehensive Review." *Journal of Solid State Electrochemistry* 23(4):997–1059. doi: 10.1007/s10008-019–04203-x.

Arya, Anil, and A. L. Sharma. 2020. "A Glimpse on All-Solid-State Li-Ion Battery (ASSLIB) Performance Based on Novel Solid Polymer Electrolytes: A Topical Review." *Journal of Materials Science* 55(15):6242–6304. doi: 10.1007/s10853-020-04434-8.

Arya, Anil, A. L. Sharma, Sweety Sharma, and Mohd Sadiq. 2016. "Role of Low Salt Concentration on Electrical Conductivity in Blend Polymeric Films." *Journal of Integrated Science and Technology* 4(1):17–20.

Borzutzki, K., J. Thienenkamp, M. Diehl, M. Winter, and G. Brunklaus. 2019. "Fluorinated Polysulfonamide Based Single Ion Conducting Room Temperature Applicable Gel-Type Polymer Electrolytes for Lithium Ion Batteries." *Journal of Materials Chemistry A* 7(1):188–201. doi: 10.1039/c8ta08391f.

Cao, Xiaoyan, Pengming Zhang, Nanping Guo, Yongfen Tong, Qiuhua Xu, Dan Zhou, and Zhijun Feng. 2021. "Self-Healing Solid Polymer Electrolyte Based on Imine Bonds for High Safety and Stable Lithium Metal Batteries." *RSC Advances* 11(5):2985–2994. doi: 10.1039/d0ra10035h.

Chen, Kai, Yuxue Sun, Chen Zhao, Yingjie Yang, Xiaorong Zhang, Jun Liu, and Haiming Xie. 2021. "A Semi-Interpenetrating Network Polymer Electrolyte Membrane Prepared from Non-Self-Polymerized Precursors for Ambient Temperature All-Solid-State Lithium-Ion Batteries." *Journal of Solid State Chemistry* 296(January):121958. doi: 10.1016/j.jssc.2021.121958.

Cheon, Sang-Eun, Ki-Seok Ko, Ji-Hoon Cho, Sun-Wook Kim, Eog-Yong Chin, and Hee-Tak Kim. 2003. "Rechargeable Lithium Sulfur Battery." *Journal of the Electrochemical Society* 150(6):A800. doi: 10.1149/1.1571533.

Cui, Jinfeng, Caixia Qing, Qingtang Zhang, Ce Su, Xiaomei Wang, Baoping Yang, and Xiaobing Huang. 2014. "Effect of the Particle Size on the Electrochemical Performance of Nano-Li_2FeSiO_4/C Composites." *Ionics* 20(1):23–28. doi: 10.1007/s11581-013-0965-3.

Das, Anurup, Madhumita Goswami, Kavya Illath, T. G. Ajithkumar, A. Arya, and M. Krishnan. 2021. "Synthesis and Characterization of LAGP-Glass-Ceramics-Based Composite Solid Polymer Electrolyte for Solid-State Li-Ion Battery Application." *Journal of Non-Crystalline Solids* 558(January):120654. doi: 10.1016/j.jnoncrysol.2021.120654.

Deng, Kuirong, Suping Zhou, Zelin Xu, Min Xiao, and Yuezhong Meng. 2022. "A High Ion-Conducting, Self-Healing and Nonflammable Polymer Electrolyte with Dynamic Imine Bonds for Dendrite-Free Lithium Metal Batteries." *Chemical Engineering Journal* 428(July 2021):131224. doi: 10.1016/j.cej.2021.131224.

Dias, Felix B., Lambertus Plomp, and Jakobert B. J. Veldhuis. 2000. "Trends in Polymer Electrolytes for Secondary Lithium Batteries." *Journal of Power Sources* 88(2):169–191. doi: 10.1016/S0378-7753(99)00529-7.

Didwal, Pravin N., Y. N. Singhbabu, Rakesh Verma, Bong Jun Sung, Gwi Hak Lee, Jong Sook Lee, Duck Rye Chang, and Chan Jin Park. 2021. "An Advanced Solid Polymer Electrolyte Composed of Poly(Propylene Carbonate) and Mesoporous Silica Nano-particles for Use in All-Solid-State Lithium-Ion Batteries." *Energy Storage Materials* 37(February):476–490. doi: 10.1016/j.ensm.2021.02.034.

Ding, Dong, Yuta Maeyoshi, Masaaki Kubota, Jungo Wakasugi, Kiyoshi Kanamura, and Hidetoshi Abe. 2021. "Non-Flammable Super-Concentrated Polymer Electrolyte with 'Solvated Ionic Liquid' for Lithium-Ion Batteries." *Journal of Power Sources* 506(March). doi: 10.1016/j.jpowsour.2021.230099.

Fang, Zhenhan, Yufeng Luo, Haitao Liu, Zixin Hong, Hengcai Wu, Fei Zhao, Peng Liu, Qunqing Li, Shoushan Fan, Wenhui Duan, and Jiaping Wang. 2021. "Boosting the Oxidative Potential of Polyethylene Glycol-Based Polymer Electrolyte to 4.36 V by

Spatially Restricting Hydroxyl Groups for High-Voltage Flexible Lithium-Ion Battery Applications." *Advanced Science* 8(16):1–13. doi: 10.1002/advs.202100736.

Fergus, Jeffrey W. 2010. "Ceramic and Polymeric Solid Electrolytes for Lithium-Ion Batteries." *Journal of Power Sources* 195(15):4554–4569. doi: 10.1016/j.jpowsour.2010.01.076.

Gao, Yunqi, Changhao Feng, Hairui Wang, Minghan Xu, Chengzhong Zong, and Qingfu Wang. 2022. "Flexible and Rigid Polyurethane Based Polymer Electrolyte for High-Performance Lithium Battery." *Journal of Applied Polymer Science* 139(5):1–10. doi: 10.1002/app.51566.

Gulzar, Umair, Subrahmanyam Goriparti, Ermanno Miele, Tao Li, Giulia Maidecchi, Andrea Toma, Francesco De Angelis, Claudio Capiglia, and Remo Proietti Zaccaria. 2016. "Next-Generation Textiles: From Embedded Supercapacitors to Lithium Ion Batteries." *Journal of Materials Chemistry A* 4(43):16771–16800. doi: 10.1039/c6ta06437j.

Guo, Huabin, Shengwen Zhong, Liping Chen, Guanghuai Peng, Fang Fang Wang, Ting Ting Yan, and Jingwei Hu. 2022. "Study on PVdF-HFP/PMMA/CMC Blended Polymer as Membrane for Lithium-Ion Batteries." *International Journal of Electrochemical Science* 17:1–15. doi: 10.20964/2022.01.47.

Guo, Jianqiang, Yapeng Chen, Yuanbin Xiao, Chenpeng Xi, Gui Xu, Borong Li, Chengkai Yang, and Yan Yu. 2021. "Flame-Retardant Composite Gel Polymer Electrolyte with a Dual Acceleration Conduction Mechanism for Lithium Ion Batteries." *Chemical Engineering Journal* 422(May):130526. doi: 10.1016/j.cej.2021.130526.

Gupta, Himani, Shalu Kataria, Liton Balo, Varun Kumar Singh, Shishir Kumar Singh, Alok Kumar Tripathi, Yogendra Lal Verma, and Rajendra Kumar Singh. 2018. "Electrochemical Study of Ionic Liquid Based Polymer Electrolyte with Graphene Oxide Coated $LiFePO_4$ Cathode for Li Battery." *Solid State Ionics* 320(March):186–192. doi: 10.1016/j.ssi.2018.03.008.

Hu, Texiong, Xiu Shen, Longqing Peng, Yizheng Liu, Xin Wang, Haoshen Ma, Peng Zhang, and Jinbao Zhao. 2021. "Preparation of Single-Ion Conductor Solid Polymer Electrolyte by Multi-Nozzle Electrospinning Process for Lithium-Ion Batteries." *Journal of Physics and Chemistry of Solids* 158(March):110229. doi: 10.1016/j.jpcs.2021.110229.

Institute of Electrical and Electronics Engineers. 2016. "SoutheastCon 2016: 30 March-3 April 2016." *Ieee* (1541108):1–6.

Jaumaux, Pauline, Qi Liu, Dong Zhou, Xiaofu Xu, Tianyi Wang, Yizhou Wang, Feiyu Kang, Baohua Li, and Guoxiu Wang. 2020. "Deep-Eutectic-Solvent-Based Self-Healing Polymer Electrolyte for Safe and Long-Life Lithium-Metal Batteries." *Angewandte Chemie: International Edition* 59(23):9134–9142. doi: 10.1002/anie.202001793.

Jiang, Yunhong, Feng Li, Yufan Mei, Yanhuai Ding, Huaji Pang, and Ping Zhang. 2021. "Gel Polymer Electrolyte Based on Hydrophilic: Lipophilic TiO_2-Modified Thermoplastic Polyurethane for High-Performance Li-Ion Batteries." *Journal of Materials Science* 56(3):2474–2485. doi: 10.1007/s10853-020-05360-5.

Jie, Jing, Yulong Liu, Lina Cong, Bohao Zhang, Wei Lu, Xinming Zhang, Jun Liu, Haiming Xie, and Liqun Sun. 2020. "High-Performance PVdF-HFP Based Gel Polymer Electrolyte with a Safe Solvent in Li Metal Polymer Battery." *Journal of Energy Chemistry* 49: 80–88. doi: 10.1016/j.jechem.2020.01.019.

Kalybekkyzy, Sandugash, Al Farabi Kopzhassar, Memet Vezir Kahraman, Almagul Mentbayeva, and Zhumabay Bakenov. 2021. "Fabrication of UV-Crosslinked Flexible Solid Polymer Electrolyte with Pdms for Li-Ion Batteries." *Polymers* 23(1):1–12. doi: 10.3390/polym13010015.

Karpagavel, K., K. Sundaramahalingam, A. Manikandan, D. Vanitha, A. Manohar, E. R. Nagarajan, and N. Nallamuthu. 2021. "Electrical Properties of Lithium-Ion Conducting Poly (Vinylidene Fluoride-Co-Hexafluoropropylene) (PVdF-HFP)/Polyvinylpyrrolidone (PVP) Solid Polymer Electrolyte." *Journal of Electronic Materials* 50(8):4415–4425. doi: 10.1007/s11664-021-08967-9.

Kim, Guk Tae, Giovanni B. Appetecchi, Fabrizio Alessandrini, and Stefano Passerini. 2007. "Solvent-Free, PYR1ATFSI Ionic Liquid-Based Ternary Polymer Electrolyte Systems. I. Electrochemical Characterization." *Journal of Power Sources* 171(2):861–869. doi: 10.1016/j.jpowsour.2007.07.020.

Kim, Jin Il, Young Gyun Choi, Yeonho Ahn, Dukjoon Kim, and Jong Hyeok Park. 2021. "Optimized Ion-Conductive Pathway in UV-Cured Solid Polymer Electrolytes for All-Solid Lithium/Sodium Ion Batteries." *Journal of Membrane Science* 619(August 2020): 118771. doi: 10.1016/j.memsci.2020.118771.

Laghari, Zubair Ahmed, Ishaque Abro, Razium Soomro, Subhan Ali Jogi, and Umair Aftab. 2020. "Polymerization and Curing Time Interactive Study of Gel-Polymer Electrolyte (GPE) for Electrical Applications." *International Journal of Electrical Engineering & Emerging Technology*, 3(1):7–11.

Lee, Tian Khoon, Nur Farahidayu Mohd Zaini, Nadharatun Naiim Mobarak, Nur Hasyareeda Hassan, Siti Aminah Mohd Noor, Shuhib Mamat, Kee Shyuan Loh, Ku Halim KuBulat, Mohd Sukor Su'ait, and Azizan Ahmad. 2019. "PEO Based Polymer Electrolyte Comprised of Epoxidized Natural Rubber Material (ENR50) for Li-Ion Polymer Battery Application." *Electrochimica Acta* 316:283–291. doi: 10.1016/j.electacta.2019.05.143.

Li, S., Zuo, C., Zhang, Y., Wang, J., Gan, H., Li, S., . . . & Xue, Z. (2020). Covalently cross-linked polymer stabilized electrolytes with self-healing performance via boronic ester bonds. *Polymer Chemistry*, 11(36), 5893–5902.

Li, Libo, Yuhang Shan, and Xueying Yang. 2021. "New Insights for Constructing Solid Polymer Electrolytes with Ideal Lithium-Ion Transfer Channels by Using Inorganic Filler." *Materials Today Communications* 26(August 2020):101910. doi: 10.1016/j.mtcomm.2020.101910.

Li, Ruiyang, Haiming Hua, Yuejing Zeng, Jin Yang, Zhiqiang Chen, Peng Zhang, and Jinbao Zhao. 2022. "Promote the Conductivity of Solid Polymer Electrolyte at Room Temperature by Constructing a Dual Range Ionic Conduction Path." *Journal of Energy Chemistry* 64:395–403. doi: 10.1016/j.jechem.2021.04.037.

Long, Man Cheng, Ting Wang, Ping Hui Duan, You Gao, Xiu Li Wang, Gang Wu, and Yu Zhong Wang. 2022. "Thermotolerant and Fireproof Gel Polymer Electrolyte toward High-Performance and Safe Lithium-Ion Battery." *Journal of Energy Chemistry* 65:9–18. doi: 10.1016/j.jechem.2021.05.027.

Luo, Kaili, Dingsheng Shao, Li Yang, Lei Liu, Xiaoyi Chen, Changfei Zou, Dong Wang, Zhigao Luo, and Xianyou Wang. 2021. "Semi-Interpenetrating Gel Polymer Electrolyte Based on PVdF-HFP for Lithium Ion Batteries." *Journal of Applied Polymer Science* 138(11):1–10. doi: 10.1002/app.49993.

Maia, Beatriz Arouca, Natália Magalhães, Eunice Cunha, Maria Helena Braga, Raquel M. Santos, and Nuno Correia. 2022. "Designing Versatile Polymers for Lithium-Ion Battery Applications: A Review." *Polymers* 14(3). doi: 10.3390/polym14030403.

Masaki Yoshio, Hideyuki Noguchi. 2009. *Lithium-Ion Batteries,* Chapter 2, DOI:https://doi.org/10.1007/978-0-387-34445-4, Springer New York, NY.

Méry, Adrien, Steeve Rousselot, David Lepage, and Mickaë Dollé. 2021. "A Critical Review for an Accurate Electrochemical Stability Composite Electrolytes." *Materials* 14(14):3840.

Nayak, Hari Chandra, Shivendra Singh Parmar, Rajendra Prasad Kumhar, and Shailendra Rajput. 2022. "Modulation in Electric Conduction of PVK and Ferrocene-Doped PVK Thin Films." *Electronic Materials* 3(1):53–62. doi: 10.3390/electronicmat3010005.

Niu, Huizhe, Le Wang, Ping Guan, Nan Zhang, Chaoren Yan, Minling Ding, Xulong Guo, Tongtong Huang, and Xiaoling Hu. 2021. "Recent Advances in Application of Ionic Liquids in Electrolyte of Lithium Ion Batteries." *Journal of Energy Storage* 40(May):102659. doi: 10.1016/j.est.2021.102659.

Sadiq, Mohd, Mohammad Moeen Hasan Raza, Sujeet Kumar Chaurasia, Mohammad Zulfequar, and Javid Ali. 2021. "Studies on Flexible and Highly Stretchable Sodium Ion Conducting Blend Polymer Electrolytes with Enhanced Structural, Thermal, Optical, and Electrochemical Properties." *Journal of Materials Science: Materials in Electronics* 32(14):19390–19411. doi: 10.1007/s10854-021-06456-7.

Sai Prasanna, Candhadai Murali, and Samuel Austin Suthanthiraraj. 2019. "PVC/PEMA-Based Blended Nanocomposite Gel Polymer Electrolytes Plasticized with Room Temperature Ionic Liquid and Dispersed with Nano-ZrO_2 for Zinc Ion Batteries." *Polymer Composites* 40(9):3402–3411. doi: 10.1002/pc.25201.

Solarajan, Arun Kumar, Vignesh Murugadoss, and Subramania Angaiah. 2016. "Montmorillonite Embedded Electrospun PVdF-HFP Nanocomposite Membrane Electrolyte for Li-Ion Capacitors." *Applied Materials Today* 5:33–40. doi: 10.1016/j.apmt.2016.09.002.

Srivastava, Nitin, Shishir Kumar Singh, Himani Gupta, Dipika Meghnani, Raghvendra Mishra, Rupesh K. Tiwari, Anupam Patel, Anurag Tiwari, and Rajendra Kumar Singh. 2020. "Electrochemical Performance of Li-Rich NMC Cathode Material Using Ionic Liquid Based Blend Polymer Electrolyte for Rechargeable Li-Ion Batteries." *Journal of Alloys and Compounds* 843:155615. doi: 10.1016/j.jallcom.2020.155615.

Swiderska-Mocek, Agnieszka, and Aleksandra Kubis. 2021. "Preparation and Electrochemical Properties of Polymer Electrolyte Containing Lithium Difluoro(Oxalato)Borate or Lithium Bis(Oxalate)Borate for Li-Ion Polymer Batteries." *Solid State Ionics* 364(March):115628. doi: 10.1016/j.ssi.2021.115628.

Tan, Li, Yuanyuan Deng, Qi Cao, Bo Jing, Xianyou Wang, and Yuewen Liu. 2019. "Gel Electrolytes Based on Polyacrylonitrile/Thermoplastic Polyurethane/Polystyrene for Lithium-Ion Batteries." *Ionics*, 25:3673–3682.

Tanwar, Shweta, Anil Arya, Anurag Gaur, and A. L. Sharma. 2021. "Transition Metal Dichalcogenide (TMDs) Electrodes for Supercapacitors: A Comprehensive Review." *Journal of Physics Condensed Matter* 33(30):303002. doi: 10.1088/1361-648X/abfb3c.

Tomaszewska, Anna, Zhengyu Chu, Xuning Feng, Simon O'Kane, Xinhua Liu, Jingyi Chen, Chenzhen Ji, Elizabeth Endler, Ruihe Li, Lishuo Liu, Yalun Li, Siqi Zheng, Sebastian Vetterlein, Ming Gao, Jiuyu Du, Michael Parkes, Minggao Ouyang, Monica Marinescu, Gregory Offer, and Billy Wu. 2019. "Lithium-Ion Battery Fast Charging: A Review." *ETransportation* 1:100011. doi: 10.1016/j.etran.2019.100011.

Tsao, Chih Hao, Hou Ming Su, Hsiang Ting Huang, Ping Lin Kuo, and Hsisheng Teng. 2019. "Immobilized Cation Functional Gel Polymer Electrolytes with High Lithium Transference Number for Lithium Ion Batteries." *Journal of Membrane Science* 572:382–389. doi: 10.1016/j.memsci.2018.11.033.

Wang, Pu, Qinghua Zhang, Yingming Li, Julius Matsiko, Ya Zhang, and Guibin Jiang. 2017. "Airborne Persistent Toxic Substances (PTSs) in China: Occurrence and Its Implication Associated with Air Pollution." *Environmental Science: Processes and Impacts* 19(8): 983–999. doi: 10.1039/c7em00187h.

Wang, Qiu Jun, Pin Zhang, Bo Wang, and Li Zhen Fan. 2021. "A Novel Gel Polymer Electrolyte Based on Trimethylolpropane Trimethylacrylate/Ionic Liquid via in Situ Thermal Polymerization for Lithium-Ion Batteries." *Electrochimica Acta* 370:137706. doi: 10.1016/j.electacta.2020.137706.

Whba, Rawdah, Mohd Sukor Su'ait, Lee TianKhoon, Salmiah Ibrahim, Nor Sabirin Mohamed, and Azizan Ahmad. 2021. "In-Situ UV Cured Acrylonitrile Grafted Epoxidized Natural Rubber (ACN-g-ENR)-LiTFSI Solid Polymer Electrolytes for Lithium-Ion Rechargeable Batteries." *Reactive and Functional Polymers* 164(May):104938. doi: 10.1016/j.reactfunctpolym.2021.104938.

Yabuuchi, Naoaki, Kazuhiro Yoshii, Seung-Taek Myung, Izumi Nakai, and Shinichi Komaba. 2011. "Detailed Studies of a High-Capacity Electrode Material for Rechargeable Batteries, Li_2MnO_3–$LiCo_{1/3}Ni_{1/3}Mn_{1/3}O_2$." *Journal of the American Chemical Society* 133(12):4404–4419.

Yu, Liping, Yong Zhang, Jirong Wang, Huihui Gan, Shaoqiao Li, Xiaolin Xie, and Zhigang Xue. 2021. "Lithium Salt-Induced in Situ Living Radical Polymerizations Enable

Polymer Electrolytes for Lithium-Ion Batteries." *Macromolecules* 54(2):874–887. doi: 10.1021/acs.macromol.0c02032.

Zhang, H. P., P. Zhang, Z. H. Li, M. Sun, Y. P. Wu, and H. Q. Wu. 2007. "A Novel Sandwiched Membrane as Polymer Electrolyte for Lithium Ion Battery." *Electrochemistry Communications* 9(7):1700–1703. doi: 10.1016/j.elecom.2007.03.021.

Zhang, Kai, Xiaopeng Han, Zhe Hu, Xiaolong Zhang, Zhanliang Tao, and Jun Chen. 2015. "Nanostructured Mn-Based Oxides for Electrochemical Energy Storage and Conversion." *Chemical Society Reviews* 44(3):699–728. doi: 10.1039/c4cs00218k.

Zhang, Pan, Rui Li, Jian Huang, Boyu Liu, Mingjiong Zhou, Bizheng Wen, Yonggao Xia, and Shigeto Okada. 2021. "Flexible Poly(Vinylidene Fluoride-Co-Hexafluoropropylene)-Based Gel Polymer Electrolyte for High-Performance Lithium-Ion Batteries." *RSC Advances* 11(20):11943–11951. doi: 10.1039/d1ra01250a.

Zhang, Yunfeng, Rupesh Rohan, Weiwei Cai, Guodong Xu, Yubao Sun, An Lin, and Hansong Cheng. 2014. "Influence of Chemical Microstructure of Single-Ion Polymeric Electrolyte Membranes on Performance of Lithium-Ion Batteries." *ACS Applied Materials and Interfaces* 6(20):17534–17542. doi: 10.1021/am503152m.

Zhao, Qing, Xiaotun Liu, Sanjuna Stalin, Kasim Khan, and Lynden A. Archer. 2019. "Solid-State Polymer Electrolytes with in-Built Fast Interfacial Transport for Secondary Lithium Batteries." *Nature Energy* 4(5):365–373. doi: 10.1038/s41560-019-0349-7.

Zhou, Binghua, Dan He, Ji Hu, Yunsheng Ye, Haiyan Peng, Xingping Zhou, Xiaolin Xie, and Zhigang Xue. 2018. "A Flexible, Self-Healing and Highly Stretchable Polymer Electrolyte: Via Quadruple Hydrogen Bonding for Lithium-Ion Batteries." *Journal of Materials Chemistry A* 6(25):11725–11733. doi: 10.1039/c8ta01907j.

Zhou, Binghua, Ye Hyang Jo, Rui Wang, Dan He, Xingping Zhou, Xiaolin Xie, and Zhigang Xue. 2019. "Self-Healing Composite Polymer Electrolyte Formed via Supramolecular Networks for High-Performance Lithium-Ion Batteries." *Journal of Materials Chemistry A* 7(17):10354–10362. doi: 10.1039/c9ta01214a.

Zhou, Da, Libo Li, Jintian Du, and Mo Zhai. 2021. "Synergistic Effect of Soy Protein Isolate and Montmorillonite on Interface Stability between Polymer Electrolyte and Electrode of All-Solid Lithium-Ion Battery." *Ionics* 27(1):137–143. doi: 10.1007/s11581-020-03803-2.

Zhou, Dong, Devaraj Shanmukaraj, Anastasia Tkacheva, Michel Armand, and Guoxiu Wang. 2019. "Polymer Electrolytes for Lithium-Based Batteries: Advances and Prospects." *Chem* 5(9):2326–2352. doi: 10.1016/j.chempr.2019.05.009.

7 Polymer Composites for Electrochromic Potential Windows

Simran Kour, Shweta Tanwar, Annu Sharma, A. L. Saroj, and A. L. Sharma

CONTENTS

7.1 INTRODUCTION

Electrochromism is the phenomenon by which a substance is able to reversibly modify its optical properties in the entire electromagnetic spectrum, such as in the near-infrared region (e.g., 1000–2000 nm) and in the visible region (400–800 nm) under the influence of an externally applied voltage (Park et al. 2016; Thakur, Ding, Ma, Lee, & Lu, 2012). This alteration in optical properties is the manifestation of the change in the electronic state of the substance due to various redox reactions occurring in the substance when an external voltage is applied. The change is reversed when the polarity of applied voltage is reversed. Electrochromic materials when integrated in the form of a device can modulate the emittance, absorbance, reflectance and transmittance of the electromagnetic radiation incident on the material (Granqvist,

DOI: 10.1201/9781003208662-9

2005; Chua et al. 2019). Currently, electrochromic devices (ECDs) are attracting the increasing attention of researchers as well as industry for exploiting their commercial applications. Some of the examples of applications of electrochromic devices include electrochromic windows, sensors, electrochromic displays, goggles, helmet visors, anti-glare mirrors, etc. (Figure 7.1). Significant benefits of ECDs such as flexible features, low power consumption, simple adjustable structures, eye-soothing display modes, etc. make them promising energy-saving candidates for smart windows and next-generation displays (Patel et al. 2017). Among all these ECDs, recently electrochromic windows (ECWs), also called smart windows, have captured great interest for energy saving and enhancing indoor comfort. ECWs are active solar control glasses capable of controlling the transmittance of solar radiation and light into vehicles and buildings. ECWs provide enhanced indoor comfort by preventing glare and thermal discomfort (Eh, Tan, Cheng, Magdassi, & Lee, 2018; Granqvist et al. 2018). They also help in saving a large amount of energy by reducing the dependence on air conditioning and artificial lighting in buildings. ECWs are more efficient in saving energy from electric lighting and cooling as compared to photochromic/thermochromics windows. Figure 7.2 shows the comparison of cooling and electric lighting energy needs for different types of windows. Clear glass with high transparency does not need too much energy for lighting, but it consumes the highest energy for cooling, while reflective glass enables better heat transfer blocking, but needs high lighting energy. It may be noted that only ECW allows minimum energy for both lighting and cooling (Gillaspie, Tenent, & Dillon, 2010; Selkowitz, 1990).

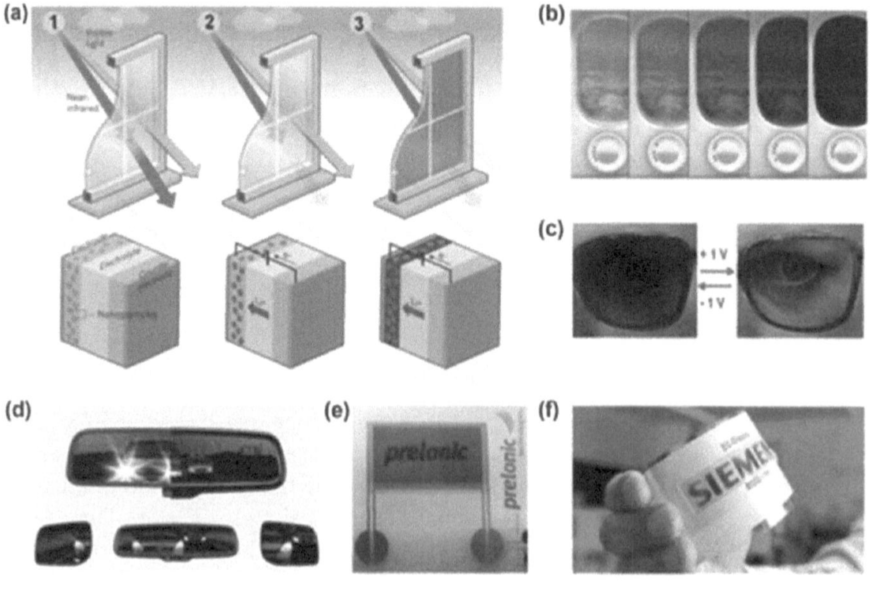

FIGURE 7.1 Various applications of ECDs. (a) EC window; (b) Smart switchable windows in an aircraft; and (c) EC lens.

Source: (a) Reproduced with permission from Korgel (2013). (b–c) Reproduced with permission from Österholm et al. (2015).

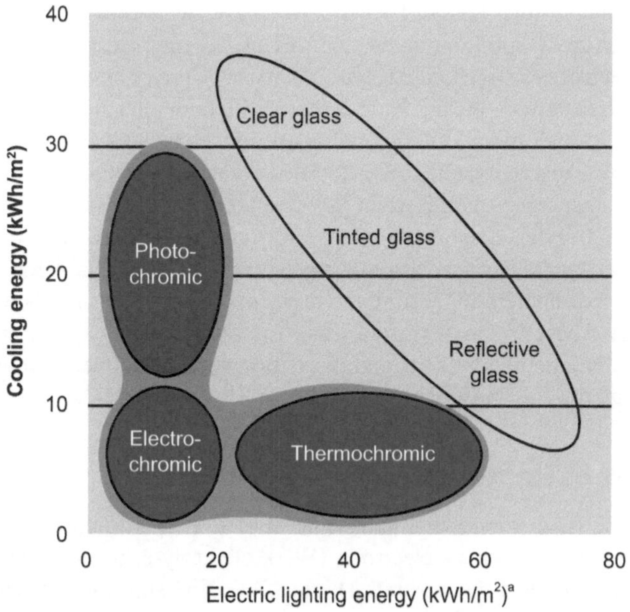

FIGURE 7.2 Comparison of electric lighting energy and cooling energy among different glazing typologies.

Source: Reproduced with permission from Granqvist et al. (2009).

ECWs can change their optical state (can be colored and bleached with application of voltage) when low voltage is applied, allowing the users to regulate the amount of heat and/or visible light entering the premise according to their preferences and comfort (Rauh, 1999; Sibilio et al. 2016). The use of electrochromic rear-view mirrors for vehicles is already well known. They help the drivers in comfortable driving during night time. Electrochromic sunglasses can be used to get protection from harmful ultraviolet (UV) radiation (Cannavale, Ayr, Fiorito, & Martellotta, 2020). Thus, ECWs have a wide potential applicability in a large number of fields. The working electrode, which is an electrochromic material, and the electrolyte are two major active components of ECWs (Sim & Pawlicka, 2020). The working efficiency of ECWs depends mainly on these two components. Electrochromic materials mainly include transition metal oxides (TMOs) such as iridium oxide, tungsten oxide, ruthenium oxide, cerium oxide, cobalt oxide, etc. and conjugated organic polymers like polyaniline, polypyrroles, polythiophenes, etc. (Yang et al. 2020). In TMOs, the transition metal ion can undergo reversible redox reactions on application of external voltage. Whereas in conjugated polymers, the delocalization of the π electrons leads to electrochromism. Electrolyte is another most important active component of ECWs. Electrolyte provides conduction of ions between the counter and working electrode. It also acts as a physical gap between the two electrodes to avoid short circuiting. The electrolyte can be in solid, liquid or gel form. Normally an electrolyte consists of one or more liquid solvents with one or more salts that provide ions on dissociation. Various kinds of electrolytes have been explored for ECWs such as liquid electrolytes, solid inorganic electrolytes, ceramic electrolytes and polymer composite electrolytes. Each electrolyte has its own pros and cons. Polymer composite electrolytes are capturing wide research attention as suitable electrolytes for ECWs, owing to their remarkable characteristics like easy processing, high mechanical strength, low reactivity, enhanced stability, wide range of operating temperature, high optical transparency, etc. Here in this chapter, we will discuss in detail the role of polymer electrolytes in enhancing the performance of ECWs. First, a brief insight of various layers and the electrochromic mechanisms of ECWs has been given. Some important key parameters suggesting the performance of ECWs has been discussed. Then, a detailed description of various kinds of polymer electrolytes used for ECWs is given. Finally a detailed results and discussion section has been provided for use of polymer electrolytes in ECWs.

7.1.1 Layers of Electrochromic Windows (ECWs)

Electrochromic (EC) windows usually consist of five layers—transparent conducting substrate (TCS), working electrode (WE), electrolyte, counter electrode (CE), and transparent conducting substrate (Figure 7.3). The substrate chosen for ECWs are usually made of glass. The substrate is coated with a conducting material (e.g., fluorine tin oxide (FTO) and indium tin oxide (ITO)). The substrate along with conductive coating makes the outermost layers of ECWs. The inner layers next to the transparent conducting substrate (TCS) are that of electroactive materials. One of them is called WE and the other is called CE. The working electrode is an

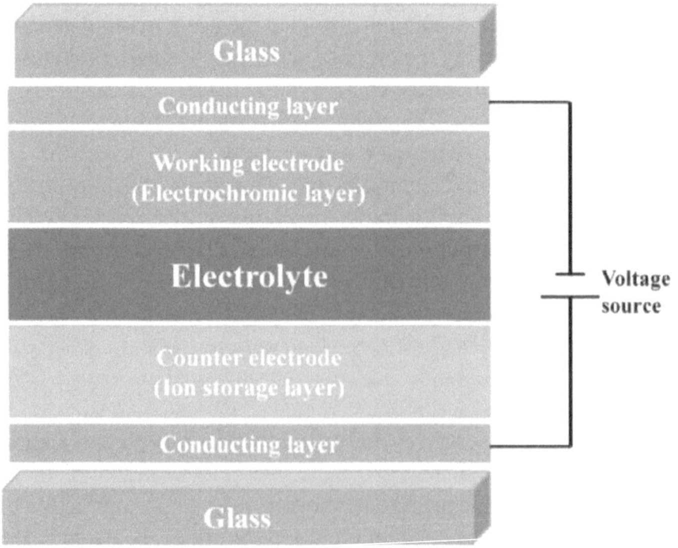

FIGURE 7.3 Schematics of various layers of ECWs.

electrochromic layer, where CE is an ion storage material that can be either an optically passive or complementary electrochromic material. The working and counter electrodes are separated by an electrolyte (solid, liquid or gel) that acts as an ionic conductor with high electronic resistance to avoid short circuiting. Optical modulation occurs when ions through the electrolyte move from the working electrode to the counter electrode and the vice versa. The movement of ions occurs through the application of potential between the two conducting layers.

7.1.2 Electrochromic Mechanisms

Electrochromism is induced in an electrochromic material by insertion of both electrolytic ions and electrons simultaneously with application of external voltage between the two transparent conducting films. Figure 7.4 represents the schematics of the electrochromic mechanism of ECWs. On applying external voltage, small electrolytic ions like H^+, K^+ and Li^+ move in between the WE and the CE (ion storage layer). Electrons from the external circuit move towards the electrochromic material and vice versa, leading to optical changes in the electrochromic material through redox reactions. The EC material undergoes a change in its color (colored state ↔ bleached state). Coloring and bleaching occurs through different redox processes. The optical state can be reversed by changing the polarity of the applied voltage. In ECWs, a very small amount of external voltage is required for switching from colored to bleached state and vice versa. The optical state (colored or bleached) remains in the same state even after removing the applied voltage. This is because of the memory effect of the EC material. Let us unfold this mechanism by taking the example of WO_3 as WE and V_2O_5 as CE. Both are EC materials. When external voltage is applied, at the cathode, WO_3 would change its color from yellow to blue after insertion of electrolytic ions M^+ and electrons. At the anode, $M_xV_2O_5$

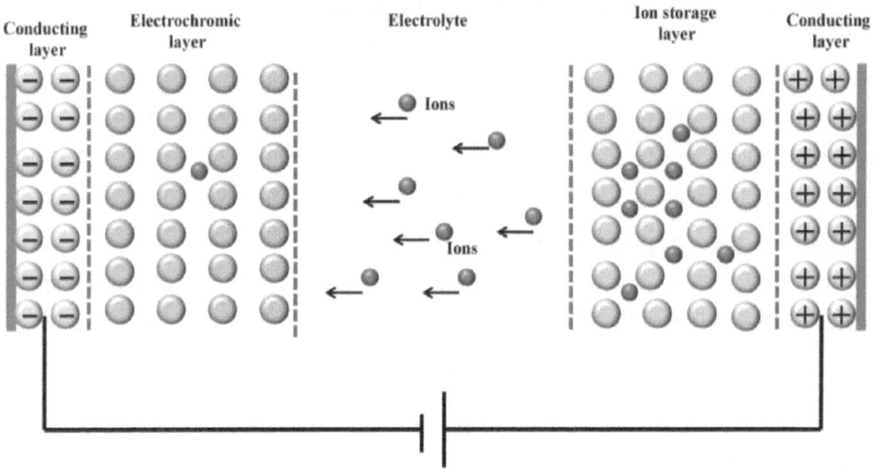

FIGURE 7.4 Schematic illustration of mechanism of an ECW.

will change its color from pale yellow to brownish yellow on removal of M^+ and electrons. The reverse reactions take place when polarity of external applied voltage is changed.

$$WO_3 + xM^+ + xe^- \rightarrow M_xWO_3$$
Yellow **Blue**

$$M_xV_2O_5 \rightarrow V_2O_5 + xM^+ + xe^-$$
Pale blue **brownish yellow**

7.1.3 Key Performance Parameters for ECWs

There are certain parameters that are used to evaluate the performance of electrochromic windows (Figure 7.5). Some of these key parameters are contrast ratio, switching time, cycling life, color, etc. Table 7.1 summarizes the requirement for large-area ECWs. Some of these parameters have been discussed here (Hassab & Padilla, 2014; Thakur et al. 2012; Yang et al. 2020).

7.1.3.1 Optical Modulation

The optical change (i.e., the change in the absorbance (ΔA) or transmittance (ΔT) between the colored and bleached state) is termed as optical modulation.

$$\Delta A = A_c - A_b$$

$$\Delta T = T_c - T_b$$

Where A_c, T_c and A_b, T_b are absorbance and transmittance in colored and bleached state, respectively.

7.1.3.2 Contrast Ratio (CR)

Optical contrast is a parameter that quantitatively determines the optical change intensity in an ECW. It is defined in terms of contrast ratio as

$$CR = \frac{I_b}{I_c}$$

Where I_c is the intensity of light reflected/transmitted from the colored state and I_b is the intensity of light reflected/transmitted from the bleached state.

7.1.3.3 Optical Density (OD)

Optical density of an electrochromic material is a measure of the optical transmittance (T) of that material corresponding to a particular wavelength. It is given as

$$OD = \log_{10}(1/T)$$

If the bleached state of the material is not completely transparent, then optical density is given as

$$OD = \log(T_b / T_c)$$

TABLE 7.1
Key performance criteria for ECWs.

Parameters	Values
Colored state solar transmittance	10–20%
Bleached state solar transmittance	50–70%
Colored state visible transmittance	≤ 10–20%
Bleached state visible transmittance	50–70%
Applied voltage (small/large area)	1–3/10–24 V
Cycling life	10^4–10^6 cycle/5–20 year
Open circuit memory	1–12 h
Operating temperature	–20 to 85 °C

Source: Lampert (1989).

7.1.3.4 Electrochromic Efficiency

Electrochromic efficiency is also known as coloration efficiency and defined as the ratio of change in optical absorbance (ΔA) for a given amount of charge injected per unit area (Q_s) with units of inverse charge density (i.e., cm^2/C).

$$\eta = \frac{\Delta A}{Q_s}$$

In terms of optical density (OD), η can be written as

$$\eta = \frac{\Delta(OD)}{Q_s} = \frac{1}{Q_s} \times \log\left(T_b / T_c\right)$$

It is a practical parameter which is used to determine the power requirements for color switching. High value of η signifies better utilization of electronic input. Electrochromic efficiency depends upon the electrochromic material used in ECWs. Inorganic EC materials with lower molar absorbance have lower efficiency in comparison to organic EC materials.

7.1.3.5 Switching Time

Switching time or response time is a key parameter to assess the performance of ECWs. It is defined as the response time required by an EC material to switch from colored to bleached state and vice versa. Defining more precisely, switching is the time taken to reach 90% of T_c from T_b and vice versa. Its value may be from few seconds to minutes depending upon the size of ECW. Switching time mainly depends upon the electronic conductivity, ionic conductivity of the electrolyte of the electrodes and electrolyte thickness. It depends also on the morphology and film thickness of the EC material, diffusion of electrolytic ions in the EC film and the applied voltage.

7.1.3.6 Write-Erase Efficiency

It is defined as the fraction of the original coloring that can be erased or bleached.

7.1.3.7 Cyclic Life

When an EC material undergoes repeated cycling between its colored and bleached state, the performance of the device degrades with time. This is due to physical and chemical changes occurring in the device. If the applied voltage exceeds the safe limit for ECW, the cyclic life of the device is reduced.

7.1.3.8 Stability

The stability of an ECW is measured by the service life of an ECW which is based on optical change after multiple switching between colored and bleached states. The applied voltage and external environmental conditions are two important factors determining the stability of an ECW. The packing of an ECW should be proper to avoid adverse impacts of external factors on the stability of the ECW. Suitable EC materials and proper choice of electrolyte with minimal leakage are encouraged for highly stable ECWs.

7.1.3.9 Memory Effect

The ability of an ECW to maintain a particular optical state (colored/bleached) during an open circuit state is termed as memory effect. The ECW device can attain a stable colored

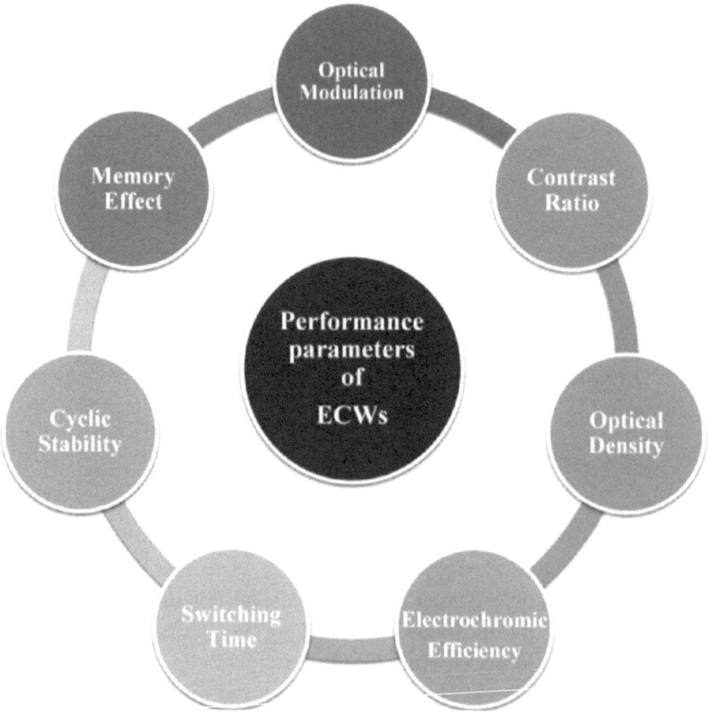

FIGURE 7.5 Performance parameters of ECWs.

as well as bleached state even after turning off the applied voltage. It does not require a continuous power supply to maintain a desired optical state as in the case of light emitting diodes (LEDs). The EC material and electrolyte play a crucial role in deciding the memory of the ECW. A solid state ECW will have high memory in comparison to liquid ECWs. This is because solid ECWs will have limited diffusion of electroactive components during the off state thus avoiding the self-erasing of the optical state.

7.2 ROLE OF ELECTROLYTES IN ECWs

The electrolyte plays a very significant role in determining the optical performance of ECWs. Electrolyte is an ionic conduction medium between WE and the CE and acts as a separator that avoids electric contact between the two electrodes. Electrolyte can be a solid, liquid or gel. For better performance of ECW, the electrolyte must possess the following useful characteristics:

- ***High ionic conductivity***
  The electrolyte should possess high ionic conductivity (σ) to facilitate the transportation of ions between the working and counter electrode. High conductivity also leads to rapid motion of electrolytic ions from the electrolyte to the electrochromic layer and the other way around leading to rapid switching between bleached and colored states. Ideally, the ionic conductivity should lie in the range of 10^{-3} to 10^{-7} S cm^{-1} (Thakur et al. 2012). The conductivity of the electrolyte should not degrade due to drying after a long time of storage. Also, the small-sized ions such as H^+, Li^+ are preferred due to their high mobility.
- ***Low/negligible electronic conductivity***
  The electrolyte should be electronically insulated. There should not be any electronic transfer between the WE and CE through electrolyte as this will lead to a short circuit condition.
- ***High optical transparency***
  The electrolyte must be optically transparent. High optical transparency of the electrolyte leads to transparency of ECW during bleaching. Electrolyte must be free from light scattering effects.
- ***Wide potential window***
  A potential window for an electrolyte is defined as the voltage range under which the electrolyte can be safely used. Exceeding the voltage beyond this limit, the electrolyte undergoes degradation. High potential window of the electrolyte means that the electrolyte can be safely used for high-voltage applications.
- ***Thermally stable***
  The electrolyte should be thermally stable. The electrolyte should be stable for a wide range of temperatures so that ECWs with such electrolytes can be used under extreme temperature conditions.
- ***Chemically stable***
  The electrolyte with high chemical and electrochemical stability is suitably preferred for ECWs. The electrolyte must be chemically stable in potential

range of switching for ECWs. Also electrolyte must be compatible with anodic and cathodic layers.

- ***Low cost and eco-friendly***
  For use of ECWs on commercial scale, it is highly desirable that different components of the device should be of low cost and environmentally safe. The electrolyte with low cost and easy availability is preferred for ECWs to reduce the cost of fabrication of ECWs. The electrolyte should also be eco-friendly.

Beside these mentioned features, the electrolyte should be compatible with both the electrodes (WE and CE), safe to humans (non-explosive, inflammable) and have long operating hours. Various kinds of electrolytes such as liquid electrolytes, polymer electrolytes, solid inorganic electrolytes, etc. are used for ECWs. Liquid electrolytes have the advantage of high ionic conductivity but there are some limitations to the use of liquid electrolytes for ECWs such as low chemical stability, leakage issue, and safety issues. Polymer electrolytes have gained wide research attention in the field of ECWs because they can be easily processed as thin films of desired thickness/shape for compact and smart packing of ECWs (Choudhury, Sampath, & Shukla, 2009; Gonçalves et al. 2012). Excellent mechanical, chemical and electrochemical stability of polymer electrolytes helps in fabrication of robust ECWs. Along with that, high ionic conductivity and low electronic conductivity of conducting polymers make them a favorable choice as electrolyte for ECWs. Use of polymer electrolytes for ECWs provides flexibility, light, and moldable shape, etc. and safety against leakage and short circuiting (Aziz et al. 2020).

7.3 POLYMER ELECTROLYTES FOR ECWs

Polymer electrolytes are used in ECWs and provide a path for ion conduction. Two principal requirements for polymer electrolyte are high conductivity and high transparency. Polymer electrolytes basically consist of molten salts dissolved in a polymer matrix. The ions of the salt are screened by the polymer host (PEO, PMMA PEI, PPO, etc.). They are widely used in a number of electrochemical devices like supercapacitors, batteries, fuel cells, etc. Over the past three decades, research has been accelerated in using polymer electrolytes for electrochromic devices such as EC displays, ECWs, anti-glare mirrors, sunroofs, etc. Polymer electrolytes offer numerous advantages over usual liquid electrolytes. Their high ionic conductivity, high electronic resistivity, wide operating windows, thermal stability, mechanical stability, chemical stability, easy processibility, flexibility, light weight, etc. (Arya & Sharma, 2019) make them a better choice as electrolyte for ECWs. Polymer electrolytes are further categorized as (i) gel polymer electrolytes (GPEs), (ii) solid polymer electrolytes (SPEs), (iii) composite polymer electrolytes (CPEs) and (iv) polyelectrolytes. The details of these electrolytes are given next:

- **Gel polymer electrolyte**
  GPEs, also called plasticized polymer electrolytes or third-generation polymer electrolytes, have gained increasing attention since their introduction by Perche and Feuillade in the year 1975 (Feuillade & Perche, 1975).

They are usually obtained by mixing a polymer matrix with an alkaline metal salt, usually a lithium salt and a solvent/liquid plasticizer capable of making a stable gel with the host polymer matrix (Krejza & Vondrák, 2009; Srivastava & Tiwari, 2009). They constitute a polymer-solvent-salt system with ionic conductivity higher than 10^{-3} S cm^{-1} (Agnihotry, Pradeep, & Sekhon, 1999). The addition of plasticizer in the polymer host matrix results in the formation of swollen polymer matrix in a suitable solvent. Based on the procedure of synthesis, gel polymer electrolytes are categorized as (i) physical gel electrolytes and (ii) chemical gel electrolytes (Arya & Sharma, 2017). Physical gel consists of a polymer matrix enclosing liquid electrolyte. For example, poly(methyl-methacrylate) matrix enclosing $LiClO_4$/EC/PC electrolyte is a physical gel electrolyte. In such polymers, there do not exist any chemical bonds between the polymer matrix and the electrolyte. On the other hand, in chemical gels, the liquid electrolyte is linked to the host polymer matrix with covalent bonds. The formation of chemical bonds can be either by direct chemical reaction of functional groups of polymer matrix with liquid electrolyte or by addition of cross-linking agents. The cohesive nature of solids and diffusive nature of liquids are simultaneously indulged in a gel polymer electrolyte (Adebahr, Byrne, Forsyth, Macfarlane, & Jacobsson, 2003). GPEs are widely used in a lot of applications (energy storage/conversion devices, chemical sensors, etc.), but recently the research has been focused to their use in electrochromic applications. Their outstanding characteristics including high ionic conductivity, low electronic conductivity, excellent thermal and chemical stability, light weight, geometric flexibility, making them ideal candidates as electrolytes for electrochromic devices including ECWs (Ngai, Ramesh, Ramesh, & Juan, 2016; Nicotera, Coppola, Oliviero, Castriota, & Cazzanelli, 2006; Uma, Mahalingam, & Stimming, 2005). The use of gel polymer electrolytes for ECWs leads to preventing leakage and internal short circuiting. Gel electrolytes also help in improving the lifetime of the device (Ramesh, Liew, & Ramesh, 2011; Song, Wang, & Wan, 1999). These unique properties of gel electrolytes make them a better replacement of liquid electrolytes in ECWs (Gonçalves et al. 2012). GPEs have certain limitations. The incorporation of liquid electrolyte inside the polymer matrix leads to poor mechanical stability of the electrolyte. Addition of plasticizers lowers the interfacial stability of the electrolyte causing poor cyclic performance of electrochromic devices (Fu, 2010).

- **Solid polymer electrolyte**

  The research for solid polymer electrolytes (SPEs) started from the year 1975 when Wright reported the electrical conductivity of poly (ethylene oxide) (PEO)-based ionic complexes (Armand, 1994; Murata, Izuchi, & Yoshihisa, 2000; Wright, 1975). SPEs are solvent-free electrolytes having an ionic conducting phase produced by dispersing an alkali salt (usually lithium salt) into a host polymer matrix. The most commonly used lithium salts are $LiClO_4$, $LiPF_4$, $LiPF_6$, $LiN(SO_2CF_3)_2$, $LiB(C_2O_4)_2$, $LiCF_3SO_3$, etc. These salts are dispersed in a suitable polymer matrix like PEO, PMMA and PVdF, etc. No organic solvent is used in SPEs. The metal ions from the metal salt are linked to the polymer chain through electrostatic forces by formation of coordination bonds. SPEs offer many advantages like light weight, wide temperature operating range, good electrode/electrolyte contact, flexibility, low volatility, excellent chemical and electrochromic stability (Ramesh & Wen, 2010). SPEs like GPEs can be easily processed in desired shape and geometry. The solvent-free nature of SPEs offers the advantage of eliminating liquid leakage and harmful gases from side reactions and in turn helps in increasing the shelf life of the device. SPEs offer excellent mechanical strength (Aziz, Woo, Kadir, & Ahmed, 2018; Ngai et al. 2016). Apart from so many advantages, SPEs have certain shortcomings also. Low ionic conductivity and high crystallinity are hampers in the wide-scale use of SPEs in various applications. Ionic conductivity of SPEs is lower than 10^{-8} S cm^{-1} at room temperature. Only at high temperatures are they able to attain practically suitable values for ionic conductivity. Usually, a polymer electrolyte consists of both crystalline and amorphous regions. But the polymer host matrix used for SPEs are usually or a crystalline or semi-crystalline nature. The amorphous regions are more prone to transportation of ions (high conductivity) as compared to crystalline regions. Thus low ionic conductivity of SPEs is related to their high crystallinity. Hence, to augment the ionic conductivity of SPEs, the crystallinity of the host polymer chains needs to be suppressed (Srivastava & Tiwari, 2009). Low crystallinity will lead to better mobility of the polymer matrix which in turn accelerates the movement of cations thereby increasing the conductivity. Various processes like copolymerization (Yağmur, Ak, & Bayrakçeken, 2013; Yu, Chen, Fu, Xu, & Nie, 2013), blending (Nguyen, Xiong, Ma, Lu, & Lee, 2011), cross-linking (Y. Wang et al. 2019), grafting (Puguan & Kim, 2017), alloying and inorganic filler blending (addition of ZnO, SiO_2, $BaTiO_3$, etc.) (Rosli, Muhammad, Chan, & Winie, 2014) of the polymer matrix can increase the ionic conductivity of SPEs to a greater extent. All these attempts to advance the performance of SPEs lead to the formation of new kinds of materials called CPEs discussed next.

- **Composite polymer electrolytes**

  Composite polymer electrolytes (CPEs) have the potential to eliminate the issues faced by SPEs. The addition of inorganic inert nano-fillers with high dielectric constant to the polymer electrolytes (low dielectric constant) leads to the formation of CPEs. The interaction of inorganic nano-fillers with the host polymer electrolytes leads to improved mechanical and electrochemical properties of the original polymer electrolyte (Kam, Liew, Lim, & Ramesh, 2014). Fillers are usually inorganic oxides such as SiO_2, ZnO, MgO, Al_2O_3, $PbTiO_3$, $BaTiO_3$, etc. (Srivastava & Tiwari, 2009). The ionic conductivity of the electrolyte depends on particle size, surface area, porosity, concentration and interaction of polymer host composite with nano-fillers (Ngai et al. 2016). Apart from high mechanical strength and ionic conductivity, CPEs also offer some other important benefits such as better interfacial contact, flexibility, thermal stability, etc. Some other ways of fabrication of CPEs include (i) copolymerization, (ii) cross-linking, (iii) blending, (iv) doping of nanomaterials and (v) addition of ionic liquids (Luo, Conrad, & Vankelecom, 2013). CPEs, owing to their enhanced optical, mechanical and electrochemical properties, have gained wide research attention as efficient electrolyte for electrochromic windows (ECWs).

- **Polyelectrolytes**
  Polyelectrolytes are polymers with a substantial portion of the repeating unit containing an electrolyte group. The electrolyte group gets dissociated in aqueous solution leaving the polymer matrix charged (Meka et al. 2017). Polyelectrolytes are thus charged polymers enjoying the features of both the electrolyte and the polymer. They are classified as (i) cationic polyelectrolytes and (ii) anionic polyelectrolytes based on the presence of a positive or negative charged ionizable group on the monomeric unit of the polymer. There is another class of polyelectrolytes called polyampholytes, which contain a mixture of both positive and negative charged monomer units. In polyelectrolytes, the anions of the salt are immobile as they are covalently linked to the host polymer matrix while the cations of the salt are mobile. Polyelectrolytes are therefore known as a single ion conductor. The ionic conductivity of polyelectrolytes results from the self ion-generating group. Perfluorosulfonic ionomers are among the most commonly used ionomer electrolytes. There exist some natural polyelectrolytes such as desoxyribonucleic acid (DNA), ribonucleic acid (RNA), proteins and polysaccharides (Thakur et al. 2012).

7.4 RESULTS AND DISCUSSION

Utilization of polymer electrolytes for the fabrication of electrochromic devices has been increasing at a rapid pace in comparison to liquid electrolytes, owing to their leak-proof ability and easy fabrication with desired shape and size. Gel electrolytes have been widely explored for electrochromic applications. For example, Lee et al. fabricated UV-cured poly(methyl methacrylate)-based gel electrolyte for their use in smart EC windows (Lee et al. 2020). The UV curing time has a huge impact on the performance of the electrolyte. An ECW based on 10 minutes of UV curing showed excellent electrochromic performance with optical transmission of 51.3% corresponding to a wavelength of 550 nm. The brilliant performance of the ECW was accredited to superior connectivity of the gel electrolyte with the electrode material. The ECW also possessed low switching time for both colored state (2 s) and bleached state (1.5 s). The device also exhibited high electrochromic efficiency of 133.1 $cm^2\,C^{-1}$ for bleaching and 178.6 $cm^2\,C^{-1}$ for coloration, respectively. The device was also able to reach high cyclic stability of 98.9% after 11,500 cycles. Further the device was tested for thermal stability at different temperatures. The devices showed thermal stability over an extensive temperature range from –20 °C to +70 °C. Thus, UV-cured gel electrolyte appeared as potential candidate for electrochromic windows.

A highly transparent gel polymer electrolyte with ionic cross-linking was developed by Chen et al. by free radical polymerization (W. Chen et al. 2019). The gel electrolyte consisted of propylene carbonate (PC), $LiClO_4$, acrylic acid (AA), acrylamide (AAm), azobisisobutylonitrile (AIBN) and was called as PADA gel polymer electrolyte. The value of σ for PADA was observed to be 1.3×10^{-2} S cm^{-1} at 25 °C. PADA was sandwiched between WO_3 and NiO electrodes coated onto FTO glass substrate to form an EC device with configuration as glass FTO/WO_3//PADA gel//NiO/FTO glass. The device showed black blue color in its colored state, corresponding to potential of −2.3 V and was transparent in its bleached state for a positive potential of +2.3 V. Transmittance for colored and bleached state was 72.4% and 11.4%, corresponding to a wavelength of 660 nm providing 61% optical modulation at 660 nm. The switching response of the device was very high with coloration time of 8.5 s and bleaching time of 7.5 s owing to superior ionic conductivity of the electrolyte. Excellent color efficiency of 78.7 cm^2 C^{-1} suggested that the device could give excellent performance corresponding to low-energy input. When tested under open circuit conditions for 24 h, the transmittance increased from 11.4% to 33.4% for the colored state, suggesting that the colored state of the fabricated device is stable to be used on daily basis without energy consumption. However, gel electrolytes suffer from limitation of poor mechanical stability owing to the incorporation of liquid electrolyte inside the polymer matrix. To overcome such limitations, solid polymer electrolytes (SPEs) have gained wide research attention. They are solvent-free electrolytes; hence the problem of leakage is solved with SPEs. Also, the solid electrolytes have excellent mechanical stability. Esin fabricated biodegradable polymer-chitosan-based SPE for electrochromic devices (Esin, 2019). The electrolyte was prepared using chitosan, LiTRIF, and PC. The effect of addition of PEDOT:PSS to the electrolyte was investigated. To test the electrochromic behavior of the electrolyte, an EC device (glass/ITO/WO_3//SPE//ITO/glass) was fabricated. Addition of PEDOT:PSS led to improved electrochromic performance of the device. The value of σ for chitosan-based electrolyte with PEDOT:PSS was calculated to be 4.2×10^{-4} S cm^{-1} which was higher than electrolyte without PEDOT:PSS (3.4×10^{-4} S cm^{-1}).

Enhanced σ led to the fast switching response of the device for coloration (0.29 s) and bleaching (3 s), respectively. The device also exhibited a contrast ratio of 22% corresponding to wavelength of 800 nm. The device exhibited high coloration efficiency of 67 cm^2 C^{-1}.

Wang et al. designed a novel PVB:PEG-based quasi-solid polymer electrolyte and tested its electrochromic performance by fabricating an ECD with PProDOT-Me_2 as WE and Li-Ti-NiO as CE (W. Wang, Guan, Li, Zheng, & Xu, 2018). The electrolyte exhibited σ value of 10^{-5} S cm^{-1} with transmittance of >70% in the visible region (400–800 nm). The fabricated ECD possessed a high contrast ratio of 43.8% corresponding to 585 nm. The device switched between its colored (deep blue color) and bleached (transparent) state on applying an external voltage of −1.8 V and +2 V, respectively. The device exhibited hasty switching between its bleached and colored state with τ_b =2.6 s and τ_C = 1.2 s. The stability of the device after repeated cycles of coloring and bleaching plays a crucial role for EC applications. It was observed that the fabricated device also showed excellent cyclic performance with 84% of optical modulation retained after 20,000 cycles. This increased stability was attributed to the highly viscous skeleton structure of PVB, making the electrolyte in a suitable structure during the transportation of ions. The results suggested that the electrolyte has wide potential applicability for ECWs and anti-glare mirrors.

Another green solid polymer electrolyte based on chitosan was developed by Eren for electrochromic applications (Eren, 2019). The electrolyte was prepared using PEDOT, PSS and chitosan polymers with LiTRF as lithium salt and PC as plasticizers. Figure 7.6 illustrates the schematics of the fabrication process of the electrolyte. The electrochromic device was fabricated using PEDOT and PMeT as EC materials with configuration as ITO PEDOT//Ch:PEDOT:PSS:LiTRIF:PC//ITO PMeT. The device exhibited an optical contrast ratio of 32.2% under an applied voltage $\pm 2V$. The device also possessed fast switching response, with switching time 0.24 s and 0.52 s for colored and bleached state, respectively. The better switching response was due to the presence of PEDOT:PSS which act as conductive binder reducing the electrolytic resistance to ease the transportation of electrolytic ions. High coloration efficiency of 228.65 $cm^2\ C^{-1}$ was observed credited to PEDOT:PSS-based percolation channels inside LiTRIF/chitosan supporting homogenous rate of coloring/bleaching process of the device. These results suggest the electrolyte has potential for fabricating efficient EC windows.

Zeng et al. fabricated a flexible and energy saving electrochromic window device using transparent solid polymer (SPE) electrolyte (Zeng et al. 2019). The electrolyte consisted of metal salt-lithium perchlorate ($LiClO_4$) and the polymers polyoxypropylene glycol (PPG), and polymethacrylate (PMMA). The electrochromic performance of the electrolyte was tested using two different working electrodes (electrochromic materials), PTCDA and 4EDOT-2B-$COOCH_3$. PTDCA was used as cathodic EC material. The device configuration was ITO PET//SPE//PTCDA

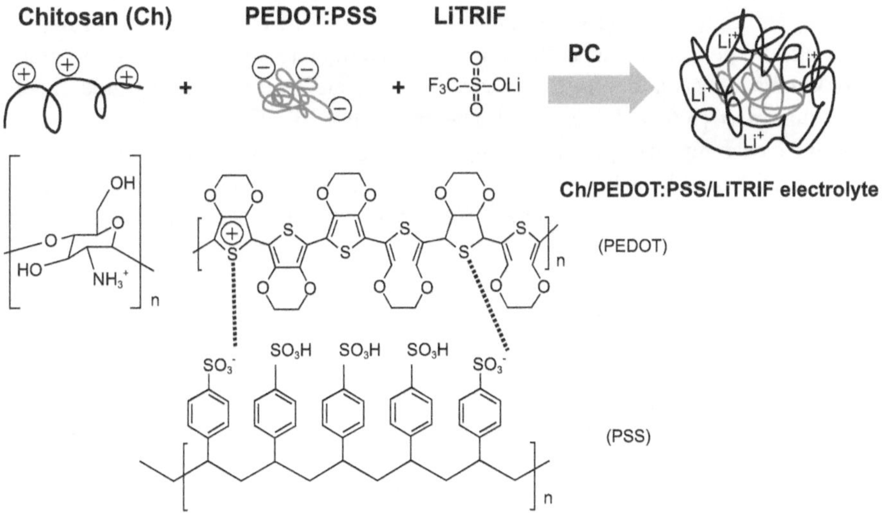

FIGURE 7.6 Schematics of fabrication of PEDOT:PSS/chitosan-based solid polymer electrolyte.

Source: Reproduced with permission from Eren (2019).

film/ITO PET (named as ECD3). 4EDOT-2B-$COOCH_3$ was used as anodic EC material with device configuration as ITO PET/SPE/4EDOT-2B$COOCH_3$ film/ ITO PET (named as ECD4). UV absorption spectroscopy was used to test the transmittance of the SPE. The electrolyte showed a high transmittance >88% in the visible region. The electrolyte exhibited high value of conductivity about 1.01×10^{-3} S cm^{-1}. The flexible device ECD3 showed a color switching from red to black with an applied voltage between −3 V and +2.7 V. The device returned from black color to red color. In a similar way, ECD4 showed color switching from red to blue by applying voltage from −2.9 V to + 2.5 V. Both the devices showed very fast switching between the colored and bleached states. ECD3 exhibited switching time of 2.4 s and 5.6 s for colored and bleached states, respectively. On the other hand, colored and bleached states of ECD4 had switching time of 5.2 s and 2.6 s, respectively. Both the devices (ECD3 and ECD4) also exhibited better cyclic stability with 79.7% and 81.4% of initial charge maintained after 600 cycles, respectively. The coloration efficiency of 186.3 cm^2 C^{-1} and 230.6 cm^2 C^{-1} was observed for ECD3 and ECD4, respectively. Open circuit memory for both the devices was also calculated. ECD3 had a transmittance of 10.2% corresponding to voltage of −3.0 V for 700 nm. When voltage was removed for 12 h, the transmittance for 700 nm increased to 22.5%, still maintaining 81.5% of initial optical contrast. In a similar way, ECD4 maintained 74.7% of the initial optical contrast. All these results suggested that the ECDs based on this kind of SPE have a wide potential applicability for electrochromic windows.

Jeong et al. investigated the use of hydroxypropyl methylcellulose (HPMC)-based SPEs for EC devices with WO_3 as WE and Prussian blue (PB) as CE (Jeong, Kubota, Chotsuwan, Wungpornpaiboon, & Tajima, 2021). HPMC, owing to its high polarity, excellent mechanical stability and environment-friendly nature, was chosen over other industrially synthesized polymers. The electrolyte was prepared using lithium perchlorate as metal salt, and polyethylene glycol (PEG) as polymer matrix. HPMC was added to the electrolyte. The electrolyte demonstrated excellent value of $\sigma\left(5.07\times10^{-3}\ \mathrm{S\ cm^{-1}}\right)$. The fabricated EC device underwent coloration and bleaching corresponding to applied voltage of −1.2 V and +1 V (Figure 7.7a). The coloration and bleaching arose from the reduction of WO_3 on intercalation of K^+ ions and oxidation of $K_4Fe_4[Fe(CN)_6]$ by deintercalation of K^+ ions and vice versa.

$$\underset{\text{Colorless}}{WO_3 + mK^+ + xe^-} \leftrightarrow \underset{\text{Blue}}{K_mWO_3}$$

$$\underset{\text{Colorless}}{K_4Fe_4[Fe(CN)_6]} \leftrightarrow \underset{\text{Blue}}{Fe_4[Fe(CN)_6]} + 4K^+ + 4e^-$$

Optical switching time for colored and bleached states was found to be 6 s and 16 s, respectively. The device also displayed high coloration efficiency of 139 cm^2 C^{-1}. Figure 7.7b represents the transmittance spectra of the device corresponding to colored and bleached state. Figure 7.7c shows the real photos of the device during coloring and bleaching. The EC performance of this solid polymer electrolyte-based device was also compared with their previously reported EC device based on a gel polymer electrolyte (Jeong, Watanabe, & Tajima, 2021; Tajima, Watanabe, Nishino, & Kawamoto, 2020a, 2020b). They observed that SPE-based ECD showed overall better electrochromic performance in comparison to GPE. The SPE solved the problem of leakage and complex manufacturing process as faced by GPEs.

Isfahani et al. fabricated an SPE with Gellan gum (Ge) host matrix doped with LiTFSI and plasticized with glycerol (Gly), and evaluated their optical and electrochromic properties (Isfahani et al. 2021). The electrolyte exhibited a non-linear temperature variation of σ with lowest conductivity of 8.69×10^{-4} S cm^{-1} at 90°C and 2.77×10^{-4} S cm^{-1} at 22°C.

An EC device using the same electrolyte with configuration ITO glass/TiO_2-CeO_2//electrolyte//PB/ITO glass was fabricated to test the electrochromic performance of the electrolyte. The device exhibited optical variation (dark blue at +0.5 V and transparent at −2.5 V), corresponding to the following redox reactions followed by intercalation/deintercalation of Li^+ ions and vice versa.

$$K(Fe[Fe(CN)_6]) + Li^+ + e^- \leftrightarrow K\left(LiFe\left[Fe(CN)_6\right]\right)$$

Dark blue Transparent

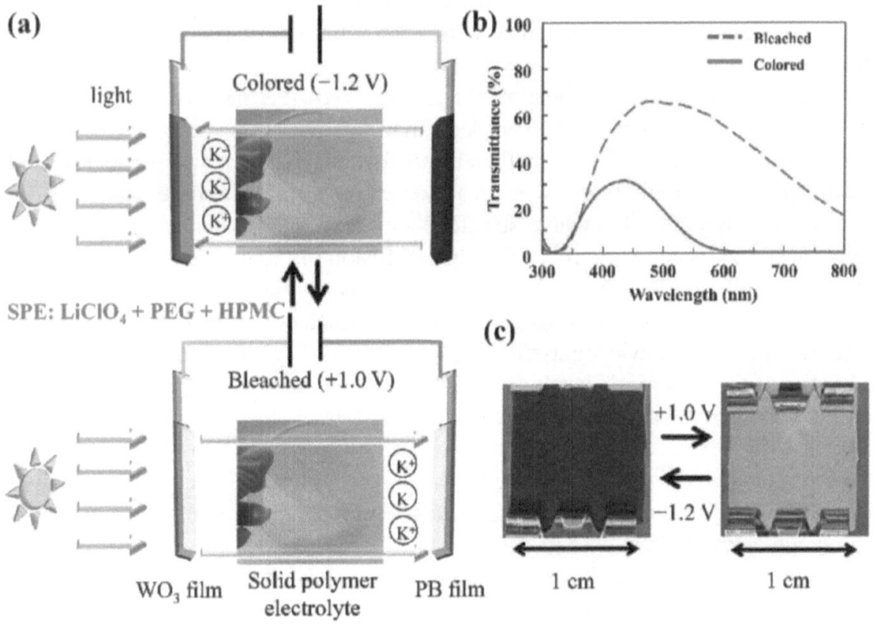

FIGURE 7.7 (a) Color and bleach state of the EC device; (b) transmission spectra of the device in color and bleach state; (c) real photos of the device in color and bleach state.

Source: Reproduced with permission from Jeong, Kubota, et al. (2021).

The device exhibited excellent electrochromic features with fast switching time, optical modulation of 37.1% at 630 nm and efficiency of 2062 $cm^2\ C^{-1}$, encouraging their potential for electrochromic applications. Although SPEs are found to exhibit excellent electrochromic performance, low ionic conductivity is a major hurdle in use of SPEs for EC applications. Composite polymer electrolytes (CPEs) had their origin in attempts to augment the ionic conductivity of SPEs. CPEs are fabricated using a variety of methods including (i) addition of nano-fillers, (ii) copolymerization, (iii) cross-linking, (iv) blending, (v) doping of nanomaterials and (vi) addition of ionic liquids. For instance, Wang et al. synthesized a self-healed dually cross-linked polymeric electrolyte for flexible EC applications (Y. Wang et al. 2019). It was synthesized using vinyl hybrid silica nanoparticles (VSN), 1-vinylimidazole (VIm) and acrylic acid (AA) by copolymerization process. The cross-linking of VSN helps in enhancing the mechanical strength of the polymer chain, and formation of a homogeneous polymeric network. The electrolyte exhibited excellent value of $\sigma \sim 1.26 \times 10^{-4}$ S cm^{-1}. The electrochromic performance of the electrolyte was tested by fabricating an EC device with viologen (1, 1′-disubstituted-4, 4′-bipyridinium) derivative (Bpy) as EC material and PANI as ion storage material. Bpy was incorporated into the electrolyte to form an EC gel that works both as an EC material and an electrolyte (Figure 7.8a). The device with an original green color turned purple during coloration and returned back to green on bleaching (Figure 7.8b). The fabricated Bpy/PANI/Fc device exhibited fast switching response for coloration (2 s) and bleaching (1.8 s) along with high color efficiency of 406.96 $cm^2\ C^{-1}$, which suggested that the device exhibited better optical modulation behavior. The device also possessed high stability with 83% retention of initial ΔT even after 5000 cycles.

Chen et al. reported the fabrication of a cross-linked GPE based on ionic liquid for EC devices (W. Chen et al. 2021). The electrolyte was synthesized using ionic liquids, methyl methacrylate (MMA), 2-diethylaminoethyl methacrylate (DEA), acrylic acid (AA) and LiTFSI. Ionic liquid was encapsulated as a uniform phase inside the 3D network of polymers. This whole structure with continuous channels available for transportation assisted rapid migration and diffusion of the ions, thereby escalating the ionic conductivity of the electrolyte. The σ value of the electrolyte was found to be 3.29×10^{-3} S cm^{-1}. The electrochromic performance of the electrolyte was tested by fabricating an EC device (glass/FTO/WO_3//CPE//FTO/glass). The devices showed coloration (blue color) and bleaching (transparent) at applied voltage of +3V and −3V, respectively. Optical contrast of 49.9% was observed corresponding to wavelength 650 nm. Also, the device exhibited fast switching time for both coloration (7s) and bleaching (4s), credited to the elevated σ for the electrolyte. High color efficiency of 96.2 $cm^2\ C^{-1}$ suggested that the device obtained high optical modulation at low input energy which can help in energy saving.

Bae et al. synthesized dual-cations-based composite polymer electrolyte (Bae, Kim, Kim, & Kim, 2021). Two ionic sources were used—LiTFSI (lithium salt) as Li ion source and PSSA as proton source. MXene was taken as a filler to boost the value of σ for the polymer electrolyte. MXene helped in fast dissociation of salts through acid-base interactions and facilitated the movement of cations between the polymer

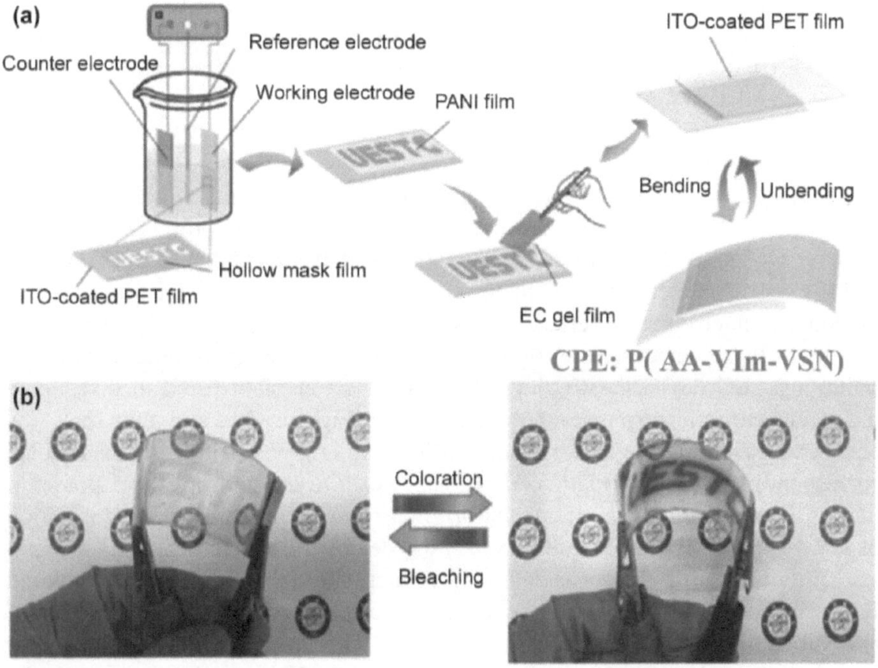

FIGURE 7.8 (a) Schematics of the fabrication process of flexible EC device on plastic sheet; (b) images of ECD during bleaching and coloring state along with bending.

Source: Reproduced with permission from Y. Wang et al. (2019).

matrix and MXene. The electrochromic behavior of the electrolyte was investigated through the fabrication of an EC device with PEDOT:PSS as WE and PANI:PSS as CE. LiTFSI provided excellent ΔT and PSSA provided very fast response. The device exhibited an excellent transmittance of 66.2% with rapid switching for coloration (8 s) and bleaching (15 s). The device also exhibited excellent color efficiency of 340.6 $cm^2\ C^{-1}$.

Gelatin, a biopolymeric material owing to its profound characteristics such as flexibility, biodegradability, renewability, intrinsic transparency, natural abundance, nontoxicity and low production cost has been widely explored as polymer electrolyte by compounding it with plasticizers and salt. Wootthikanokkhan et al. fabricated a transparent adhesive polymer electrolyte based on gelatin for the fabrication of EC glass (Wootthikanokkhan, Jaruphan, Azarian, & Yosthisud, 2020). The electrolyte was prepared by blending EAA copolymer-based ionomer resin with gelatin. $LiClO_4$ was used as a lithium source. The optical performance of the electrolyte was tested by fabricating an ECD with configuration ITO/WO_3//electrolyte// NiO/ITO. The optical contrast (ΔT)% for gelatin-based electrolyte was 43.35% while it was 30.07% for EAA (2.5% wt.)/gelatin-based electrolyte. The switching time for both gelatin and EAA/gelatin electrolyte was same for both coloration (10 s) and bleaching (50 s). The coloration efficiency of EAA/gelatin-based electrolyte was higher (60.34 $cm^2 C^{-1}$) in comparison to that for gelatin-based electrolyte (53.90 $cm^2 C^{-1}$). Higher efficiency suggested that less energy is required to obtain larger optical modulation for EAA/gelatin electrolyte in comparison to gelatin electrolyte. Thus, EAA ionomers lead to perking up the overall optical performance of the electrochromic device.

Sydam et al. synthesized a hybrid kind of polymer electrolyte called as ionogel which is a blend of a polymer matrix entrapping an ionic liquid (IL) (Sydam, Ojha, & Deepa, 2021). Heptyl viologen (HV) was used as cathodic EC material and PB was used as anodic EC material. An ionic additive, EDTA was added to the ionogel electrolyte to perk up the write-erase efficiency of the HV-based EC device, as these devices were found to become permanently colored after a few switchings. EC devices with and without EDTA were fabricated to investigate the electrochromic response of the electrolyte. Figures 7.9a and 7.9b show the color change for EC devices without EDTA. At −1.8 V, the device turned blue corresponding to the formation of HV^+ species. On applying positive potential of +1.5 V, the device returned to its original transparent state due to formation of HV^{2+} species. But after a few cycles, the device did not return to its original state after bleaching. This was due to the reduction of HV^+ radical to pale colored HV^0. The addition of EDTA in the ionogel electrolyte prevented the reduction of HV^+ radical to pale colored HV^0 by interacting with the radical cation and hence, improved the EC performance of the device. Figures 7.9c and 7.9d–f represent the bleach and color states of the device at various potentials with a clear transparent bleached state.

To enhance the conductivity of polymer electrolytes, Zhang et al. focused on synthesizing a hybrid organic-inorganic gel electrolyte for EC applications (W. Zhang, Zhang, & Shen, 2021). The electrolyte was fabricated by blending methoxypolyethylene glycol poly(propylene glycol)-based precursor with bis(2-aminopropylene glycol)-based precursor followed by hydrosilylation and addition of LiTFSI-based liquid electrolyte. The electrolyte possessed excellent $\sigma \sim 1.67 \times 10^{-3}$ S cm^{-1}. A TGA test was conducted to find the weight loss of the electrolyte with temperature. It was observed that the electrolyte did not undergo any appreciable weight loss for temperature change from 25 °C to 315 °C suggesting excellent thermal stability of the electrolyte over an extensive temperature range. To explore the electrochromic behavior of the electrolyte, an EC device (ITO glass/WO_3//CPE//ITO glass) was made up. The device exhibited excellent EC behavior with high optical contrast (38%), and high color efficiency of 282.9 $cm^2 C^{-1}$.

There is another class of polymer electrolytes called polyelectrolytes that are widely explored for electrochromic applications. For instance, Puguan et al. fabricated polyelectrolyte based on 1,2,3-triazole with a pentaoxyethylene spacer for EC devices (Puguan, Boton, & Kim, 2018). The electrolyte was developed by cyclic addition of copper-catalyzed alkyne-azide- terminal monomer quaternized using alkyl halides followed by anion exchange with several fluorinated salts. An EC device was made up by sandwiching the synthesized polyelectrolyte between two ITO glass-based PEDOT:PSS electrodes. The device exhibited an optical contrast of 18% corresponding to wavelength 648 nm with switching time of 4.75 s for

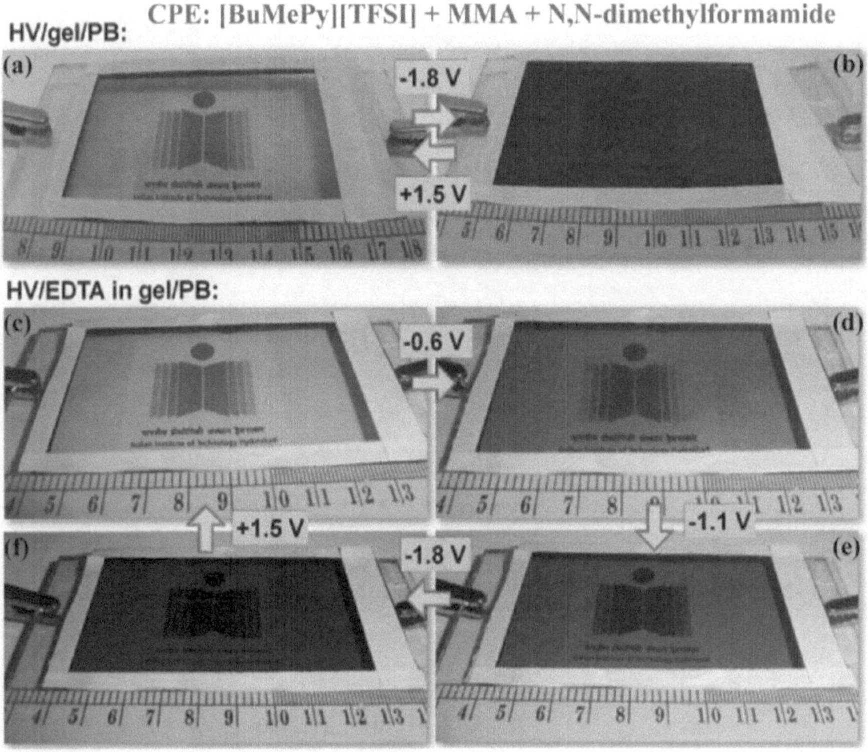

FIGURE 7.9 (a–b) Images of HV/gel/PB ECD in bleached and colored state, respectively; (c) in bleached state; (d–f) HV/EDTA in gel/PB ECD in progressive colored states under the different applied potentials.

Source: Reproduced with permission from Sydam et al. (2021).

coloration and 11.8 s for bleaching. The color efficiency was evaluated to be 356 $cm^2\ C^{-1}$. Wang et al. fabricated a viologens-based EC device with poly ionic liquids (PIL) as electrolyte (X. Wang, Guo, Cao, & Zhao, 2020). For fabrication of the EC device, various electrochromic chromophore viologens—($DHV(PF_6)_2$, $HBV(PF_6)_2$ and $PHBV(PF_6)_2$)—were blended with ferrocence (Fc) and poly(VBImBr) electrolyte gel, and the obtained gel was sandwiched between two ITO glass substrates. PIL electrolyte helped in suppressing the dimerization of viologen radical cation during the process of switching from colored to bleached state and vice versa, thereby enhancing the electrochromic performance of the device. The optical contrast for DHV-, HBV- and PHBV-based ECDs was found to be 47.7%, 45.4% and 41.1% respectively. The optical efficiency for the three ECDs was 109.8 $cm^2\ C^{-1}$, 106.2 $cm^2\ C^{-1}$ and 105.7 $cm^2\ C^{-1}$, in that order. DHV-based ECD exhibited maximum cyclic stability of 96% after 4000 cycles. Figure 7.10 represents the transmission spectra of different viologen-based ECDs along with images representing their colored

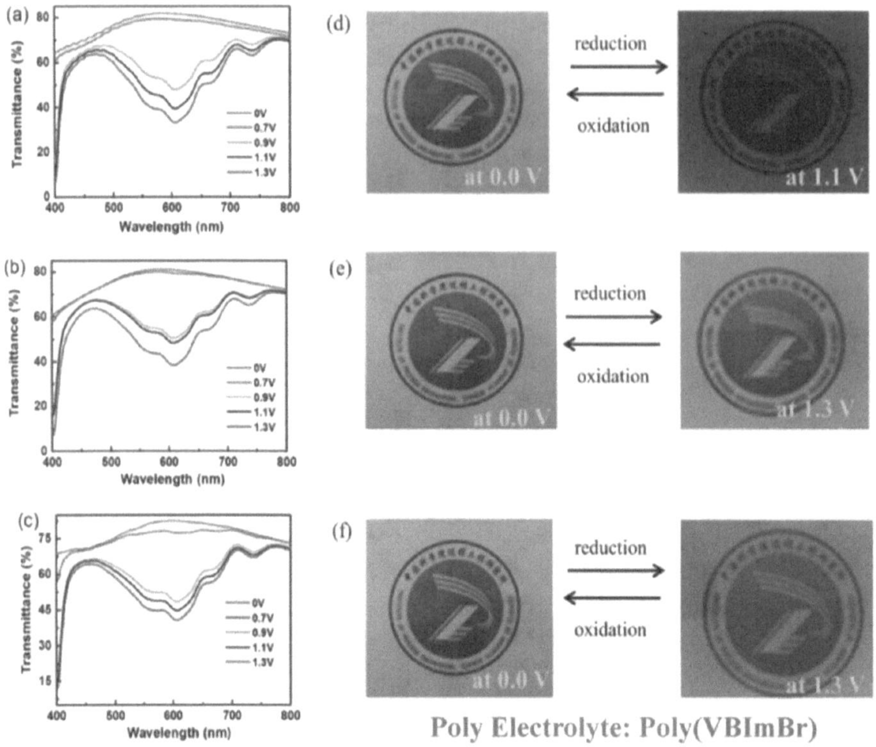

FIGURE 7.10 The transmission spectra at different applied voltages and the images showing colored and bleached states of (a) DHV, (b) HBV and (c) PHBV-based ECDs.

Source: Reproduced with permission from X. Wang et al. (2020).

and bleached states. Table 7.2 presents the performance parameters of some of the recently fabricated EC devices using polymer electrolytes.

7.5 CONCLUSIONS

In the present age, electrochromic windows (ECWs) have gained huge research interest in the field of academics and industries owing to their controllable transmission. In the energy sector, ECWs are proving to be revolutionary energy saving technology. Using ECDs or smart windows in houses and buildings, the amount of sunlight entering the building can be controlled reducing the cost and energy consumption for air conditioning and artificial lightning. Use of ECWs helps in building a healthy environment for future generations by enhancing the energy efficiency of the solar energy and reducing the dependence of the present generation on the artificial sources of energy. The fabrication of an ECW consists of five layers—transparent conducting substrate (TCS), working electrode (WE), electrolyte, counter electrode (CE), and transparent conducting substrate. EC material

TABLE 7.2
Electrochromic Parameters of Various EC Devices Using Polymer Electrolytes.

EC Device	Electrolyte Host	Ionic Conductivity (S cm^{-1})	Optical Modulation ΔT(%)	V_c/V_b (V)	ΔOD	π (cm^2 C^{-1})	τ_C (s)	τ_b (s)	Stability% (Cycles)	Reference
Gel Polymer Electrolytes (GPE) 1.3×10^{-2}										
ITO/WO_3//GPE//ITO	2-APPG	1.01×10^{-3} at room temp.	–	−1/+1	0.38	193.3 (550 nm)	–	–	–	(Zhou, Wang, Li, & Shen, 2018b)
ITO glass/WO_3//gel electrolyte/ITO	PMMA	–	51.3(550 nm)	−1.2/0	–	–	2.0	1.5	98.9 (11,500)	(Lee et al. 2020)
FTO WO_3//GPE//NiO FTO	PADA	at 25 °C	61 (660 nm)	−2.3/+2.3	–	78.7 (660 nm)	7.5	8.5	98.5 (25), 93.8 (1 month)	(W. Chen et al. 2019)
ITO/WO_3//GPE//ITO	PMDA+2-APPG	3.1×10^{-4} at 25 °C	81.9 (550 nm)	−2.5/2.0	0.19	198.9	–	–	–	(D. Zhang, Zhou, & Shen, 2021)
PProDOT-Me_2//GPE//ITO PET	GMA+AAm		49 (582 nm)	−1/+1	–	—	1.6	2.0	96 (1000)	(Q. Chen, Shi, Sheng, Zheng, & Xu, 2021)
ITO glass/WO_3//CPE// NiO_x/ITO glass	Acrylate-based UV curable glue	–	53.2 (550 nm)	–	–	62.2 (550 nm)	11.1	4	84 (2000)	(Mengying, Hang, Xu, & Xungang, 2021)
Solid Polymer Electrolyte (SPE)										
ITO glass/WO_3//SPE// CeO_2-TiO_2/ITO glass	Chitosan	1.68×10^{-5} at 30 °C	4.1 (550 nm), 9.2 (633 nm)	−2.8/+2.8	0.03 (550 nm), 0.04 (633 nm)	–	15	15	–	(Alves et al. 2018)
ITO glass/WO_3//SPE//ITO glass	Chitosan+PEDOT+PSS	4.2×10^{-4}	22.0 (800 nm)	−3/+3	–	67 (800 nm)	0.29	3	–	(Esin, 2019)
FTO glass/WO_3//SPE// ATO/ITO glass	PEGDMA+PEGMA	1.31×10^{-5} at 30 °C	35 (800 nm)	−1.5/+1.0	–	–	60	300	–	(Choi et al. 2019)

(Continued)

TABLE 7.2 *(Continued)*
Electrochromic Parameters of Various EC Devices Using Polymer Electrolytes.

EC Device	Electrolyte Host	Ionic Conductivity (S cm⁻¹)	Optical Modulation ΔT(%)	V_c/V_b (V)	ΔOD	π (cm^2 C^{-1})	τ_C (s)	τ_b (s)	Stability% (Cycles)	Reference
ITO glass/PProDOT-Me_2// electrolyte//Li-Ti-NiO/ ITO glass	PVB+PEG	~ 1.0×10^{-5}	43.8 (585 nm)	−1.8/+2 1.01×10^{-3}	–	106.0	1.2	2.6	84 (20,000)	(W. Wang et al. 2018)
ITO PEDOT//SPE//ITO PMeT	Chitosan+PEDOT+PSS	–	32.2 (650 nm)	−2/+2	–	228.65 (700 nm)	0.24	0.52	74.0 (50)	(Eren, 2019)
ITO PET/PTCDA//SPE// ITO PET	PPG+PMMA		61.0 (700 nm)	−3/+2.7	–	186.3 (700 nm)	2.4	5.6	79.7 (600)	(Zeng et al. 2019)
ITO PET//SPE//4EDOT-2BCOOCH$_3$/ITO PET	PPG+PMMA	1.01×10^{-3}	48.6 (700 nm)	−2.9/+2.5	–	230.6 (700 nm)	5.2	2.6	81.4 (600)	(Zeng et al. 2019)
WO_3//SPE//PB	PEG	5.07×10^{-3}	49 (633 nm)	−1.2/+1.0	2.349	139 (633 nm)	6	16	–	(Jeong, Kubota, et al. 2021)
ITO glass/TiO_2-CeO_2// electrolyte//PB/ITO glass	Gellan gum	2.77×10^{-3}	37.1 (630 nm)	−2.5/+0.5	1.34	2062 (630 nm)	–	–	–	(Isfahani et al. 2021)
Composite Polymer Electrolyte (CPE)										
ITO glass/WO_3//CPE//ITO glass	PPG-PEG–PPG triblock copolymer	1.1×10^{-4} at 30 °C	18 (550 nm)	–	0.56	675 (550 nm)	9	30	–	(Liu, Saikia, Wu, Fang, & Kao, 2017)
ITO glass/WO_3//CPE//ITO glass	2-APPG+ICS	1.43×10^{-3} at 30 °C	–	−3/+2	–	267.4	13	19	–	(Zhou, Wang, Li, & Shen, 2018a)
ITO glass/WO_3//PAEG// ITO glass	PMHS+PPG-PEG-PPG	6.5×10^{-4} at 30 °C	49 (550 nm)	−2/+2	0.39	–	–	–	–	(Deka et al. 2019)
Bpy/PANI/Fc	PAA	1.26×10^{-4} at room temp.	45 (581 nm)	−1.2/+0.3	–	406.9 (581 nm)	2	1.8	83.0 (5000)	(Y. Wang et al. 2019)

glass/FTO/WO_3//CPE// FTO/glass	PMMA+PAA+DEA	3.29×10^{-3}	49.9 (650 nm)	−3/+3	–	96.2 (650 nm)	7	4	–	(W. Chen et al. 2021)
PEDOT:PSS//Electrolyte// PANI:PSS	PEGDA	–	66.7 (600 nm)	−1.2/+0.5	–	340.6 (600 nm)	8	15		(Bae et al. 2021)
ITO/WO_3//CPE//ITO NiO	Gelatin	4.46×10^{-6}	30.1 (600 nm)	−3/+3	0.35	60.34 (600 nm)	10	50	–	(Wootthikanokkhan et al. 2020)
HV/EDTA in gel/PB	PMMA	–	73.1 (606 nm)	−0.6/+1.5	–	346.2 (606 nm)	16	35	~ 93 (10,000), 68 (2 years)	(Sydam et al. 2021)
ITO glass/WO_3//CPE//ITO glass	2-APPG	1.67×10^{-3} at 25 °C	38	−3/+3	0.279	282.9	–	–	–	(W. Zhang et al. 2021)
Polyelectrolytes										
ITO glass/PEDOT:PSS// PIL//PEDOT:PSS/ITO glass	PEG	1.20×10^{-4}	22 (648 nm)	–	–	–	2.5	3.2	–	(Puguan, Jadhav, Boton, & Kim, 2018)
ITO glass/PEDOT:PSS// PIL//PEDOT:PSS/ITO glass	poly(3-alkyl-4-PEG1,2,3-triazolium)	1.16×10^{-4} at 30 °C	18 (648 nm)	–	–	356 (648 nm)	4.75	11.8	–	(Puguan, Boton, et al. 2018)
ITO glass//DHV^+ EC-based gel//ITO glass	poly(VBImBr)	–	46.8 (606 nm)	1.1/0	–	109.8 (606 nm)	11.1	19.9	96 (4000)	(X. Wang et al. 2020)

and electrolyte are the most important components of ECWs. The choice of suitable EC material and the electrolyte leads to the development of ECWs with excellent performance parameters. Some important performance parameters of ECWs have been discussed herein. The performance of ECWs highly depends on the chosen electrolyte. The electrolyte with high ionic conductivity, high electronic resistivity, high thermal/mechanical/chemical stability and high compatibility with electrode material is highly recommended for ECWs. Among various kinds of electrolytes, polymer electrolytes have gained wide research focus for ECWs owing to their leakage resistance, high ionic conductivity and chemical stability. Various kinds of polymer electrolytes such as GPEs, SPEs, CPEs and poly electrolytes have been discussed in detail in this chapter. In recent years, a lot of research has been focused on using these polymer electrolytes in ECWs. Among them SPEs and CPEs are found to have superior mechanical stability in comparison to GPEs. SPEs possess high mechanical stability but their ionic conductivity is low. CPEs, on the other hand, possess high ionic conductivity and high mechanical stability along with excellent electrochromic performance. Recently the research has been focused on fabricating new kinds of SPEs and CPEs with excellent electrochromic performance to be used in ECW applications.

REFERENCES

Adebahr, J., Byrne, N., Forsyth, M., Macfarlane, D. R., & Jacobsson, P. (2003). Enhancement of ion dynamics in PMMA-based gels with addition of TiO_2 nano-particles. *Electrochimica Acta*, *48*(14–16), 2099–2103.

Agnihotry, S., Pradeep, P., & Sekhon, S. (1999). PMMA based gel electrolyte for EC smart windows. *Electrochimica Acta*, *44*(18), 3121–3126.

Alves, R., Sentanin, F., Sabadini, R., Fernandes, M., de Zea Bermudez, V., Pawlicka, A., & Silva, M. M. (2018). Samarium (III) triflate-doped chitosan electrolyte for solid state electrochromic devices. *Electrochimica Acta*, *267*, 51–62.

Armand, M. (1994). The history of polymer electrolytes. *Solid State Ionics*, *69*(3–4), 309–319.

Arya, A., & Sharma, A. (2017). Polymer electrolytes for lithium ion batteries: A critical study. *Ionics*, *23*(3), 497–540.

Arya, A., & Sharma, A. L. (2019). Dielectric relaxations and transport properties parameter analysis of novel blended solid polymer electrolyte for sodium-ion rechargeable batteries. *Journal of Materials Science*, *54*(9), 7131–7155.

Aziz, S. B., Brza, M., Nofal, M. M., Abdulwahid, R. T., Hussen, S. A., Hussein, A. M., & Karim, W. O. (2020). A comprehensive review on optical properties of polymer electrolytes and composites. *Materials*, *13*(17), 3675.

Aziz, S. B., Woo, T. J., Kadir, M., & Ahmed, H. M. (2018). A conceptual review on polymer electrolytes and ion transport models. *Journal of Science: Advanced Materials and Devices*, *3*(1), 1–17.

Bae, S., Kim, Y., Kim, J. M., & Kim, J. H. (2021). Dual-cation electrolytes crosslinked with MXene for high-performance electrochromic devices. *Nanomaterials*, *11*(4), 874.

Cannavale, A., Ayr, U., Fiorito, F., & Martellotta, F. (2020). Smart electrochromic windows to enhance building energy efficiency and visual comfort. *Energies*, *13*(6), 1449.

Chen, Q., Shi, Y., Sheng, K., Zheng, J., & Xu, C. (2021). Dynamically cross-linked hydrogel electrolyte with remarkable stretchability and self-healing capability for flexible electrochromic devices. *ACS Applied Materials & Interfaces*, *13*(47), 56544–56553.

Chen, W., Liu, S., Guo, L., Zhang, G., Zhang, H., Cao, M., . . . Peng, Y. (2021). A self-healing ionic liquid-based ionically cross-linked gel polymer electrolyte for electrochromic devices. *Polymers*, *13*(5), 742.

Chen, W., Zhu, C., Guo, L., Yan, M., Wu, L., Zhu, B., . . . Peng, Y. (2019). A novel ionically crosslinked gel polymer electrolyte as an ion transport layer for high-performance electrochromic devices. *Journal of Materials Chemistry C*, *7*(13), 3744–3750.

Choi, D., Kim, H., Lee, M., Son, M., Ahn, S.-h., & Lee, C. S. (2019). Low-voltage modulated inorganic smart windows using solid polymer electrolyte. *Solar Energy Materials and Solar Cells*, *200*, 109966.

Choudhury, N., Sampath, S., & Shukla, A. (2009). Hydrogel-polymer electrolytes for electrochemical capacitors: An overview. *Energy & Environmental Science*, *2*(1), 55–67.

Chua, M. H., Tang, T., Ong, K. H., Neo, W. T., & Xu, J. W. (2019). Introduction to Electrochromism , in *Electrochromic Smart Materials: Fabrication and Applications*, 2019, pp. 1–21. DOI: 10.1039/9781788016667-00001

Deka, J. R., Saikia, D., Lou, G.-W., Lin, C.-H., Fang, J., Yang, Y.-C., & Kao, H.-M. (2019). Design, synthesis and characterization of polysiloxane and polyetherdiamine based comb-shaped hybrid solid polymer electrolytes for applications in electrochemical devices. *Materials Research Bulletin*, *109*, 72–81.

Eh, A. L. S., Tan, A. W. M., Cheng, X., Magdassi, S., & Lee, P. S. (2018). Recent advances in flexible electrochromic devices: Prerequisites, challenges, and prospects. *Energy Technology*, *6*(1), 33–45.

Eren, E. (2019). Improved performance and stability of solid state electrochromic devices with eco-friendly chitosan-based electrolytes. *Solid State Ionics*, *334*, 152–159.

Esin, E. (2019). Li^+ doped chitosan-based solid polymer electrolyte incorporated with PEDOT: PSS for electrochromic device. *Journal of the Turkish Chemical Society Section A: Chemistry*, *5*(3), 1413–1422.

Feuillade, G., & Perche, P. (1975). Ion-conductive macromolecular gels and membranes for solid lithium cells. *Journal of Applied Electrochemistry*, *5*(1), 63–69.

Fu, X. (2010). Polymer electrolytes for electrochromic devices. In *Polymer Electrolytes* (pp. 471–523). Elsevier, The Netherlands.

Gillaspie, D. T., Tenent, R. C., & Dillon, A. C. (2010). Metal-oxide films for electrochromic applications: Present technology and future directions. *Journal of Materials Chemistry*, *20*(43), 9585–9592.

Gonçalves, A., Costa, C., Pereira, S., Correia, N., Silva, M., Barbosa, P., . . . Fortunato, E. (2012). Study of electrochromic devices with nanocomposites polymethacrylate hydroxyethylene resin based electrolyte. *Polymers for Advanced Technologies*, *23*(4), 791–795.

Granqvist, C. G. (2005). Electrochromic devices. *Journal of the European Ceramic Society*, *25*(12), 2907–2912.

Granqvist, C. G., Arvizu, M. A., Pehlivan, İ. B., Qu, H.-Y., Wen, R.-T., & Niklasson, G. A. (2018). Electrochromic materials and devices for energy efficiency and human comfort in buildings: A critical review. *Electrochimica Acta*, *259*, 1170–1182.

Granqvist, C. G., Lansåker, P. C., Mlyuka, N. R., Niklasson, G. A., & Avendano, E. (2009). Progress in chromogenics: New results for electrochromic and thermochromic materials and devices. *Solar Energy Materials and Solar Cells*, *93*(12), 2032–2039.

Hassab, S., & Padilla, J. (2014). Use of ionic liquids in electrochromic devices. In *Ionic Liquids in Separation Technology* (pp. 301–333). Elsevier, The Netherlands.

Isfahani, V. B., Pereira, R. F., Fernandes, M., Sabadini, R. C., Pereira, S., Dizaji, H. R., . . . Rego, R. (2021). Gellan-Gum and LiTFSI-based solid polymer electrolytes for electrochromic devices. *ChemistrySelect*, *6*(20), 5110–5119.

Jeong, C. Y., Kubota, T., Chotsuwan, C., Wungpornpaiboon, V., & Tajima, K. (2021). All-solid-state electrochromic device using polymer electrolytes with a wet-coated electrochromic layer. *Journal of Electroanalytical Chemistry*, *897*, 115614.

Jeong, C. Y., Watanabe, H., & Tajima, K. (2021). Adhesive electrochromic WO_3 thin films fabricated using a WO_3 nanoparticle-based ink. *Electrochimica Acta*, 138764.

Kam, W., Liew, C.-W., Lim, J., & Ramesh, S. (2014). Electrical, structural, and thermal studies of antimony trioxide-doped poly (acrylic acid)-based composite polymer electrolytes. *Ionics*, *20*(5), 665–674.

Korgel, B. A. (2013). Composite for smarter windows. *Nature*, *500*(7462), 278–279.

Krejza, O., & Vondrák, J. (2009). *Gel polymer electrolytes for electrochromic devices*. Brno: Brno University of Technology, Faculty of Electrical Engineering and Communications.

Lampert, C. M. (1989). IEA solar R&D. Task 10: Solar materials R&D. In *Failure and Degradation Modes in Selected Solar Materials: A review, Lawrence Berkeley National Laboratory,* Berkeley.

Lee, H. J., Lee, C., Song, J., Yun, Y. J., Jun, Y., & Ah, C. S. (2020). Electrochromic devices based on ultraviolet-cured poly (methyl methacrylate) gel electrolytes and their utilisation in smart window applications. *Journal of Materials Chemistry C*, *8*(26), 8747–8754.

Liu, H.-M., Saikia, D., Wu, C.-G., Fang, J., & Kao, H.-M. (2017). Solid polymer electrolytes based on coupling of polyetheramine and organosilane for applications in electrochromic devices. *Solid State Ionics*, *303*, 144–153.

Luo, J., Conrad, O., & Vankelecom, I. F. (2013). Imidazolium methanesulfonate as a high temperature proton conductor. *Journal of Materials Chemistry A*, *1*(6), 2238–2247.

Meka, V. S., Sing, M. K., Pichika, M. R., Nali, S. R., Kolapalli, V. R., & Kesharwani, P. (2017). A comprehensive review on polyelectrolyte complexes. *Drug Discovery Today*, *22*(11), 1697–1706.

Mengying, W., Hang, Y., Xu, W., & Xungang, D. (2021). High-performance of quasi-solid-state complementary electrochromic devices based on Al^{3+}/Li^{+} dual-ion electrolyte. *Solar Energy Materials and Solar Cells*, *230*, 111196.

Murata, K., Izuchi, S., & Yoshihisa, Y. (2000). An overview of the research and development of solid polymer electrolyte batteries. *Electrochimica Acta*, *45*(8–9), 1501–1508.

Ngai, K. S., Ramesh, S., Ramesh, K., & Juan, J. C. (2016). A review of polymer electrolytes: Fundamental, approaches and applications. *Ionics*, *22*(8), 1259–1279.

Nguyen, C. A., Xiong, S., Ma, J., Lu, X., & Lee, P. S. (2011). High ionic conductivity P (VDF-TrFE)/PEO blended polymer electrolytes for solid electrochromic devices. *Physical Chemistry Chemical Physics*, *13*(29), 13319–13326.

Nicotera, I., Coppola, L., Oliviero, C., Castriota, M., & Cazzanelli, E. (2006). Investigation of ionic conduction and mechanical properties of PMMA—PVdF blend-based polymer electrolytes. *Solid State Ionics*, *177*(5–6), 581–588.

Österholm, A. M., Shen, D. E., Kerszulis, J. A., Bulloch, R. H., Kuepfert, M., Dyer, A. L., & Reynolds, J. R. (2015). Four shades of brown: Tuning of electrochromic polymer blends toward high-contrast eyewear. *ACS Applied Materials & Interfaces*, *7*(3), 1413–1421.

Park, S.-I., Quan, Y.-J., Kim, S.-H., Kim, H., Kim, S., Chun, D.-M., . . . Ahn, S.-H. (2016). A review on fabrication processes for electrochromic devices. *International Journal of Precision Engineering and Manufacturing-Green Technology*, *3*(4), 397–421.

Patel, K., Bhatt, G., Ray, J., Suryavanshi, P., & Panchal, C. (2017). All-inorganic solid-state electrochromic devices: a review. *Journal of Solid State Electrochemistry*, *21*(2), 337–347.

Puguan, J. M. C., Boton, L. B., & Kim, H. (2018). Triazole-based ionene exhibiting tunable structure and ionic conductivity obtained via cycloaddition reaction: A new polyelectrolyte for electrochromic devices. *Solar Energy Materials and Solar Cells*, *188*, 210–218.

Puguan, J. M. C., Jadhav, A. R., Boton, L. B., & Kim, H. (2018). Fast-switching all-solid state electrochromic device having main-chain 1, 2, 3-triazolium-based polyelectrolyte with extended oxyethylene spacer obtained via click chemistry. *Solar Energy Materials and Solar Cells*, *179*, 409–416.

Puguan, J. M. C., & Kim, H. (2017). ZrO_2-silane-graft-PVdFHFP hybrid polymer electrolyte: Synthesis, properties and its application on electrochromic devices. *Electrochimica Acta*, *230*, 39–48.
Ramesh, S., Liew, C.-W., & Ramesh, K. (2011). Evaluation and investigation on the effect of ionic liquid onto PMMA-PVC gel polymer blend electrolytes. *Journal of Non-Crystalline Solids*, *357*(10), 2132–2138.
Ramesh, S., & Wen, L. C. (2010). Investigation on the effects of addition of SiO_2 nanoparticles on ionic conductivity, FTIR, and thermal properties of nanocomposite PMMA—$LiCF_3SO_3$—SiO_2. *Ionics*, *16*(3), 255–262.
Rauh, R. D. (1999). Electrochromic windows: An overview. *Electrochimica Acta*, *44*(18), 3165–3176.
Rosli, N. H. A., Muhammad, F. H., Chan, C. H., & Winie, T. (2014). *Effect of filler type on the electrical properties of hexanoyl chitosan-based polymer electrolytes*. Paper presented at the Advanced Materials Research.
Selkowitz, S. E. (1990). *Application of large-area chromogenics to architectural glazings*. Paper presented at the Large-Area Chromogenics: Materials and Devices for Transmittance Control.
Sibilio, S., Rosato, A., Scorpio, M., Iuliano, G., Ciampi, G., Vanoli, G. P., & De Rossi, F. (2016). A review of electrochromic windows for residential applications. *Int. J. Heat Technol*, *34*, S481–S488.
Sim, L. N., & Pawlicka, A. (2020). Polymer electrolytes for electrochromic windows. *Polymer Electrolytes: Characterization Techniques and Energy Applications*, 365–389.
Song, J., Wang, Y., & Wan, C. C. (1999). Review of gel-type polymer electrolytes for lithium-ion batteries. *Journal of Power Sources*, *77*(2), 183–197.
Srivastava, N., & Tiwari, T. (2009). New trends in polymer electrolytes: A review. *e-Polymers*, *9*(1).
Sydam, R., Ojha, M., & Deepa, M. (2021). Ionic additive in an ionogel for a large area long lived high contrast electrochromic device. *Solar Energy Materials and Solar Cells*, *220*, 110835.
Tajima, K., Watanabe, H., Nishino, M., & Kawamoto, T. (2020a). Electrochromic properties of WO_3 thin films fabricated by magnetron sputtering, ion plating, and spin coating: A comparative investigation. *Journal of the Ceramic Society of Japan*, *128*(7), 381–386.
Tajima, K., Watanabe, H., Nishino, M., & Kawamoto, T. (2020b). Green fabrication of a complementary electrochromic device using water-based ink containing nanoparticles of WO_3 and Prussian blue. *RSC Advances*, *10*(5), 2562–2565.
Thakur, V. K., Ding, G., Ma, J., Lee, P. S., & Lu, X. (2012). Hybrid materials and polymer electrolytes for electrochromic device applications. *Advanced Materials*, *24*(30), 4071–4096.
Uma, T., Mahalingam, T., & Stimming, U. (2005). Conductivity studies on poly (methyl methacrylate)—Li2SO4 polymer electrolyte systems. *Materials Chemistry and Physics*, *90*(2–3), 245–249.
Wang, W., Guan, S., Li, M., Zheng, J., & Xu, C. (2018). A novel hybrid quasi-solid polymer electrolyte based on porous PVB and modified PEG for electrochromic application. *Organic Electronics*, *56*, 268–275.
Wang, X., Guo, L., Cao, S., & Zhao, W. (2020). Highly stable viologens-based electrochromic devices with low operational voltages utilizing polymeric ionic liquids. *Chemical Physics Letters*, *749*, 137434.
Wang, Y., Zheng, R., Luo, J., Malik, H. A., Wan, Z., Jia, C., . . . Yao, X. (2019). Self-healing dynamically cross linked versatile polymer electrolyte: A novel approach towards high performance, flexible electrochromic devices. *Electrochimica Acta*, *320*, 134489.

Wootthikanokkhan, J., Jaruphan, P., Azarian, M. H., & Yosthisud, J. (2020). Effects of ethylene-acrylic acid ionomer on thermomechanical and electrochromic properties of electrochromic devices using gelatin-based electrolytes. *Journal of Applied Polymer Science*, *137*(44), 49362.

Wright, P. V. (1975). Electrical conductivity in ionic complexes of poly (ethylene oxide). *British Polymer Journal*, *7*(5), 319–327.

Yağmur, İ., Ak, M., & Bayrakçeken, A. (2013). Fabricating multicolored electrochromic devices using conducting copolymers. *Smart Materials and Structures*, *22*(11), 115022.

Yang, G., Zhang, Y.-M., Cai, Y., Yang, B., Gu, C., & Zhang, S. X.-A. (2020). Advances in nanomaterials for electrochromic devices. *Chemical Society Reviews*, *49*(23), 8687–8720.

Yu, W., Chen, J., Fu, Y., Xu, J., & Nie, G. (2013). Electrochromic property of a copolymer based on 5-cyanoindole and 3, 4-ethylenedioxythiophene and its application in electrochromic devices. *Journal of Electroanalytical Chemistry*, *700*, 17–23.

Zeng, J., Wan, Z., Zhu, M., Ai, L., Liu, P., & Deng, W. (2019). Flexible electrochromic energy-saving windows with fast switching and bistability based on a transparent solid-state electrolyte. *Materials Chemistry Frontiers*, *3*(11), 2514–2520.

Zhang, D., Zhou, J., & Shen, F. (2021). A hybrid gel polymer electrolyte with imide groups modified by the coupling agent and its application in electrochromic devices. *Journal of Sol-Gel Science and Technology*, *97*(2), 393–403.

Zhang, W., Zhang, D., & Shen, F. (2021). Synthesis of a highly conductive organic-inorganic hybrid gel electrolyte and its characterization and application in electrochromic devices. *Journal of Materials Science: Materials in Electronics*, *32*(18), 23500–23512.

Zhou, J., Wang, J., Li, H., & Shen, F. (2018a). Hybrid gel polymer electrolyte with good stability and its application in electrochromic device. *Journal of Materials Science: Materials in Electronics*, *29*(7), 6068–6076.

Zhou, J., Wang, J., Li, H., & Shen, F. (2018b). A novel imide-based hybrid gel polymer electrolyte: synthesis and its application in electrochromic device. *Organic Electronics*, *62*, 516–523.

8 Polymer Composites for Fuel Cells

Soubhagya Ranjan Bisoi, Naresh Kumar Sahoo, Ankur Soam, and Prasanta Kumar Sahoo

CONTENTS

8.1 INTRODUCTION

Climate change, population increase, and rising energy consumption are just a few of the major worldwide problems driving contemporary efforts to create and expand renewable, sustainable, and clean energy options (Boboescu et al. 2016). The need to create alternative energy sources and ways for regenerating waste materials and energy sources has resulted in a slew of large-scale research projects that focus not

DOI: 10.1201/9781003208662-10

only on waste treatment but also on waste harvesting for energy and value-added goods. The discovery of electrochemically active microorganisms that employ transfer of electrons through a direct way in their physiology has lately spawned the area of bio-electrochemical waste remediation (Kim et al. 1999). One example of this technology is microbial fuel cells (MFCs). MFCs use live catalysts to transform energy into electricity from industrial, sustainable, and home waste sources. MFCs may also be used to collect bio-electrochemical power from carbohydrates, in addition to removing organic materials in wastewater (He, Minteer and Angenent 2005). Microbial fuel cells (MFCs) are devices that oxidize organic and inorganic substances and generate electricity using microorganisms as catalysts. H-shaped MFCs are generally made up of two parts: the anodic and cathodic chambers, which include the two electrodes joined by a tube carrying a half-cell separator, including a salt bridge, ceramic or PEM, that is commonly a CEM (cation exchange membrane; e.g. Nafion). Electrons created by bacteria on these substrates are transmitted to the anode (negative terminal) and flow to the cathode (positive terminal) via a conductive substance including a resistor, or the system is operated under load (i.e. producing electricity that runs a device) (Figure 8.1). A positive current flows in the opposite direction of electron flow, from the positive to the negative terminal, by convention. The device must be able to replenish the substrate oxidized at the anode on a continuous or intermittent basis; otherwise, the system is termed a biobattery.

Electrons can be transmitted to the anode via electron mediators or shuttles (Rabaey et al. 2005), direct membrane associated electron transfer (Bond and Lovley 2003), so-called nanowires (Reguera et al. 2005) created by bacteria, or perhaps additional as yet unknown mechanisms. The separator permits protons to readily migrate to the cathode owing to a potential gradient while preventing O_2 (or the electron acceptor used in the cathodic compartment) from diffusing to the anode in a way that might harm the bacteria present. The biocatalyst is also significant, and it is widely available, since it may be derived from a variety of wastewater resources. The electrode materials show a major role among all MFC components in the generation of electricity for MFCs. The cost of the electrodes will be a major element in selecting whether or not to use MFC technology on a wide scale.

8.1.1 Electrode Material

Electrode materials are important in MFCs because electron transport is crucial to their function. Materials of metals, carbonaceous, or their mixtures with polymers, are commonly employed as electrode materials in MFCs. The durability, biocompatibility, and chemical/environmental inertness of carbon-polymer composite materials have piqued curiosity, whereas metal electrodes are unstable and corrosive in aquatic environments (Narayanasamy and Jayaprakash 2020). There are some common requirements for anode and cathode materials.

8.1.1.1 Anode

Low cost, non-fouling, non-corrosive, non-toxic to microorganisms, big surface area, and highly conductive are the most essential parameters for an anode material. The

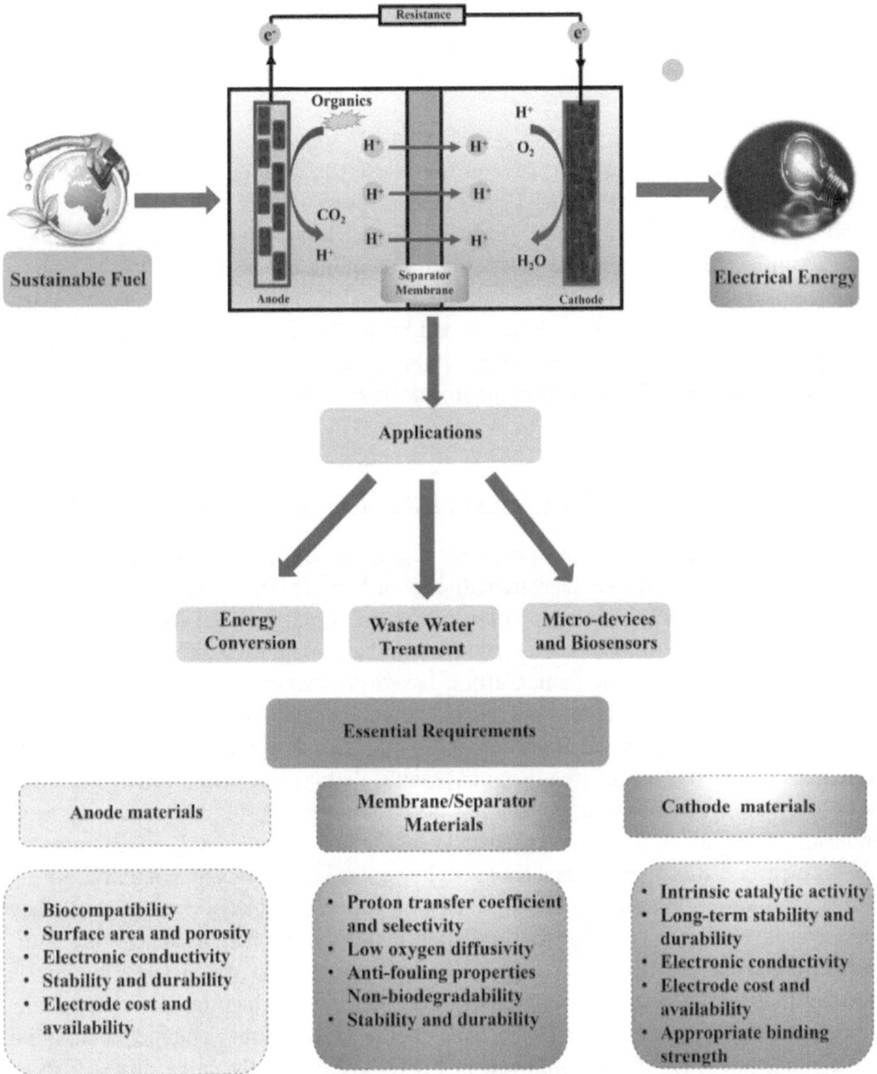

FIGURE 8.1 General set up of a typical two-chamber MFC highlighting the potential applications and essential requirements for anodes, separator/membranes, and cathode materials in terms of energy and environment aspects.

anode chamber's produced electrons must be capable of moving from the solution to the anode material's surface. Microorganisms must have the ability to adhere to the surface and create biofilm in order to achieve effective electron transfer (Chatterjee et al. 2019). As a result, despite their strong conducting qualities, some materials, like stainless steel, are not appropriate for the production of power in MFCs. Carbon-based electrodes are a cost-effective alternative to metal-based anodes. Carbon-based materials are thus preferred for the treatment of vast amounts of wastewater at a low cost and the increase of transfers of electrons since they are inexpensive and stimulate the requisite bacterial growth.

8.1.1.2 Cathode

Carbon-based materials, which are commonly used as cathodes, have the same fundamental substance as an anode. However, processes that need a catalyst (like Pt) to actually occur on a cathode's surface, like ORR (oxygen reduction reaction), pose a significant obstacle. The most common application of platinum is as a cathode catalyst. However, using these cathode catalysts is neither sustainable nor cost-effective. As a result, when used on a broad basis, Pt-based catalysts are necessary for replacement by less expensive compounds of metal like TMO (transitional metal oxides) and Ni, Co, Fe metal centre macro cyclic organic compounds (Santoro et al. 2018).

8.2 CARBON-POLYMER-BASED NANOCOMPOSITES ELECTRODE

Electrode modification is an appealing technique for improving the catalytic activity (cathode) and the surface area (anode) of MFCs and, as a result, their performance. Improved performance of electrodes in microbial fuel cells is mostly dependent on lowering internal resistance, but it should also be aimed on lowering costs to promote widespread deployment. Carbon-based products like carbon cloth, graphite rod, rod of carbon, carbon brush, carbon sheet, carbon mesh, carbon felt, granular graphite, granular activated carbon, activated carbon power, graphite plate, and reticulated vitreous carbon are all typical materials for electrodes in MFCs (Huang et al. 2016). Carbon-based materials are potentials for scaling-up because of their superior chemical stability and electrical conductivity (Figure 8.2a–l). 3D materials, like carbon mesh and brush, yield higher current densities as compared to 2D materials, like graphite rod or plate and carbon paper, because of their large surface area. Researchers have previously used chemical or physical approaches, like addition of electroactive coating or highly conductive coating, to construct numerous bulk and surface-modified carbon electrodes. Adding conducting polymers to carbon electrodes is the most commonly used chemical technique. The most beneficial conducting polymers (CPs) employed in MFCs are generally polyaniline (PANI), polypyrrole (PPy), poly(3, 4-ethylenedioxythiophene) (PEDOT), and polythiophene (PTh). There were several different types of conducting polymers that were often employed in microbial fuel cell applications. In conjugated chains, an electron phonon cloud pair or an electron deformation pair is known to exist in the structure of conducting polymers (i.e. double and single bonds). As shown in Figure 8.2m, the conductivity of CPs is triggered by the molecule's constant migration of double bonds, which stabilizes the charge by surrounding atoms. By including conductive fillers like carbon fiber, carbon black, and some metallic species, the electrical conductivity of certain insulating polymers can be increased while maintaining their polymeric features. Some nanoscale conductive fillers are presently available like graphene materials (Sahoo et al. 2021) and metal nanoparticles (Aepuru et al. 2020), and carbon nanotubes (CNTs) (Chou et al. 2014) have also been researched

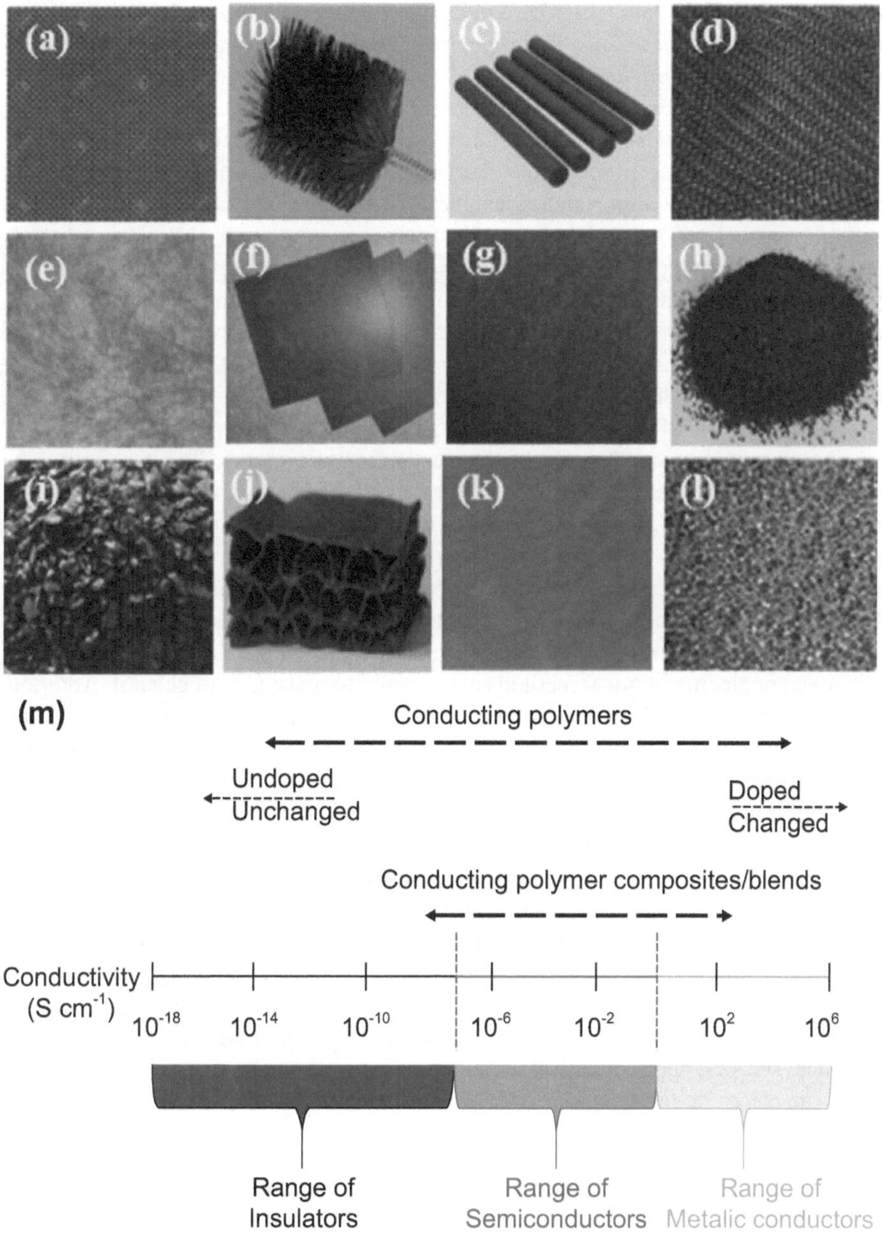

FIGURE 8.2 Digital images of various carbon electrodes normally used in MFCs: (a) carbon cloth, (b) carbon brush, (c) carbon rod, (d) carbon mesh, (e) carbon veil, (f) carbon paper, (g) carbon felt, (h) granular activated carbon, (i) granular graphite, (j) carbonized cardboard, (k) graphite plate, (l) reticulated vitreous carbon; (m) conductivity range of the conducting polymers.

Source: (a–l) Santoro et al. (2017). (m) Ramanujam and Annamalai (2017).

in general for the manufacture of conductive polymer composites for application in MFCs. Filler particles inside the polymer matrix have conductive channels; as a result of the conductive paths, the conductivity of polymer composites improves, which is influenced by filler matrix interactions, including a number of factors: the intrinsic characteristics, geometry and dispersion state of nanofillers. As a result, selecting a composite preparation process of filler distribution with the proper amount is of primary importance since the concentration of conductive filler affects a composite's electrical conductivity. The filler concentration is sufficient to generate a continuous conductive network throughout the composite when it reaches the percolation threshold. The composite achieves a percolation threshold at a certain critical loading, when the conductivity begins to increase by several orders of magnitude with only a minor increase in filler quantity. As a result, it's been identified that as the aspect ratio (length to diameter) of the filler increases, the amount of the percolation threshold drops.

The electrical characteristics of the conducting polymers (CPs) are similar to those of metals and semiconductors. These polymers have good electro kinetic characteristics, as well as high electron mobility, biocompatibility, stability, and anticorrosion qualities. They are more appealing than traditional polymers because of their ease of synthesis process, which could be because of the possibility of chemical surface alteration with species that are physiologically active to improve the functionality and biocompatibility of the electrodes that are essential for electrodes of MFC. Chemical or electrochemical methods can be used to make CPs in general. Addition polymerization or condensation polymerization are commonly used in chemical synthesis. Chemical synthesis allows for the creation of a wide range of CPs using a variety of processes (e.g. condensation or addition polymerization) and is thus the most often used approach for CP production. Chemical polymerization is the only means to make certain unique CPs with changed monomers, while electrochemical synthesis is confined to particular polymeric materials. Because, in some situations, when a potential for polymerization is applied, the monomer may be oxidized, forming reactive radical ion intermediates. Due to its interesting properties, such as environmental stability, non-biodegradability, ease of fabrication, and high conductivity at room temperature, PANI acts as a potential candidate for different applications in the synthesis of biosensors, batteries, supercapacitors, actuators, electronic devices, chemical sensors, and fuel cell applications.

In the electrode fabrication for MFCs, various other criteria like redox potential appropriate to bacterial metabolism, biocompatibility, and surface area are taken into account in addition to conductivity of electrodes. As a result, over the last few years, there have been lots of researches into bulk modification or surface modification of conducting, semiconducting, or insulating polymers. This chapter examines the current state of knowledge and data on MFCs with electrodes of carbon-polymer-based composites. To support the selection of materials for this study, a comparison of their performances has been offered. MFCs commonly use carbon-polymer-based composite electrodes, which are divided into two types:

- Bulk-modified carbon-based electrodes (BMEs)
- Surface-modified carbon-based electrodes (SMEs)

8.2.1 Bulk-Modified Electrodes (BMEs) for Microbial Fuel Cells

Carbonization, doping, electroactive coatings, or chemical changes are used to modify carbon-based electrodes in bulk. There are two types of bulk-modified carbon-based electrodes:

1. Polymer-derived carbon-based bulk-modified electrodes (PBMEs)
2. Polymer/carbon composite-based bulk-modified electrodes (PCBMEs)

8.2.1.1 Polymer-Derived Carbon-Based Bulk-Modified Electrodes (PBMEs)

Graphite is mechanically stiff, which is critical to create precise shapes with large surface areas. Converting polymeric materials to carbonaceous structures can be a good way to make a low-cost carbon electrode, which is doped by heteroatoms containing high carbon with large surface area. The polymer precursor is carbonized (graphitized) or pyrolyzed to increase the amount of carbon and doped with a heteroatom like F and N to keeps the porous structure intact to make polymer-derived carbon electrodes. Carbonization, which needs a long residence period because of pyrolysis, and stepwise heating, which needs a short residence period because of the rapid rate of heating, are the two procedures most commonly utilized for fabricating PBMEs. Both of these processes add to the carbon concentration. Nitrogen-containing polymers are commonly used in this carbonization procedure. Polymers containing N, F such as polytetrafluoroethylene, polyacrylonitrile, and polyaniline Produce heteroatom-doped carbon compounds when thermally treated. It has been claimed that adding a second heteroatom to N-doped carbon materials, such as B, S, or P, controls the electronic and surface polarities, increasing the carbon material's ORR catalytic activity. The polymers that are commonly utilized as PBMEs are polyacrylonitrile (PAN), polyacrylamide (PAM), polydopamine (PDA), polyaniline (PANI), and polytetrafluoroethylene (PTFE) (Zeng et al. 2018). The benefits of utilizing polymer-derived carbon-based BMEs include (i) because of heteroatom doping, catalytic activity of ORR increases, and (ii) the porous shape and large surface area is suitable to biofilm adhesion. The following sections address different PBMEs utilized in MFC investigations.

8.2.1.1.1 PBMEs as Anode Material

Pre-oxidation, hot-pressing, and carbonization (up to 1050 °C) of polyacrylonitrile (PAN)-based carbon electrodes were obtained employing an activating agent. PAN carbon fibers are ionically and electrically conducting, porous, active electrochemically, and also have large surface areas, making them ideal for use as electrodes in MFCs. Electrospinning and solution-blowing of a 3D carbon fiber anode resulted in a bio electrocatalytic anode current density of up to 30 A/m^2 (Chen et al. 2011). With such an ultrahigh porosity of 98.5%, it was the highest value ever recorded for electroactive microbial biofilms. Wang et al. (2015) and Chen et al. (2011) developed a unique open-cell scaffold (CS) anode having 3D configuration using supercritical CO_2 as the physical foaming agent and low-cost PAN as precursor (Wang et al. 2015; Chen et al. 2011). The maximal power density of the MFC with the CS anode is 30.7 mW m^{-2}, that is 28.5 percent greater than a commercially available carbon felt anode (23.9 mW m^{-2}). As a result, the MFC's performance is enhanced by the 3D open-cell scaffold (CS) anode (Wang et al. 2015). How fiber diameter affects anodic performance was investigated by He et al. (2011) and they found that moderate diameter and high porosity of fiber anodes (size of microorganism range between about diameter of 0.5 μm and length of 1 μm) are also necessary for greater current density (He et al. 2011). Carbon black (CB) deposition and APS/H_2SO_4-based chemical oxidation resulted in the material of a three-dimensional network structure that serves as a solid substrate for the development of bio films. For a thickness of 3 mm, the capital expense of ACS was just $2.5 per m^2, which is like an order of magnitude lower than the cost of most marketed carbon electrodes. The output of the composite anode produced 2.2 times that of the CC anode (926 mW m^{-2}) and had a faster start-up period (23 h).

8.2.1.1.2 PBMEs as Cathode Material

Because the heteroatomic dopants have a synergistic effect on carbon, catalysts for heteroatom-doped carbon-based cathodes with varied dopants, like Fe, Co, graphene, and N, have been used to obtain improved the activity ORR (Guo et al. 2019). Figure 8.3a shows how to make Co_2NX-T@NC by copolymerizing the Co^{2+}-Aniline-Pyrrole system and heating it to a higher temperature and pyrolyzing it, resulting in efficient oxygen reduction electrocatalysts made of Co_2N nanoparticles embedded in N-doped mesoporous carbon. Similarly, phytic acid-doped polyaniline was pyrolyzed onto AC to produce N and P co-doped carbon-modified activated carbon (NPC@AC) (Lv, Zhang and Chen 2018). In an air-cathode MFC, NPC@AC-0.7 had mesoporous properties, a greater degree of graphitization and bigger surface area, resulting in 2 times greater power density of 1223 mW m^{-2} and higher ORR electrocatalytic activity compared to a pristine AC catalyst with power density of 595 mW m^{-2}. In a similar way, PANI-Fe-C, a new type of carbon-nitrogen-metal catalyst that was produced using calcination technique and obtained a max power density about 10.17 W m^{-3} in the MFC, which is marginally greater than Pt/C-based catalyst (9.56 W m^{-3}) (Lai et al. 2013).

Graphite oxide-polyaniline hybrid (GO-PANI) carbonization with activation of KOH (PNCN) was also observed by Wen et al. (2014), resulting in a greater value of power density about 1159.34 mW m^{-2} in MFC, which was greater as compared to Pt/C-based catalyst (858.49 mW m^{-2}) (Wen et al. 2014). Melamine, like polyaniline, has been proven to be effective as a heteroatom-doped cathodic catalyst due to its nitrogen and carbon content (Bi et al. 2018; Chan et al. 2015; Figure 8.3b). Sawant et al. (2018) made NCFs (N-doped carbon foams) with a 3D configuration out of RF resin (i.e. resorcinol-formaldehyde resin) and showed a greater power density about 35.74 Wm^{-3}, which was greater about 1.15 times as compared to commercial Pt/C-based catalyst (Sawant et al. 2018; Figure 8.3c). Meng et al. (2015) analyzed the ORR catalytic effect in carbon black (BP-NF) which is N- and F-co-doped, and was synthesized through pyrolysis method of BP-2000 mixture and PTFE (polytetrafluoroethylene) under ammonium atmosphere, which show as extremely proficient ORR electrocatalysis in the air cathode, with a max power density (672 mA cm^{-2} in the MFC) that was greater as compared to marketable Pt/C-based catalyst (572 mA cm^{-2}) (Meng et al. 2015). Ghasemi et al. (2011) applied an alternative ORR catalyst which is activated carbon nanofibers from PAN by applying chemically activated electrospun

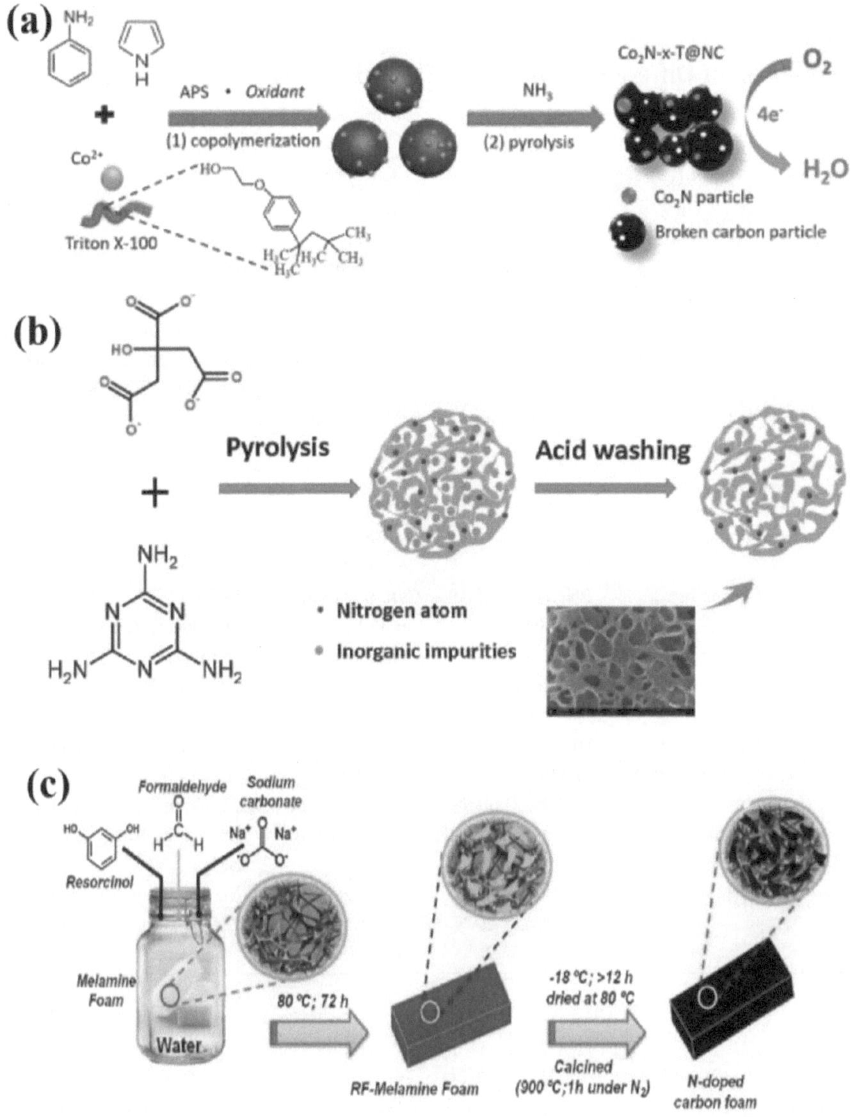

FIGURE 8.3 (a) Schematic representation of the synthesis of Co_2N-X-T@NC through the combination of copolymerization and pyrolysis processes at 600–800 °C in presence of NH_3 gas; (b) graphical representation of synthesis of heteroatom-doped porous carbon (N/PCs) from sodium citrate (C source) and melamine (N source); and (c) schematic illustration of the preparation process of NCFs.

Source: (a) Guo et al. (2019). (b) Bi et al. (2018).

carbon nanofibers (ACNFs) with 8 M KOH, which could generate more power (78 %) as compared to physically and 16 % more power than chemically activated ACNFs with 4 M KOH and plain carbon paper, respectively (Ghasemi et al. 2011).

TABLE 8.1
Comparison of Different PBMEs for MFCs Applications.

Electrode Materials	Anode: Polymer Source	Cathode: Polymer Source	Carbonization/ Pyrolysis Condition	Power Density/ Increased by	References
PBMEs as anode material	PAN—GR: Polyacrylonitrile	0.3 mg/cm² Pt—catalyst (20wt% Pt/C) on CP	Carbonization at 1000 °C in N_2 (4.5 h)	30.7 mW/m² (28.5% higher than CF)	Wang et al. (2015)
	N/PCs: Melamine on carbon cloth	Carbon brush	Pyrolysis at 800 °C in Ar (5 h)	2777.7 mW/m² (98% higher than CC)	Bi et al. (2018)
	NCFs:RF sol—gels	Pt-carbon paper	Carbonization at 900 °C in N_2 (1 h)	35.74 W/m³ (15% higher than Pt-CP)	Sawant et al. (2018)
	NC@CCT: Polydopamine	0.5 mg/cm² Pt on CC	Carbonization at 1000 °C in Ar (1 h)	931 ± 61 mW/m² (80.5% higher than CF)	Zeng et al. (2018)
PBMEs as cathode material	Carbon brush	PNCN coated SS net: PANI	Carbonization at 850 °C (1 h)	1159.34 mW/m² (35% w.r.t. Pt/C)	Wen et al. (2014)
	Carbon fiber brush	NPGC on SS mesh: Melamine	Carbonization at 620–700 °C in N_2 (4 h)	1323 mW/m² (11% w.r.t. Pt/C)	Chan et al. (2015)
	Heat treated carbon mesh	BPNF on CC: PTFE	Pyrolysis at 950 °C in NH_3	672 mW/m² (17% w.r.t. Pt/C)	Meng et al. (2015)
	Heat treated graphite felt brush	NPC@AC on SS mesh: PA-doped PANI	Pyrolysis at 950 °C in Ar (2 h)	1223 mW/m² (105% w.r.t. AC/ SSM)	Lv et al. (2018)
	Carbon felt	CNTs/CNFs electrode: PAN	Carbonization at 1050 °C in N_2 (1 h)	306 ± 14 mW/m² (39.09% w.r.t. Pt/CCC)	Cai et al. (2019)

The main disadvantage was that ACNFs with 8 M KOH were 2.65 times more expensive per unit power as compared to typical platinum cathodes. According to Cai et al. (2019), interconnected fiber aggregated into the structure like thorn on carbon nanotube (CNT)/(CNF) electrodes led to strong ORR catalytic activity exhibiting low internal resistance (0.18 Ω cm^{-2}) and excellent exchange current density (13.68 A m^{-2}) (Cai et al. 2019). With these electrodes, MFCs with a maximum power density of 306 ± 14 mWm^{-2} were achieved, which is 140 percent greater than with Pt/C. Garcia-Gomez et al. (2015) created anode mats made of TiO_2 and carbon dual electrospun nanofibers (TiO_2-PVP-PANI), with a high current density of 8 A m^{-2} (0.8 mA cm^{-2}) and a low resistance of 3.149 Ω in a half microbial fuel cell (Garcia-Gomez et al. 2015). Table 8.1 compares the significant improvements in power performance of different polymeric materials utilized as anodes and cathodes in PBMEs.

8.2.1.2 Polymer/Carbon Composite-Based BMEs (PCBMEs)

While synthesizing cathodic catalysts for MFCs, to bind the catalyst/dopants to the carbon materials, polymeric materials are commonly used as binder like carbon black, activated carbon and graphite. They're also employed as anodic modifiers to make carbon materials more biocompatible and hydrophilic. The price of the binder has a considerable impact on the overall value of MFC. The importance of the binder cannot be overstated for the modified electrodes' excellent performance and stability. Here's a more in-depth look at the topic of several polymers employed as binders and their price use in carbon electrodes which is bulk modified.

8.2.1.2.1 PCBMEs as Anode Material

A simple and environmentally friendly PPy-CMC-CNTs/CB composite anode was prepared to obtain a high energy storage and good performance for MFCs (Wang, Zhu and An 2020; Figure 8.4a). The power density of the PPy-CMC-CNTs/CB composite anode (2970 mW/m^2) was 4.34 times higher than the bare anode (683 mW m^{-2}) due to its excellent biocompatibility and presence of selective electrogenic bacteria. PVA (Polyvinyl alcohol) (Chen et al. 2015), which is hydrophilic due to oxygen-containing groups, has been suggested for use as an anode binder in MFCs, because of its properties like biocompatibility and hydrophilicity. As a anode binder for electrocatalysts applied in MFCs, PVA has indeed been proposed as a substitute for PTFE (Chen et al. 2015). The maximal output power of an E. coli-based MFC employing PVA as an electrocatalyst in that CNT anode, was 1.631 W m^{-2}, which is 97.9% higher than the MFC using PTFE as the binder (0.824 W m^{-2}). However, because PVA also has the qualities of an electronic insulator, it may have a negative impact on electron transmission in between the anode and bacterium. Roh (2015) used in situ chemical polymerization for coat polypyrrole on activated PAN/CNT nanofibers, resulting in a 40% increase in maximum power density over carbon cloth (CC) anodes that is unmodified (Roh 2015). GPF (graphite phenol formaldehyde) was tested as an anode by Navaneeth et al. (2015; Figure 8.4b). It was constructed by mixing 25% (w/v) natural GP (graphite powder) with 75% (by volume) industrial-grade novolac-type phenol-formaldehyde (PF) resin. The efficiency of the GPF was somewhat lower than that of a graphite electrode (GE) oriented photo-bio electro-catalytic fuel cell (PhFC) (Navaneeth et al. 2015). Polytetrafluoroethylene (PTFE) in a composite with graphite electrodes at 24 to 36% (w/w) could have a substantial impact on current generation efficiency in MFCs. The power density of an E. coli-catalyzed MFC with a composite anode containing 30% PTFE and a conventional air cathode was 760 mW m^{-2}. AC cathodes produced with various amounts of the binder (10 to 40% PTFE) (Wei et al. 2012) did not indicate any consistent pattern for current produced, implying that the PTFE binder content was not a necessary factor in the performance of AC and that applying of binder can be minimized to actually reduce the cost of the cathode. Wang et al. (2010) employed a cheap mixture of PTFE and Nafion to bind Pt/C catalyst to air cathodes, and with the quantity of Nafion in the binder, the maximum power density changed linearly, with values of 844 and 685 mW m^{-2} for 67% and 33% respectively. Given the high expense of large-scale MFCs, this suggests that Nafion-PTFE mixtures could be utilized as a replacement for pure Nafion. Using a diethylamine-functionalized Nafion polymer

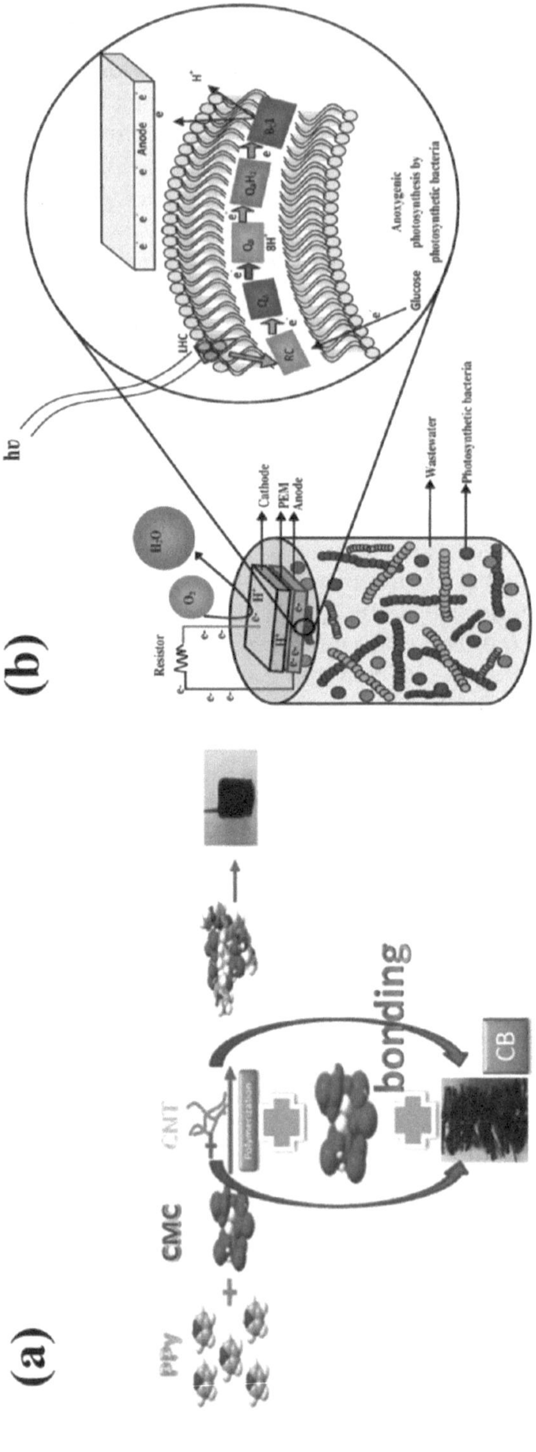
(a)
PPy
CMC
CNT
bonding
CB
(b)
Resistor
Cathode
PEM
Anode
Wastewater
Photosynthetic bacteria
hυ
Anode
Glucose
Anoxygenic photosynthesis by photosynthetic bacteria

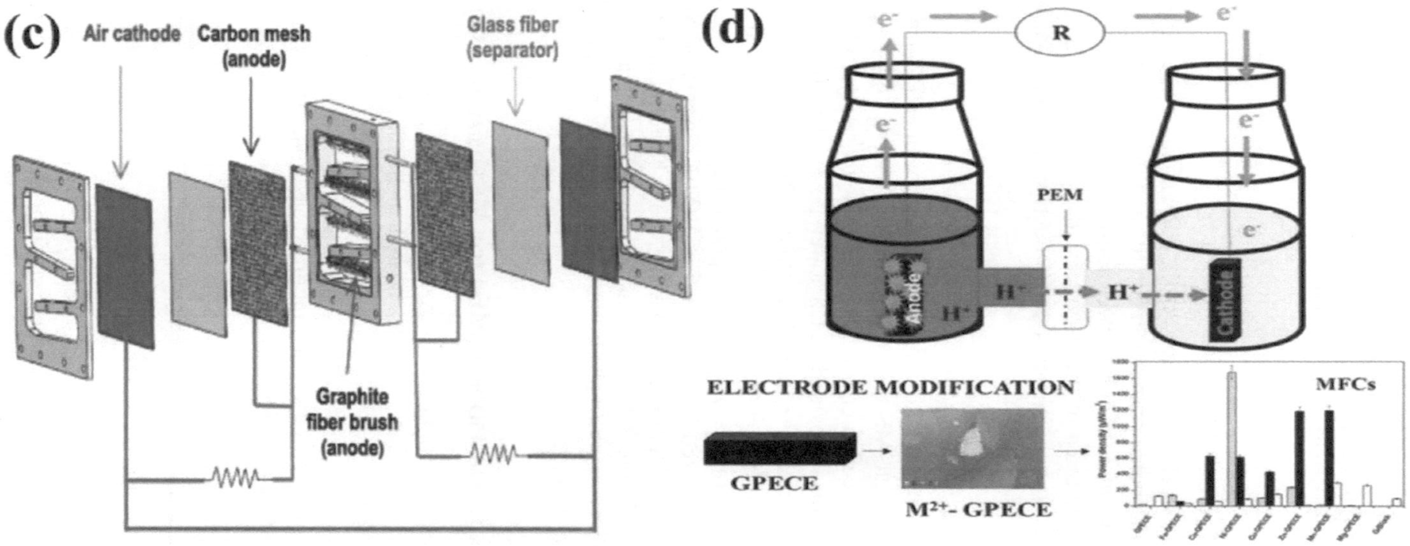

FIGURE 8.4 (a) Preparation and integration of PPy-CMC-CNTs/CB composite material to obtain a high-capacitance 3D anode for MFCs; (b) occurring of cellular mechanism in the photosynthetic bacterial membrane; (c) separator assembly component of the air-cathode MFC; (d) schematic illustration of electrode modification in a dual-chambered MFC setup and their performance comparison.

Source: (a) Wang et al. (2020). (b) Navaneeth et al. (2015). (c) Joel Koffi and Okabe (2020). (d) Narayanasamy and Jayaprakash (2018).

as the catalyst binder improved the growth of nitrifying bacterial biofilm on the surface of cathode. Furthermore, changing the chemical functionality of the catalyst binder, which is the outer part of an air cathode's solution facing side, does not add another layer to the cathode surface, and hence does not raise the internal resistance (Yang et al. 2019).

8.2.1.2.2 PCBMEs as Cathode Material

In MFC air cathode (Joel Koffi and Okabe 2020), polyvinylidene fluoride (PVdF) was utilized as a binder through a phase inversion method instead of the traditional "mix and paste" technique, and the max power density was somewhat lower (286 mW m^2) (Figure 8.4c). Low price of the polymer and the simplicity of the phase inversion technique may encourage its use in the production of MFC cathodes (Yang et al. 2014).

8.2.1.2.3 PCBMEs as Both Anode and Cathode Materials

Graphite powder (GP) was mixed with commercial epoxy resin and doped with different MS (metal salts) to create MS-GECE (graphite-epoxy composite electrodes), that were verified utilizing Pseudomonas aeruginosa as the biocatalyst. Other metal-doped electrode combinations provided much lower power density in the MFC (1679.9 ± 98.04 μW m^{-2}) than the Fe^{3+} doped graphite cathode. Casting was used to make bulk improved GPECE (graphite polyester composite electrodes) doped with MS-GPECE (metal salts) (Narayanasamy and Jayaprakash 2018). A laboratory-scale H-shaped Pseudomonas catalyzed microbial fuel cell (MFC) was used to investigate the impact of redox behaviour of electrodes suitable to bacterial metabolism on improved biofilm growth (Figure 8.4d). With cathode material such as Ni- GPECE and anode material such as graphite block, the maximum power density (1575 ± 223.26 μW m^{-2}) has been attained.

8.2.2 Surface-Modified Electrodes (SMEs) for Microbial Fuel Cells

Surface-modified polymer/carbon-based electrodes are categorized into three kinds based on whether they are doped, electroactive coated, or chemically treated just on the active or inactive surface:

1. Polymer-derived carbon-based surface-modified electrodes (PSMEs)
2. Polymer/carbon composite-based surface-modified electrodes (PCSMEs)

8.2.2.1 Polymer-Derived Carbon-Based SMEs (PSMEs)

8.2.2.1.1 PSMEs as Anode and Cathode Materials

Three-dimensional (3D) electrodes provide the enormous active surface areas for bacterial adhesion, making them suitable anodes in MFCs (Chen et al. 2019). The power density of a carbon-encapsulated metallic nanoparticle-based cathode with a core shell shape developed for MFCMBR by carbonization of melamine foam was found to be 38 times greater than that of unmodified electrodes. When analyzed as a high-performance cathode in a MFC, carbon nanofiber-skinned 3D Ni/carbon micropillars modified by carbonization and activation of the phenol-melamine precursor-based polymeric film produced a maximum power density of ~ 2496 mW m^{-2}, 10 times greater than that obtained by pristine carbon film electrodes.

8.2.2.2 Polymer/Carbon Composite-Based SMEs (PCSMEs)

There are two forms of polymer/carbon composite-based SMEs (PCSMEs): polymer coated on carbon and carbon coated on polymer.

8.2.2.2.1 Polymer Coated on Carbon as Anode and Cathode Materials

Conducting polymers like polyaniline, polythiophene, and polypyrrole have been produced in various shapes using various techniques, and their morphology has a significant impact on MFC power output. Several groups (Zhao et al. 2018; Zhao et al. 2019) have developed conducting polymers with various morphologies on the surface of carbon cloth that might be employed as anode materials. Due to the large specific area that provided enough surface for growth of microbial and transport of charged species, electrochemical measurements demonstrated that both charge transfer resistance and electron transfer resistance were low. Interfacial polymerization was employed to create conductive polyaniline nanofibers (PANInf) that were applied in electrode composites using CB (carbon black). Commonly, two varieties of process, namely electro-polymerization (Lv et al. 2014) and chemical polymerization (Papiya et al. 2018) are employed in synthesizing conductive polymers (Figure 8.5a; Table 8.2). Polyaniline's anodic performance is affected by doping with metal or non-metal dopants. As a result, polyaniline anode modification is an effective method for increasing the performance of MFCs (Liao et al. 2015). To enhance power output, carbon nanotube/polyaniline composites have been tried as anodes in MFCs (Yellappa et al. 2019). The findings showed that a 20 wt% carbon nanotube composite anode with Escherichia coli as the microbial catalyst had the best electrochemical activity, with a higher power density about 42mW/m^2. Electro-polymerized PANI and CNTs deposited on the surface (CNT/PANI/GF) achieved an output voltage about 342 mV across an external resistance of 1.96 kΩ constant load, and the highest power density (257 mW m^{-2}) rose by 343 and 186%, respectively, compared to MFCs with pure GF and PANI/GF. During MFC operation, a larger number of bacteria was associated to the CNT/PANI/GF anode compared to the PANI/GF anode. As a result, the PANI/CNT-based composites performed well as both anode and cathode (Kashyap et al. 2015). In the absence of external electron mediators, an MFC using CNT composites (i.e. polypyrrole (PPy)-coated as the anode and Escherichia coli as the biocatalyst) displayed a higher power density about 228 mW m^{-2}. Various researchers created conductive (PPy)/(rGO) composites using environmentally friendly, simple, cost- and time-effective bio-reduction and in situ polymerization processes (Rikame, Mungray and Mungray 2018). Researchers found that the highest power (output) of the MWCNTs/PANI composite anode was 527.0 mW m^{-2}, four times greater as compared to carbon felt which is unmodified (Fu et al. 2016). Metal-oxide or metals dopants, like V_2O_5 (Ghoreishi et al. 2014), MnO_2 (Wang et al. 2017), TiO_2 (Yin et al. 2019), Fe_2O_3 (Prakash et al. 2020), and SnO_2 and WO_3 (Wang et al. 2013), are thought to be influenced by the conducting polymer electrode's catalytic activity by a variety of factors. In benthic MFCs, electrochemical behaviour of surface-modified electrodes was investigated with Fe_2O_3 (FP), MnO_2-Fe_2O_3 (MFP), and polypyrrole (PPy)-coated MnO_2 (MP) nanocomposites. Carbon felt composite with polyaniline

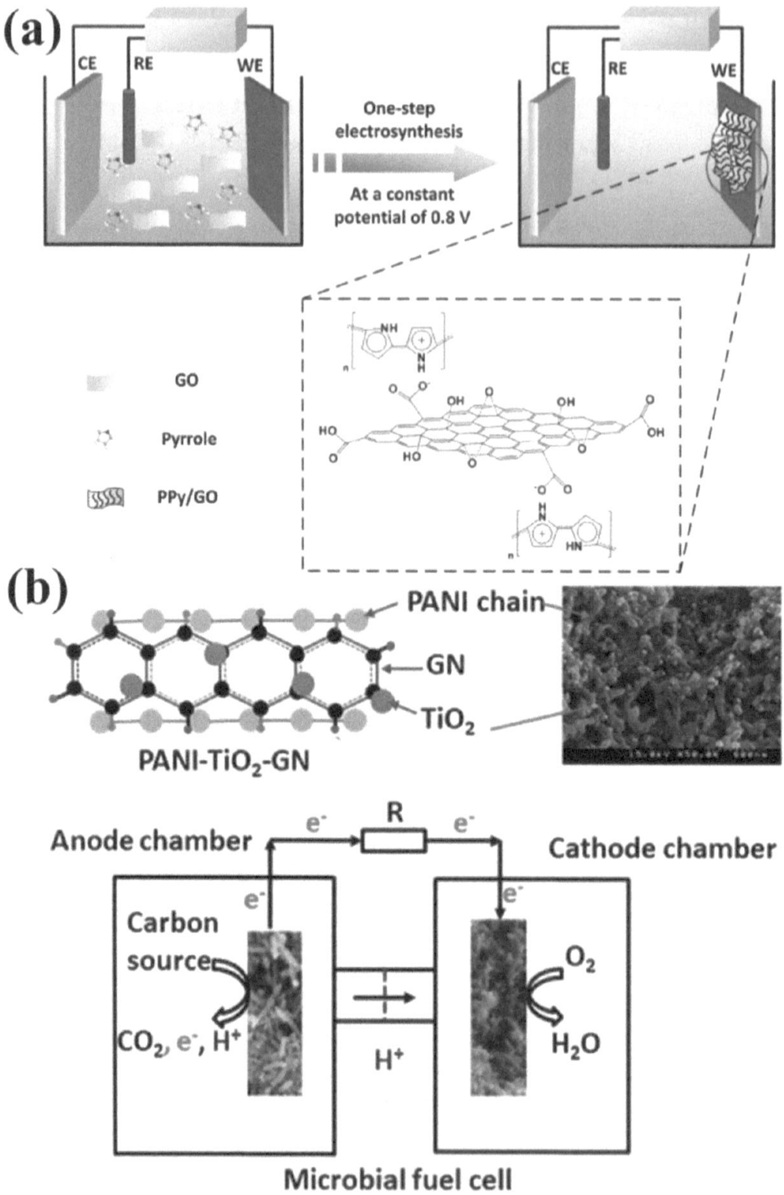

FIGURE 8.5 (a) Electrosynthesis of polypyrrole/graphene oxide composites though electropolymerization method, (b) improved performance of PANI-TiO_2-GN nanocomposite as a bifunctional catalyst in both the cathode and anode for low-cost MFCs.

Source: (a) Lv et al. (2013). (b) Han et al. (2018).

TABLE 8.2
Comparison of Different PCBMEs in Microbial Fuel Cell Applications.

Electrode Materials	Anode	Cathode	Percentage of Polymer Used as Binder	Power Density/ Increased By	References
PCBMEs as anode material	PVA coated carbon felt (3.0 cm × 3.0 cm)	Pt/C on Cp (0.2 mg cm^{-2})	10% PVA	1.631 W/m^2 (97.9% w.r.t. PTFE)	Chen et al. (2015)
	PPy-PAN/CNT	CC/Pt (0.5 mg cm^{-2})	10 wt% PAN	455 mW/m^2 (40% w.r.t. CC)	Roh (2015)
	Graphite polymer composite electrode (GPF) (4 × 4 × 0.5 cm)	Gr plate	75% PF resin with natural graphite powder (25% w/v)	114 mW/m^2 (−57% w.r.t. graphite plate)	Navaneeth et al. (2015)
PCBMEs as cathode material	Graphite fiber brushes	AC and PTFE on SSM	10 wt% PTFE	1100 mW/m^2 (12% w.r.t. Pt/C)	Wei et al. (2012)
	Graphite block	Ni-doped graphite polyester composite electrode (Ni-GPECE)	50% (unsaturated polyester) with graphite	1575 ± 223.26 $\mu W/m^2$ (1557% w.r.t. Gr block)	Narayanasamy and Jayaprakash (2018)
	Carbon mesh	PVdF-based activated carbon on SSM	10% (w/v) PVdF solution containing 26.5 mg/cm^2 of AC and 8.8 mg/cm^2 of carbon black (CB)	3.96 ± 3.01 W/m^3 (118% w.r.t. Pt/Cp)	Joel Koffi and Okabe (2020)

(PANI) and petaline NiO (NiO@PANI-CF) was fabricated by in situ polymerization technique and growth. Zhong et al. (2018) noted that the combined high capacitive properties of NiO and high conductivity of PANI effectively improved the capacity of the generation of electricity in NiO@PANI-CF. The MFC's highest output power density and charge transfer resistance were 563% (1078.8 mW m^{-2}) greater and 68% (10.4 Ω) lower than CF-MFC's. The hybrid bioanode (PEDOT/MnO_2/(CF) had the greatest electron transfer efficiency about 6.3×10^{-9} mol cm^{-2} $s^{-1/2}$ and a maximum power density of 1534 ± 13 mW m^{-2}, which is about 57.7% greater as compared to a bare carbon felt anode that is about 972 ± 21 mW m^{-2} (Liu et al. 2019). In the applications of MFC, regarding anode materials, Table 8.2 analyzes numerous polymer/carbon composite-based SMEs (PCSMEs).

With a 75 wt% PANI/MWNT composite cathode, Jiang et al. (2014) achieved a greater power density about 476 mW m^{-2}, that was greater as compared to pure MWNT cathode (367 mW m^{-2}) but lower as compared to Pt/C cathode (541 mW m^{-2}). Some of THE ternary composites have been proposed as both anode and cathode materials in various articles. At the cathode, a ternary PANI-TiO_2-GN catalyst showed higher activity of ORR, allowing for increased extracellular electron transport to the anode (Han et al. 2018; Figure 8.5b). PANI-TiO_2-GN outperformed PANI-TiO_2 and PANI as an ORR catalyst, with greater catalytic activity, stability, and power output. The immobilization of bacteria on the catalyst matrix boosted the EET substantially at the anode, hence improving the MFC's performance. The MFC with a S. oneidensis (S. o)-immobilized PANI-TiO_2-GN anode generated a high-power density of 79.3 mW m^{-2}, which was 1.3 and 2.7 times greater than that of PANI-TiO_2-GN without S. o and plain carbon paper anode with immobilized S. o, respectively. Likewise, Khilari et al. (2015) showed a noble metal-free MFC composite material employing a manganese ferrite ($MnFe_2O_4$)/polyaniline (PANI) as both anode and cathode. The researchers discovered that adding $MnFe_2O_4$ NPs to Vulcan XC or PANI increased catalytic activity at both the anode and cathode. In a single-chambered MFC, the $MnFe_2O_4$ nanoparticles (NPs) and $MnFe_2O_4$ NPs/PANI hybrid composite not just outperformed the air cathode in terms of ORR activity, the redesigned carbon cloth anode's half-cell potential was also enhanced. The $MnFe_2O_4$ NPs/PANI composite's ORR activity has been found to be equivalent to that of ordinary Pt/C. The anode treated with $MnFe_2O_4$ NPs had a higher half-cell potential than the bare CC anode, resulting in more power being generated in the MFC. The presence of multivalence cations and capacitive characteristic of $MnFe_2O_4$ NPs were attributed to the exoelectrogens' enhanced extracellular electron transport. Furthermore, electrochemically active PANI was shown to offer a superior catalytic support than traditional Vulcan XC. The PANI/C/FePc cathode's maximum power density (630.5 mW m^{-2}) was shown to be greater as compared to C/FePc cathode power density of 336.6 mW m^{-2}, and even the Pt cathode has a power density of 575.6 mW m^{-2}. The PANI/C/FePc cathode, on the other hand, has 7.5 times the power per cost as the Pt cathode. As a consequence, PANI/C/FePc might be a suitable replacement for platinum in MFCs (Yuan, Ahmed and Kim 2011). MFCs using mechanically mixed-$MnCo_2O_4$ NRs/PPy, in situ-manganese cobaltite nanorods ($MnCo_2O_4$ NRs/PPy), catalyst-free (just Vulcan XC) and $MnCo_2O_4$ NRs/Vulcan XC cathodes had maximum volumetric power densities of 4.22, 6.11, 1.77, and 5.05 W m^{-3}, respectively.

This indicated not only that the conducting polymer PPy is superior to the commonly used Vulcan XC as a conducting support, and also that the cathode composite production technique is critical for improved electrode performance. Electrochemical and chemical polymerization processes were used to synthesize PEDOT (Poly(3,4-ethylenedioxythiophene))-Polystyrene sulfonate that is characterized for its electrochemical characteristics on various carbon substrates, and was effective including anodes and cathodes. Air-cathode rGO/PEDOT/Fe_3O_4/CC composite had the highest MFC power density about 3525 mW m^{-2}.

8.2.2.2.2 Carbon Coated on Polymer as Anode and Cathode Materials

Carbon anode materials having a porous structure and large specific surface area, such as graphite granules, graphite foam, polyurethane-sponges, and graphite fiber, offer a lot of promise for boosting MFC power generation. The tiny holes within the anode design, on the other hand, provide high transfer resistance and also create a limited region for bacterial development, limiting bacterial colonization and bacterial blockage of the pores (Xu et al. 2018). Electrochemical behaviour, biomass distribution, MFC performance, and active biomass content are all used to evaluate electrode optimization (Li, Cheng and Thomas 2017; Zhou et al. 2018). Polyurethane foam (PUF) containing macropores has lower toxicity and greater mechanical strength. Perez-Rodriguez et al. (2016) employed polyurethane (PU) foam/graphite/polypyrrole (PU/Graph/PPy) composites to grow anaerobic bacteria and test them as anodes in microbial fuel cells (MFCs) utilizing municipal wastewater as fuel.

The anode efficiency of the low-cost MFC utilizing PU/Graph/PPy-$FeCl_3$ was 2.13 times higher than that of the pure PU foam MFC. Wang et al. (2019) mentioned that a multilayer porous sponge 3D material that is coated with nitrogen-doped carbon nanotube/polyaniline/manganese dioxide (S/N-CNT/PANI/MnO_2) as a bioanode (capacitive bioanode) in an MFC attained a greater power density (1019.5mW/m^2), which about 2.2 and 5.8 times greater as compared to the bioanode (S/N-CNT/MnO_2) (Wang et al. 2019). In the manufacture of electrodes for MFCs, polyester has been used as a support material (Pang, Gao and Choi 2018). Xie et al. (2011b) made CNT-textile anodes and found that the highest current density was 7.2 A/m^2, which was 2.6 times greater than carbon cloth anodes (2.8 A/m^2), indicating that the carbon nanotube covering acted as a key electron carrier in between textile substrate and biofilm (Xie et al. 2011a). Pt nanoparticles were placed on this material by Xie et al. (2011b), and the maximum power density of the CNT-textile-Pt cathode was about 559 mW/m^2, 2.14 times greater than the CNT-Pt cathode (391 mW m^{-2}), trying to imply that the CNT-textile-Pt cathode had a larger surface area and porous network as compared to the CNT-Pt cathode (Xie et al. 2011b).

8.3 CONCLUSIONS AND OUTLOOK

Modified electrode materials with improved electron transport, biocompatibility, and conductivity are now a hot topic in MFC research. As a result, the recent progress in the development of electrode materials for microbial fuel cells based on carbon-polymer nanocomposites has been discussed in this chapter. The advantages of enhancing the performance of microbial fuel cells have also been underlined, as have the recently discovered surface- and bulk-modified composite electrodes. Simple pyrolysis/graphitization (carbonization) of the polymer precursor improved the carbon content while keeping the porous structure, making doping with heteroatoms like F and N easier. Polymeric materials, such as polyvinylidene fluoride (PVdF), polyvinyl alcohol (PVA), polyester and epoxy resin, on the other hand, when used to attach metal ions/metal oxide to carbon compounds as binders, like carbon black, graphite, and activated carbon, effectively enhanced the rate of oxygen reduction reaction and anode/biofilm electron transfer. The kind of polymer used in design of electrodes has a significant influence on the overall cost of an MFC. The trend of progress in MFCs reveal that carbon-polymer nanocomposites can have a major role in cost and performance of the MFC's electrodes. However, this synergic effect of the carbon-polymer nanocomposites alone cannot fulfil the requirement for electrodes in MFCs. However, various combinations of carbon-polymer nanocomposite electrodes can be very efficient alternative electrodes in MFCs in the near future, with reference to energy production as well as sustainable environmental application.

REFERENCES

Aepuru R, Viswanathan M, Rao BV B, Panda H S, Sahu S, Sahoo P K (2020) Tailoring the performance of mechanically robust highly conducting Silver/3D graphene aerogels with superior electromagnetic shielding effectiveness. *Diam. Relat. Mater* 109:108043. https://doi.org/10.1016/j.diamond.2020.108043

Bi L, Ci S, Cai P, Li H, Wen Z (2018) One-step pyrolysis route to three dimensional nitrogen-doped porous carbon as anode materials for microbial fuel cells. *Appl Surf Sci* 427: 10–16. https://doi.org/10.1016/j.apsusc.2017.08.030

Boboescu IZ, Gherman V. D, Lakatos G, Pap B, Biro T, Maroti G (2016) Surpassing the current limitations of biohydrogen production systems: The case for a novel hybrid approach. *Bioresource Technology* 204:192–201. https://doi.org/10.1016/j.biortech.2015.12.083

Bond DR, Lovley DR (2003) Electricity production by Geobacter sulfurreducens attached to electrodes. *Applied and Environmental Microbiology* 69(3):1548–1555. https://doi.org/10.1128/AEM.69.3.1548-1555.2003

Cai T, Huang Y, Huang M, Xi Y, Pang D, Zhang W (2019) Enhancing oxygen reduction reaction of supercapacitor microbial fuel cells with electrospun carbon nanofibers composite cathode. *Chem Eng J* 371:544–553. https://doi.org/10.1016/j.cej.2019.04.025

Chatterjee P, Dessi P, Kokko M, Lakaniemi AM, Lens P (2019) Selective enrichment of biocatalysts for bioelectrochemical systems: A critical review. *Renewable and Sustainable Energy Reviews* 109:10–23. https://doi.org/10.1016/j.rser.2019.04.012

Chen L, Li Y, Yao J, Wu G, Yang B, Lei L, Hou Y, Li Z (2019) Fast expansion of graphite into superior three-dimensional anode for microbial fuel cells. *J Power Sources* 412:86–92. https://doi.org/10.1016/j.jpowsour.2018.11.033

Chen S, Hou H, Harnisch F, Patil SA, Carmona-Martinez AA, Agarwal S, Zhang Y, Sinha-Ray S, Yarin AL, Greiner A, Schröder U (2011) Electrospun and solution blown three dimensional carbon fiber nonwovens for application as electrodes in microbial fuel cells. *Energy Environ Sci* 4:1417–1421. https://doi.org/10.1039/c0ee00446d

Chen XF, Wang XS, Liao KT, Zeng LZ, Xing LD, Zhou XW, Zheng XW, Li WS (2015) Improved power output by incorporating polyvinyl alcohol into the anode of amicrobial fuel cell. *J Mater Chem A* 3:19402–19409.https://doi.org/10.1039/C5TA03318G

Chan YZ, Dai Y, Li R, Zou JL, Tian GH, Fu HG (2015) Low temperature synthesized nitrogen-doped iron/iron carbide/partly-graphitized carbon as stable cathode catalysts for enhancing bioelectricity generation. *Carbon N Y* 89:8–19. https://doi.org/10.1016/j.carbon.2015.03.026

Chou HT, Lee HJ, Lee CY, Tai NH, Chang HY (2014) Highly durable anodes of microbial fuel cells using a reduced graphene oxide/carbon nanotube-coated scaffold. *Bioresour Technol* 169:532–536. https://doi.org/10.1016/j.biortech.2014.07.027

Fu YB, Liu ZH, Su G, Zai XR, Ying M, Yu J (2016) Modified carbon anode by MWCNTs/PANI used in marine sediment microbial fuel cell and its electrochemical performance. *Fuel Cells* 16:377–383. https://doi.org/10.1002/fuce.201500103

Garcia-Gomez NA, Balderas-Renteria I, Garcia-Gutierrez DI, Mosqueda HA, Sánchez EM (2015) Development of mats composed by TiO_2 and carbon dual electrospun nanofibers: A possible anode material in microbial fuel cells. *Mater SciEng B* 193:130–136. https://doi.org/10.1016/j.mseb.2014.12.003

Ghasemi, M., Shahgaldi, S., Ismail, M., Kim, B. H., Yaakob, Z., & Daud, W. R. W. (2011). Activated carbon nanofibers as an alternative cathode catalyst to platinum in a two-chamber microbial fuel cell. *International Journal of Hydrogen Energy*, 36(21), 13746–13752.

Ghoreishi KB, Ghasemi M, Rahimnejad M, Yarmo MA, Daud WRW, Asim N, Ismail M (2014) Development and application of vanadium oxide/polyaniline composite as a novel cathode catalyst in microbial fuel cell. *Int J Energy Res* 38:70–77. https://doi.org/10.1002/er.3082

Guo D, Tian Z, Wang J, Ke X, Zhu Y (2019) Co_2N nanoparticles embedded N-doped mesoporous carbon as efficient electrocatalysts for oxygen reduction reaction. *Appl Surf Sci* 473:555–563. https://doi.org/10.1016/j.apsusc.2018.12.204

Han TH, Parveen N, Shim JH, Nguyen ATN, Mahato N, Cho MH (2018) Ternary composite of polyaniline graphene andTiO_2 as a bifunctional catalyst to enhance the performance of both the bioanode and cathode of a microbial fuel cell. *Ind Eng Chem Res* 57:6705–6713. https://doi.org/10.1021/acs.iecr.7b05314

He G, Gu Y, He S, Schroder U, Chen S, Hou H (2011) Effect of fiber diameter on the behavior of biofilm and anodic performance of fiber electrodes in microbial fuel cells. *Bioresour Technol* 102:10763–10766. https://doi.org/10.1016/j.biortech.2011.09.006

He Z, Minteer SD, Angenent LT (2005) Electricity generation from artificial wastewater using an upflow microbial fuel cell. *Environmental Science & Technology* 39(14):5262–5267. https://doi.org/10.1021/es0502876

Huang L, Li X, Ren Y, Wang X (2016) In-situ modified carbon cloth with polyaniline/graphene as anode to enhance performance of microbial fuel cell. *Int J Hydrogen Energy* 41:11369–11379. https://doi.org/10.1016/j.ijhydene.2016.05.048

Jiang Y, Xu Y, Yang Q, Chen Y, Zhu S, Shen S (2014) Power generation using polyaniline/multi-walled carbon nanotubes as an alternative cathode catalyst in microbial fuel cells. *Int J Energy Res* 38:1416–1423. https://doi.org/10.1002/er.3155

Joel Koffi N, Okabe S (2020) Domestic wastewater treatment and energy harvesting by serpentine up-flow MFCs equipped with PVdF-based activated carbon air—cathodes and a low voltage booster. *Chem Eng J* 380:122443. https://doi.org/10.1016/j.cej.2019.122443

Kashyap D, Kim C, Kim SY, Kim YH, Kim GM, Dwivedi PK, Sharma A, Goel S (2015) Multi walled carbon nanotube and polyaniline coated pencil graphite based bio-cathode for enzymatic biofuel cell. *Int J Hydrogen Energy* 40:9515–9522. https://doi.org/10.1016/j.ijhydene.2015.05.120

Khilari S, Pandit S, Varanasi JL, Das D, Pradhan D (2015) Bifunctional manganese ferrite/polyaniline hybrid as electrode material for enhanced energy recovery in microbial fuel cell. *ACS Appl Mater Interfaces* 7:20657–20666. https://doi.org/10.1021/acsami.5b05273

Kim BH, Kim HJ, Hyun MS, Park DH (1999) Direct electrode reaction of Fe (III)-reducing bacterium, Shewanella putrefaciens. *Journal of Microbiology and Biotechnology* 9(2): 127–131.

Lai, B., Wang, P., Li, H., Du, Z., Wang, L., & Bi, S. (2013). Calcined polyaniline–iron composite as a high efficient cathodic catalyst in microbial fuel cells. *Bioresource technology*, 131, 321–324.

Liao ZH, Sun JZ, Sun DZ, Si RW, Yong YC (2015) Enhancement of power production with tartaric acid doped polyaniline nanowire network modified anode in microbial fuel cells. *Bioresour Technol* 192:831–834. https://doi.org/10.1016/j.biortech.2015.05.105

Liu P, Zhang C, Liang P, Jiang Y, Zhang X, Huang X (2019) Enhancing extracellular electron transfer efficiency and bioelectricity production by vapor polymerization Poly(3,4-ethylenedioxythiophene)/MnO2 hybrid anode. *Bioelectrochemistry* 126:72–78. https://doi.org/10.1016/j.bioelechem.2018.07.011

Lv K, Zhang H, Chen S (2018) Nitrogen and phosphorus co-doped carbon modified activated carbon as an efficient oxygen reduction catalyst for microbial fuel cells. *RSC Adv* 8:848–855. https://doi.org/10.1039/C7RA12907F

Li S, Cheng C, Thomas A (2017) Carbon-based microbial-fuel cell electrodes: From conductive supports to active catalysts. *Adv Mater* 29:1602547. https://doi.org/10.1002/adma.201602547

Lv Z, Chen Y, Wei H, Li F, Hu Y, Wei C, Feng C (2013) One step electrosynthesis of polypyrrole/graphene oxide composites for microbial fuel cell application. *Electrochim Acta* 111:366–373. https://doi.org/10.1016/j.electacta.2013.08.022

Lv Z, Xie D, Li F, Hu Y, Wei C, Feng C (2014) Microbial fuel cell as a biocapacitor by using pseudo-capacitive anode materials. *J Power Sources* 246:642–649. https://doi.org/10.1016/j.jpowsour.2013.08.014

Meng K, Liu Q, Huang Y, Wang Y (2015) Facile synthesis of nitrogen and fluorine co-doped carbon materials as efficient electrocatalysts for oxygen reduction reactions in air—cathode microbial fuel cells. *J Mater Chem A* 3:6873–6877. https://doi.org/10.1039/c4ta06500j

Narayanasamy S, Jayaprakash J (2018) Improved performance of pseudomonas aeruginosa catalyzed MFCs with graphite/polyester composite electrodes doped with metal ions for azo dye degradation. *Chem Eng J* 343:258–269. https://doi.org/10.1016/j.cej.2018.02.123

Narayanasamy S, Jayaprakash J (2020) Application of carbon-polymer based composite electrodes for Microbial fuel cells. *Reviews in Environmental Science and Bio/Technology*: 1–26. https://doi.org/10.1007/s11157-020-09545-x

Navaneeth B, Hari Prasad R, Chiranjeevi P, Chandra R, Sarkar O, Verma A, Subudhi S, Lal B, Venkata Mohan S (2015) Implication of composite electrode on the functioning of photo-bioelectrocatalytic fuel cell operated with heterotrophic-anoxygenic condition. *Bioresour Technol* 185:331–340. https://doi.org/10.1016/j.biortech.2015.02.065

Pang S, Gao Y, Choi S (2018) Flexible and stretchable biobatteries: monolithic integration of membrane-free microbial fuel cells in a single textile layer. *Adv Energy Mater* 8:1702261. https://doi.org/10.1002/aenm.201702261

Papiya F, Pattanayak P, Kumar P, Kumar V, Kundu PP (2018) Development of highly efficient bimetallic nanocomposite cathode catalyst, composed of Ni: Co supported sulfonated polyaniline for application in microbial fuel cells. *Electrochim Acta* 282:931–945. https://doi.org/10.1016/j.electacta.2018.07.024

Pérez-Rodríguez P, Ovando-Medina VM, Martínez-Amador SY, Rodríguez-de la Garza JA (2016) Bioanode of polyurethane/graphite/polypyrrole composite in microbial fuel cells. *Biotechnol Bioprocess Eng* 21:305–313. https://doi.org/10.1007/s12257-015-0628-5

Prakash O, Mungray A, Chongdar S, Mungray AK, Kailasa S (2020) Performance of polypyrrole coated metal oxide composite electrodes for benthic microbial fuel cell(BMFC). *J Environ Chem Eng* 8:102757. https://doi.org/10.1016/j.jece.2018.11.002

Rabaey K, Boon N, Hofte M, Verstraete W (2005) Microbial phenazine production enhances electron transfer in biofuel cells. *Environmental Science & Technology* 39(9):3401–3408. https://doi.org/10.1021/es048563o

Ramanujam BTS, Annamalai PK (2017) Condúcting polymer graphite binary and hybrid composites: Structure, properties, and applications. *Applications. Elsevier Ltd., Hybrid Polymer Composite Materials*. https://doi.org/10.1016/B978-0-08-100785-3.00001-2

Reguera G, McCarthy KD, Mehta T, Nicoll JS, Tuominen MT, Lovley DR (2005) Extracellular electron transfer via microbial nanowires. *Nature* 435(7045):1098–1101. https://doi.org/10.1038/nature03661

Rikame SS, Mungray AA, Mungray AK (2018) Modification of anode electrode in microbial fuel cell for electrochemical recovery of energy and copper metal. *Electrochim Acta* 275:8–17. https://doi.org/10.1016/j.electacta.2018.04.141

Roh SH (2015) Electricity generation from microbial fuel cell with polypyrrole-coated carbon nanofiber composite. *J Nanosci Nanotechnol* 15:1700–1703. https://doi.org/10.1166/jnn.2015.9317

Sahoo BB, Biswal K, Jena A, Naik B, Nayak NC, Dash BP, Mahanto BS, Soam A, Sahoo PK (2021) Pd supported on 3D graphene aerogel as potential electrocatalyst for alkaline direct methanol fuel cells. *Mater Today: Proc* 41:150–155. https://doi.org/10.1016/j.matpr.2020.08.446

Santoro C, Arbizzani C, Erable B, Ieropoulos I (2017) Microbial fuel cells: From fundamentals to applications: A review. *J Power Sources* 356:225–244. https://doi.org/10.1016/j.jpowsour.2017.03.109

Santoro C, Kodali M, Herrera S, Serov A, Ieropoulos I, Atanassov P (2018) Power generation in microbial fuel cells using platinum group metal-free cathode catalyst: Effect of the catalyst loading on performance and costs. *Journal of Power Sources* 378:169–175. https://doi.org/10.1016/j.jpowsour.2017.12.017

Sawant SY, Han TH, Ansari SA, Shim JH, Nguyen ATN, Shim JJ, Cho MH (2018) A metal-free and non-precious multifunctional3D carbon foam for high-energy density supercapacitors and enhanced power generation in microbial fuel cells. *J Ind Eng Chem* 60:431–440. https://doi.org/10.1016/j.jiec.2017.11.030

Wang X, Feng Y, Liu J, Shi X, Lee H, Li N, Ren N (2010) Power generation using adjustable Nafion/PTFE mixed binders in air—cathode microbial fuel cells. *Biosens Bioelectron* 26:946–948. https://doi.org/10.1016/j.bios.2010.06.026

Wang Y, Li B, Zeng L, Cui D, Xiang X, Li W (2013) Polyaniline/mesoporous tungsten trioxide composite as anode electrocatalyst for high-performance microbial fuel cells. *Biosens Bioelectron* 41:582–588. https://doi.org/10.1016/j.bios.2012.09.054

Wang Y, Wen Q, Chen Y, Qi L (2017) A novel polyaniline interlayer manganese dioxide composite anode for high performance microbial fuel cell. *J Taiwan Inst Chem Eng* 75:112–118. https://doi.org/10.1016/j.jtice.2017.03.006

Wang Y, Zheng H, Chen Y, Wen Q, Wu J (2019) Macroporous composite capacitive bioanode applied in microbial fuel cells. *Chin Chem Lett* 31:205–209. https://doi.org/10.1016/j.cclet.2019.05.052X

Wang Y, Zhu L, An L (2020) Electricity generation and storage in microbial fuel cells with porous polypyrrole-base composite modified carbon brush anodes. *Renew. Energy* 162:2220–2226. https://doi.org/10.1016/j.renene.2020.10.032

Wang Y-Q, Huang H-X, Li B, Li W-S (2015) Novelly developed three-dimensional carbon scaffold anodes from polyacrylonitrile for microbial fuel cells. *J Mater Chem A* 3:5110–5118. https://doi.org/10.1039/C4TA06007E

Wei B, Tokash JC, Chen G, Hickner MA, Logan BE (2012) Development and evaluation of carbon and binder loading in low-cost activated carbon cathodes for air-cathode microbial fuel cells. *Rsc Adv* 2:12751–12758. https://doi.org/10.1039/c2ra21572a

Wen Q, Wang S, Yan J, Cong L, Chen Y, Xi H (2014) Porous nitrogen-doped carbon nanosheet on graphene as metal free catalyst for oxygen reduction reaction in air—cathode microbial fuel cells. *Bioelectrochemistry* 95:23–28. https://doi.org/10.1016/j.bioelechem.2013.10.007

Xie X, Hu L, Pasta M, Wells GF, Kong D, Criddle CS, Cui Y (2011a) Three-dimensional carbon nanotube—textile anode for high-performance microbial fuel cells. *Nano Lett* 11:291–296. https://doi.org/10.1021/nl103905t

Xie X, Pasta M, Hu L, Yang Y, McDonough J, Cha J, Criddle CS, Cui Y (2011b) Nano-structured textiles as high-performance aqueous cathodes for microbial fuel cells. *Energy Environ Sci* 4:1293. https://doi.org/10.1039/c0ee00793e

Xu H, Wu J, Qi L, Chen Y, Wen Q, Duan T, Wang Y (2018) Preparation and microbial fuel cell application of sponge structured hierarchical polyaniline-texture bioanode with an integration of electricity generation and energy storage. *J Appl Electrochem* 48:1285–1295. https://doi.org/10.1007/s10800-018-1252-9

Yang W, He W, Zhang F, Hickner MA, Logan BE (2014) Single-step fabrication using a phase inversion method of poly(vinylidene fluoride) (PVdF) activated carbon air cathodes for microbial fuel cells. *Environ Sci Technol Lett* 1:416–420. https://doi.org/10.1021/ez5002769

Yang Y, Choi C, Xie G, Park JD, Ke S, Yu JS, Zhou J, Lim B (2019) Electron transfer interpretation of the biofilm coated anode of a microbial fuel cell and the cathode modification effects on its power. *Bioelectrochemistry* 127:94–103. https://doi.org/10.1016/j.bioelechem.2019.02.004

Yellappa M, Sravan JS, Sarkar O, Reddy YVR, Mohan SV (2019) Modified conductive polyaniline-carbon nanotube composite electrodes for bioelectricity generation and waste remediation. *Bioresour Technol* 284:148–154. https://doi.org/10.1016/j.biortech.2019.03.085

Yin T, Zhang H, Yang G, Wang L (2019) Polyaniline compositeTiO2 nanosheets modified carbon paper electrode as a high performance bioanode for microbial fuel cells. *Synth Met* 252:8–14. https://doi.org/10.1016/j.synthmet.2019.03.027

Yuan Y, Ahmed J, Kim S (2011) Polyaniline/carbon black composite-supported iron phthalocyanine as an oxygen reduction catalyst for microbial fuel cells. *J Power Sources* 196:1103–1106. https://doi.org/10.1016/j.jpowsour.2010.08.112

Zeng L, Zhao S, He M (2018) Macroscale porous carbonized polydopamine-modified cotton textile for application as electrode in microbial fuel cells. *J Power Sources* 376:33–40. https://doi.org/10.1016/j.jpowsour.2017.11.071

Zhao N, Ma Z, Song H, Wang D, Xie Y (2018) Polyaniline/reduced graphene oxide-modified carbon fiber brush anode for high-performance microbial fuel cells. *Int J Hydrogen Energy* 43:17867–17872. https://doi.org/10.1016/j.ijhydene.2018.08.007

Zhao N, Ma Z, Song H, Xie Y, Zhang M (2019) Enhancement of bioelectricity generation by synergistic modification of vertical carbon nanotubes/polypyrrole for the carbon fibers anode in microbial fuel cell. *Electrochim Acta* 296:69–74. https://doi.org/10.1016/j.electacta.2018.11.039

Zhong D, Liao X, Liu Y, Zhong N, Xu Y (2018) Enhanced electricity generation performance and dye wastewater degradation of microbial fuel cell by using a petaline NiO@ polyaniline-carbon felt anode. *Bioresour Technol* 258:125–134. https://doi.org/10.1016/j.biortech.2018.01.117

Zhou Y, Lu H, Wang J, Zhou J, Leng X, Liu G (2018) Catalytic performance of quinone and graphene-modified polyurethane foam on the decolorization of azo dye Acid Red18 by Shewanella sp. RQs-106. *J Hazard Mater* 356:82–90. https://doi.org/10.1016/j.jhazmat.2018.05.043

9 Polymer Composites for Dye-Sensitized Solar Cells

A. L. Saroj, Pooja Rawat, and A. L. Sharma

CONTENTS

9.1 INTRODUCTION

The increase in pollution of the earth and shortage of safe, sustainable, and environmentally friendly energy resources become a serious threat in the world. The tremendous increase in the use of non-renewable energy resources like oil, coal, fossil fuel,

DOI: 10.1201/9781003208662-11

and natural gas causes environmental issues such as air pollution, CO_2 emissions, and the greenhouse effect and also leads to a shortage of available energy resources. The efforts in resolving these problems have forced scientists of the world to search the alternative energy resources like wind energy, solar energy, thermal energy, geothermal energy, etc. Among various renewable energy resources, solar energy is considered one of the most powerful alternative energy resources due to its clean, safe, and abundant energy supply (Gong et al. 2017). The fraction of energy consumed by humans in a year to the total energy released by the sun in an hour is very small; therefore solar energy is expected to meet the world's energy demand (Lewis 2007). To fulfill environmental security and energy demand, photovoltaic (PV) solar cells, which convert solar energy into electrical energy, have been developed. Moreover, these PV solar cells generate electricity without any movable mechanical components. The evolution of PV solar cells consists of three generations, namely, first-, second-, and third-generation solar cells (Chebrolu and Kim 2019). First-generation solar cells were based on single-crystalline and multi-crystalline silicon cells which governed the market and are very popular due to their high efficiencies but their high manufacturing costs become an obstacle in their worldwide applications (Pi et al. 2012; Dekkers et al. 2006). Second-generation solar cells are comparatively much cheaper to produce but have lower efficiencies than first-generation solar cells. Second-generation solar cells include amorphous silicon (a-Si) solar cells, Cadmium Telluride (CdTe) solar cells, and Copper Indium Gallium Selenide (CIGS) solar cells. The arising third-generation solar cells include organic tandem solar cells (Meng et al. 2018; Ameri et al. 2013), inorganic solar cells (Miles et al. 2007; McCandless and Sites 2011), organic and inorganic perovskite solar cells (Zhang et al. 2016; Tzounis et al. 2017), quantum dots solar cells (Tvrdy and Kamat 2011; Ning et al. 2015), and dye-sensitized solar cells (DSSCs) (Kavan 2017; Feldt et al. 2010; Dette et al. 2014), which are still in the development phase of production for commercialization. Nowadays, silicon-based solar cells of the first-generation, whose light-to-electricity conversion efficiency reaches up to 15–20%, dominate the market. However, the requirement of highly purified silicon and its high cost of production has paved the way for research for environmentally friendly and low-cost dye-sensitized solar cells.

In 1991, Brian O'Regan and Michael Grätzel designed a photovoltaic solar cell known as the Grätzel cell or dye-sensitized solar cell whose workings were based on plant photosynthesis. The incident photon to current conversion efficiency (IPCE) for this cell was reported as 7.1–7.9%. The excellent stability of about five million turnovers without decomposition, a large short circuit current density (J_{SC} > 12mAcm^{-2}), and the low cost of fabrication had established the practical and feasible utilization of this cell. The current energy conversion efficiency of lab-sized DSSC improves from 7.1% achieved in 1991 to 14.3% reported in 2015 (Regan and Grätzel 1991; Kakiage et al. 2015). This improvement arises due to optimization of the device, use of transition metal redox coupled with desirable dye, and a solvent having low viscosity like acrylonitrile (ACN). A new DSSC configuration has been designed using copper-based electrolytes with the highest energy conversion efficiency of about 32% under low light intensity conditions (Cao et al. 2018). These spectacular improvements in conversion efficiency of DSSC have resulted in developing large-scale DSSCs and also the small module that focuses on convenient electronics

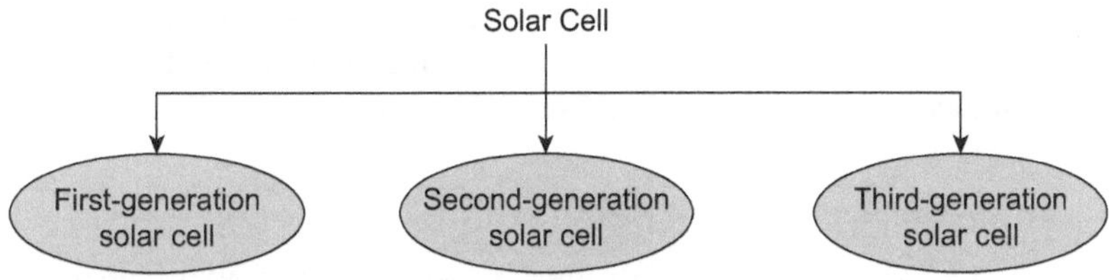

FIGURE 9.1 Classification of solar cells/photovoltaic (PV) cells.

(Pradhan et al. 2018). A dye-sensitized solar cell (DSSC) consists of a photoanode (working electrode) and cathode (counter electrode) with an electrolyte sandwiched between them.

9.2 CLASSIFICATION OF SOLAR CELLS

Solar cells (or PV cells) are generally made up of two types of semiconducting materials, namely n-type and p-type, and these materials have some important features to absorb incident sunlight. Solar cells can be fabricated by using only a single layer of light-absorbing materials (i.e. single-junction solar cells) or by using multi-layered light-absorbing materials that form multi-junctions to take the advantage of various absorption and charge separation mechanisms. Based on these properties of materials, solar cells can be classified into different classes as shown in Figure 9.1.

9.2.1 First-Generation Solar Cell

This generation of solar cells is the oldest and most commonly used solar cell due to their high efficiency. These cells are manufactured on wafers that can provide a 2–3 Watt power supply per wafer. To enhance the power, an assembly of such solar cells (known as solar modules) is constructed. Based on crystallization level, first-generation solar cells are divided into two categories: (i) single crystal solar cells and (ii) multi-crystal solar cells. Single crystal solar cells are produced by only one crystal (i.e. the whole wafer consists of a single crystal). On the other hand, multi-crystal solar cell wafers consist of crystal grains. The multi-crystal-based solar cell has a lower efficiency in comparison with a single crystal solar cell, but its easier and cheaper cost of fabrication made them good competition.

9.2.2 Second-Generation Solar Cell

The second-generation solar cell involves amorphous Si (a-Si)-based thin-film solar cell, CIGS-solar cells, and CdTe/cadmium sulfide (CdS)-solar cells (Carlson and Wronski 1976; Choubey et al. 2012). Second-generation solar cells are based on thin-film technology to produce a cheaper solar cell. This generation of solar cells uses less material and has a low manufacturing cost as compared to the first-generation solar cell. Other advantages of thin-film solar cells are their flexibility, and they can be extended up to large areas of nearly 6 m^2, while first-generation solar cells can be grown up to wafer dimensions (Sharma et al. 2015; Fakharuddin et al. 2014). However, the use of toxic materials and lower overall efficiency than first-generation solar cells put restrictions in the way of second-generation solar cells.

9.2.3 Third-Generation Solar Cell

The main motive for the evolution of the third-generation solar cell is to increase the electrical performance of second-generation solar cells with very low manufacturing costs. Third-generation solar cells are based on nanostructured materials and consist of pure organic components or a mixture of organic and inorganic components which allow for a huge and limitless choice of materials. Among various types of third-generation solar cells are nano-crystal-based solar cells, polymer solar cells, DSSCs, and concentrated solar cells. DSSCs have attracted many researchers for the development of next-generation renewable and sustainable energy devices because of their feasible properties such as low production cost, flexibility, eco-friendliness, ease of fabrication, long life, mechanical robustness, and they have the potential for both indoors and outdoors application and can work under the low intensity of incident light (Choubey et al. 2012; Sharma et al. 2015; Fakharuddin et al. 2014).

9.3 DYE-SENSITIZED SOLAR CELLS (DSSCs)

DSSC was introduced by Grätzel and co-workers two decades ago and is a photo electrochemical cell that converts solar energy into electricity. DSSC is assembled by sandwiching a thin film of electrolyte between a working electrode soaked with a dye (or sensitizer) and a counter electrode (Regan and Grätzel 1991; Pradhan et al. 2018).

9.3.1 Working Principle of DSSC

The construction and working principle of DSSC is shown in Figure 9.2. The working mechanism of DSSC follows a series of processes such as absorption of the incident photon, electron injection, and transportation of electrons through external circuit and collection of current. These basic processes which DSSC goes through are explained next (Grätzel 2004). Absorption of the photon: initially light incident on working electrode and electrons in the ground state (D) of dye absorbs photos and gets transferred to the excited state (D^*) of dye.

$$D + hu \rightarrow D^*$$

1. Electron injection process: the excited electron of dye is injected into the conduction band of the semiconductor (TiO_2) which results in the oxidation of dye.

$$D^* \rightarrow D^+ + e^-$$

2. Transportation of electrons through the external circuit: these injected electrons after traveling through semiconductor nanoparticles diffuse towards transparent conducting FTO glass plate and enter the external circuit. Through the external circuit, electrons reach the counter electrode.

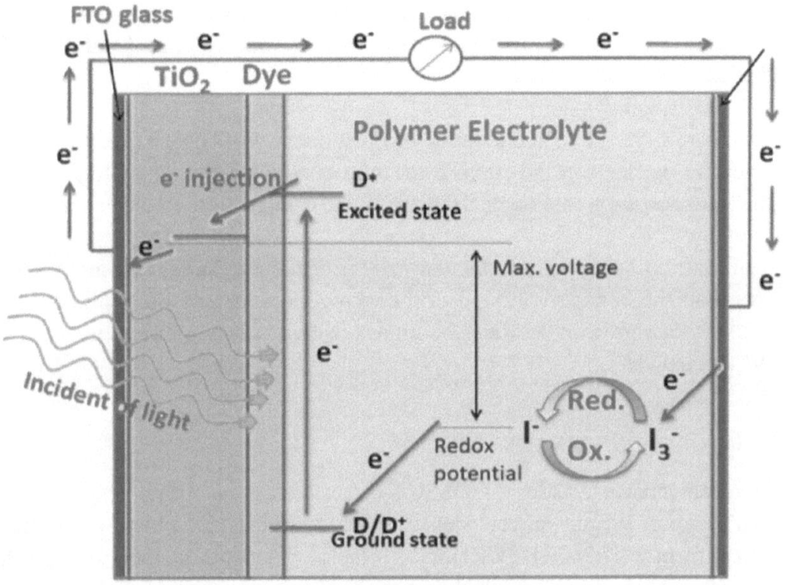

FIGURE 9.2 Construction and working principle of DSSC.

3. Regeneration of dye: the electron deficiency of cationic dye is required by accepting an electron from the I^- ion on the electrolyte and giving rise to the initial state (D) of the dye, and also oxidation of the I^- ion into I_3^- ion occurs.

$$D^+ + 3I^- \rightarrow D + I_3^-.$$

4. The I_3^- ions diffuse towards the counter electrode and reduce into I^- ion by accepting electrons at the counter electrode.

$$I_3^- + 2e^- \rightarrow 3I^-$$

Thus, the circuit is completed via transportation of electrons through the external load and I^- ion being regenerated by the reduction of the I_3^- ion at the counter electrode. Therefore, in the DSSC, by following the previous steps the conversion of an incident photon to electricity takes place. Along with the forward charge transportation processes, backward charge transportation processes also take place in one complete cycle, which results in a sharp decrease in the efficiency of DSSC. The following processes lead to the backward charge transportation mechanism in DSSC:

1. Extraction of electrons by the oxidized dye from the semiconductor
2. Transfer of electrons from the excited state to the ground state of dye
3. Production of dark current by recombining injected electrons with the electrolyte

To minimize the previously mentioned effects of the backward charge transportation processes, the following steps should be followed (Andersen and Lian 2005; Hara et al. 2003; Gong et al. 2012):

1. Transfer of charge to semiconductor must occur with a high quantum yield
2. The rate of electron injection to the semiconductor should be higher than the transportation rate of electrons from the excited state to the ground state of dye
3. The lowest unoccupied molecular orbital of the dye should be more negative than the conduction band of the semiconductor and the highest occupied molecular orbital should be more positive than the redox potential of the electrolyte

9.3.2 Important Parameters of DSSC

The important characteristics of DSSCs—like overall electrical conversion efficiency (η%), short circuit current density (J_{sc}), open-circuit voltage (V_{oc}), incident photon to current efficiency (IPCE), and fill factor—depend on spectroscopic properties of dyes, surface morphological properties of semiconductors, and the electrical properties of electrolytes. The production of electrical power on receiving incident light by the DSSC shows the capability of the DSSC to generate a current through the external load and voltage over an external load simultaneously. The characteristic plots of current density vs. voltage (*J-V*) and power vs. voltage (*P-V*) curve of the DSSC are shown in Figure 9.3.

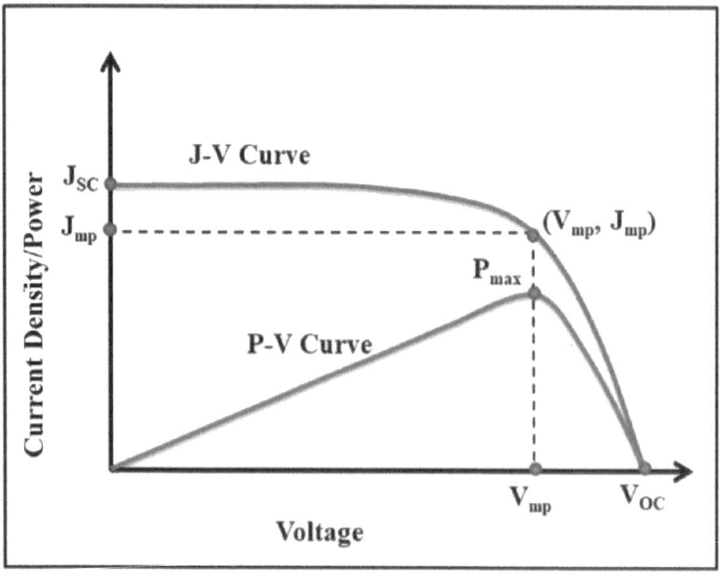

FIGURE 9.3 Short circuit current density, J_{SC}, and power, *P*, vs. voltage plot of DSSC.

The maximum current (short circuit current, I_{sc}, or short circuit current density, J_{sc}) is produced when the cell is short-circuited under illumination (i.e. at zero voltage). Thus, the electrical power generated is also zero. On increasing the intensity of incident light, the short-circuit current reached its maximum value (J_{max}) expressed as:

$$J_{sc} = J_{max}\left(V = 0\right) \tag{9.1}$$

The maximum voltage (open-circuit voltage, V_{oc}) is generated under open-circuit conditions when no current flows through the circuit. As the incident light intensity increases, the open-circuit voltage increases (V_{max}) and is written as:

$$V_{oc} = V_{max}(I = 0) \tag{9.2}$$

The maximum efficiency of the DSSC required to convert incident solar energy into electricity illustrates the maximum power output (P_{max}) of the DSSC. The maximum output power point can be determined by evaluating the maxima of output power (p_{max}) which is calculated by the product of V_{mp} and J_{mp} expressed as:

$$P_{max} = V_{max} \times J_{max} \tag{9.3}$$

It can be concluded from the plot of power versus voltage (as shown in Figure 9.3) that on increasing the current density and voltage, initially the value of output power increases and reaches its optimal value at maximum voltage and maximum current density and then gradually starts decreasing up to zero value at an open-circuit voltage (V_{oc}). The fill factor (*FF*) is an important parameter of DSSC which signifies the overall capability of the cell and indicates the electric generation loss. It is a measure of the quality and idealness of the solar cell. The value of the fill factor can be obtained by comparing the maximum short circuit current density (J_{sc}) and maximum open-circuit voltage (V_{oc}) to the maximum power output (P_{max}).

$$FF = \frac{P_{max}}{V_{OC} \times J_{SC}} = \frac{V_{mp} \times J_{mp}}{V_{OC} \times J_{SC}} \tag{9.4}$$

The $\eta(\%)$ of DSSC is the ratio of maximum electrical output power (P_{max}) to the input power (P_{in}) of incident light. It describes the percentage of incident light energy converted into electricity, and the overall conversion efficiency of the cell is measured by using the following equations:

$$\eta(\%) = \frac{P_{max}}{P_{in}} \times 100 = \frac{J_{mp} \times V_{mp}}{P_{in}} \times 100 \tag{9.5}$$

$$\text{and } \eta(\%) = \frac{V_{OC} \times J_{SC} \times FF}{P_{in}} \times 100 \tag{9.6}$$

Incident photon to current conversion efficiency, *IPCE*(%), is one of the most significant parameters of DSSC. It has been used to analyze different mechanism processes in DSSC such as current generation, recombination, and diffusion mechanism. It is calculated by the ratio of the number of electrons flowing through the external circuit to the total number of photons incident on the surface of DSSC at any wavelength, λ:

$$IPCE(\%) = \frac{P_{in} \times q \times \lambda}{J_{SC} \times hc} \times 100 \tag{9.7}$$

where q is the electron charge, h is Planck's constant, and c is the speed of light.

The ideal output power, P_{ideal}, of DSSC is obtained as a product of short-circuit current density, J_{sc} and open-circuit voltage, V_{oc}.

$$P_{ideal} = J_{SC} \times V_{OC} \tag{9.8}$$

On the other hand, the product of V_{mp} and J_{mp} gives the value of maximum output power, P_{max}. Equations (9.5) and (9.6) show the dependence of the overall conversion efficiency of DSSC on both the maximum output power and ideal output power. The difference between the ideal output power and maximum output power occurs due to the resistances of the cell, electrolyte, electrode, and other components of DSSC.

9.3.3 Components of DSSC

9.3.3.1 Transparent and Conductive Substrate (FTO/ or ITO Coated Glass Sheet)

The construction of DSSCs requires two glass sheets coated with a transparent conductive substrate which provides support for the deposition of semiconductor and catalyst and also acts as a current collector in DSSCs. The substrate must have the following two properties to achieve effective DSSC performance: (i) the transparency of substrate must be greater than 80% for focusing maximum sunlight to the active area of the DSSCs and (ii) the electrical conductivity of the substrate must also be high to produce an efficient transfer of charges and to decrease energy loss in DSSCs (Sima et al. 2010). FTO (Fluorine tin oxide, SnO_2:F) and ITO (Indium tin oxide, In_2O_3:Sn), which consist of a soda-lime glass sheet coated with fluorine tin oxide and indium tin oxide layer respectively, are generally employed as the transparent conductive substrates. ITO film exhibits an average light transmittance >80% and FTO film has a transmittance of nearly 70% in the visible region. On sintering for 2 h at 450° C in the oxygen atmosphere, the sheet resistance of ITO film enhanced from 18 Ω cm^{-2} to 52 Ω cm^{-2} while that of FTO film remained constant (i.e. 8.5 Ω/cm^2). This property of low and temperature-independent sheet resistance made FTO-based photoanodes highly suitable for DSSCs (Murakami et al. 2004). Due to being less expensive and flexible, polymers can be treated as an alternative to the glass substrate. Although polyethylene terephthalate (PET) and polyethylene napthalate (PEN) coated with ITO have reported 3.8% and 7.8% efficiencies, respectively (Ito et al. 2006; Weerasinghea et al. 2013), the restrictions in the accessible temperature range prohibited the use of polymers as substrates in DSSCs (Jun et al. 2007). Metals like tungsten, titanium, and stainless steel can also be used as substrates (Lupan et al. 2010). On using stainless steel as substrate, an efficiency of 6.1%

has been observed. However, corrosion caused by electrolytes and their high cost limit the use of these metals as substrates in DSSCs.

9.3.3.2 Working Electrode

The working electrode is developed by applying a thin paste of semiconducting materials such as titanium dioxide (TiO_2), zinc oxide (ZnO), niobium pentaoxide (Nb_2O_5), silicon dioxide (SiO_2), and tin dioxide/stannic dioxide (SnO_2) (Han et al. 2009; Fukai et al. 2007), etc. on a transparent conducting glass sheet usually made of FTO or ITO. The electron transport rate which affects the efficiency of DSSC depends on the morphological, crystallinity, and surface area of semiconducting materials. Among several semiconducting materials, TiO_2 is considered an ideal semiconductor due to its good morphological and photovoltaic features, non-toxic nature, low cost, easy availability, and greater efficiency than ZnO and SnO_2 based DSSC (Park et al. 2000). Out of two allotropic forms of TiO_2, viz. anatase and rutile, anatase-based DSSCs are more efficient than the rutile form due to some interesting features of the anatase form such as high energy band gap of 3.2 eV, better chemical stability, fast electron transport rate, and more short-circuit current density (Li et al. 2005). The semiconducting materials assign a large surface area for the deposition of dye molecules. The semiconductors receive electrons from the excited state of the dye and transfer them to the external circuit to conduct electric current (Dai and Rabani 2002).

9.3.3.3 Dye (or Photosensitizer)

Dye or photosensitizer is one of the important components of DSSC that absorb maximum photos from the incident light. The semiconducting layer deposited on a transparent conducting glass plate absorbs only a small fraction of light. Thus, to enhance the absorption process at the surface of the semiconductor, the working electrode is immersed in a dye solution. Because of the highly porous structure and large surface area provided by the semiconductor, a large number of dye molecules get attached to its surface. Until now synthetic dyes have been widely used in the fabrication of DSSC; the overall power conversion efficiency of DSSC using Ruthenium complexes dye was reported as 11 to 12%. Although Ruthenium complex is the most stable and effective dye, it is expensive, toxic, and difficult to produce. Therefore, natural dyes, which can be extracted from flowers, roots, plants, and fruit have been developed. Despite their lower stability and efficiency than synthetic dyes, several features of natural dyes, like their abundance in nature, easy production, non-toxicity, and low production cost (Neale et al. 2005), attract research, and various modifications have been performed to improve the efficiency of DSSC. A photosensitizers or dye should have the following characteristics (Kusama and Arakawa 2005):

- Dye should have the property of luminescence.
- It should absorb the light in ultraviolet-visible (UV-vis) and near-infrared (NIR) regions.
- It should easily transfer the electron from the lowest unoccupied molecular orbit to the semiconductor (i.e. TiO_2).
- Dye should be compatible with the electrolyte (i.e. the highest unoccupied molecular orbital should lie lower).

- Dye should have high chemical stability which enhances the durability (long-term stability) of cells.
- Dye should have the minimum aggregation in between dye and semiconductor surfaces. The aggregation of dye molecules near the dye semiconductor interface limits the recombination reaction (lowers the cell efficiency).

9.3.3.4 Counter Electrode

The function of the counter electrode in DSSC is to regenerate the electrolyte. The oxidized electrolyte is propagated towards the counter electrode where it receives the electrons from the external circuit. A catalyst is deposited on the transparent conducting glass plate which speeds up the reduction process of electrolyte. Platinum (Pt) is generally used as a catalyst for counter electrodes due to its high catalytic ability, high exchange current density, high efficiency, and transparency. However, Pt is very expensive and less abundant. Other materials which can be used as a catalyst for counter electrodes are graphene and conductive polymers (Grätzel 2004). But they show lower efficiency as compared to platinum catalysts.

9.3.3.5 Electrolyte with Redox Couple

The DSSCs are thin-layer PV cells formed by sandwiching the soluble redox couple [(iodide/triiodide(I^-/I_3^-) or Co^{2+}/Co^{3+}]-based electrolyte in between two electrodes, photoanode and photocathode (photoanode—a mesoporous TiO_2-coated semiconducting layer with a photosensitizer; photocathode—platinum/graphite-coated counter electrode). The electrolyte having redox couple is one of the key components of DSSC which greatly affect its performance and durability. An electrolyte consisting of redox couple should have the following properties as follows (Nogueira et al. 2004; Andrade et al. 2011):

- Redox couples present in the electrolyte should be able to regenerate the oxidized dye efficiently
- Electrolyte should have high chemical, thermal, mechanical, and electrochemical stabilities
- It should have a non-corrosive property with the other components of DSSCs
- Electrolyte should have a high electrical conductivity ($\sim 10^{-4} - 10^{-2}$S/cm), a proper electrode-dye interface that also permits the fast diffusion of charge emission
- Redox couples should have low volatility and minimum corrosive property
- The compatibility of electrolyte (high chemical stability) with dye is required to prevent the degradation of dye molecules, hence electrolyte must be compatible with the dye

The essential role of an electrolyte in DSSC is to regenerate the dye molecules after the electrons from the excited state of dye are injected into the conduction band of the semiconductor. The long-term stability of the DSSC is influenced by the properties of the electrolyte. In the last few decades, the developments of DSSCs are quite good in terms of efficiency, but still, they need further modification in terms of durability, performance, and conversion efficiency, etc. Thus, to enhance the conversion efficiency and other features of these cells, researchers have to focus on the development of electrolytes, electrode materials, and sensitizers. To improve the stability and performance of the DSSCs different electrolytes such as gel electrolytes, quasi-solid state electrolytes, ionic liquid as electrolytes, ionic liquid-based polymer electrolytes, composite polymer electrolytes, etc. have been used.

Liquid electrolytes (i.e. organic solvent/ionic liquid-based electrolytes) were widely used and investigated in DSSCs due to the following properties like low viscosity, high conversion efficiency, and fast ion diffusion easy to design (O'Regan and Grätzel 1991; Dresselhaus and Thomas 2001; Nazeeruddin et al. 1993). DSSC based on liquid electrolyte has a high conversion efficiency of the order of 13% (Mathew et al. 2014). The constituents of the electrolytes include organic solvent/ionic liquid, redox couple, and some additives. Organic solvents may be nitrile like acetonitrile, valeronitrile, 3-methoxy propionitrile and some esters like propylene carbonate (PC), ethylene carbonate (EC), and γ-butyrolactone (GBL), etc. In place of redox couple generally, I^-/I_3^- couple is used. Whereas some other redox couple was also used like Br^-/Br_2, $SCN^-/(SCN)_2$, Co^{2+}/Co^{3+}, and $SeCN^-/(SeCN)_2$ (Bergeron et al. 2005; Sapp et al. 2002; Oskam et al. 2001). But the performance of the couple I_3^-/I^- is much better than the others. The commonly used additives are 4-tert-butyl pyridine (TBP) and N-methyl benzimidazole (NMBI). The addition of these additives in the electrolyte may suppress the dark current and improve the conversion efficiency. But the DSSCs based on organic electrolytes have several disadvantages like less durability, leakage problems, volatility in nature, chemical instability (Oskam et al. 2001), etc.

9.4 IONIC LIQUIDS (ILs) AS ELECTROLYTES IN DSSCs

Ionic liquids, also known as room-temperature molten salts, are very exciting materials to use as additives in the electrolyte for DSSC fabrication. ILs have several advantages over organic solvents such as high chemical and thermal stability, high ionic conductivity, negligible vapor pressure, non-flammability, wide electrochemical stability window, and ions (charge carriers) are present in dissociated form. The DSSCs have high potential applications to be commercialized due to their simplicity in fabrication and low cost. The use of nanometer-sized TiO_2 nanofiller having efficiency ~7–8% with the electrode material was a breakthrough in the development of DSSCs in 1991 (O'Regan and Grätzel 1991). An organic solvent like acetonitrile was used to form an electrolyte to develop DSSCs. But it has a low boiling temperature and easily vaporizes at room temperature. Later on, organic solvents having high boiling temperatures like ethylene carbonate (EC), propylene carbonate (PC), valeronitrile, 3-methoxy propionitrile (MPN), and N-methyl-2-pyrrolidone (NMP) were used as stable electrolytes (Wu et al. 2015). Nevertheless, DSSCs still face several problems related to the fabrication of DSSCs at a large scale due to their leakage problem, chemical instability problem, sealing problem (Harikisun and Desilvestro 2011), etc.

As a result, several substitutes for organic solvent were discovered as additives in the electrolyte. In all the solvents, ionic liquids (ILs) were identified as ideal materials. ILs have several advantages over the other organic solvents stated as follows (Eftekhari et al. 2016; Hosseinnezhad et al. 2017; Gorlov and Kloo 2008):

- High ionic conductivity
- Almost negligible vapor pressure
- Large liquids range
- High thermal and chemical stability
- Large chemical stability window
- Non-flammability
- Poor coordination between cations and anions and no need for solvent for dissociation of its cations and anions

Due to their excellent properties, ILs have quickly gained popularity and are accepted as green solvents for several potential applications as well as industrial applications (Omara et al. 2017). For the development of an electrolyte, imidazolium-based ionic liquids are commonly used, and by using ILs several solid and gel-based electrolytes have been discovered. The ILs are also known as designer solvents because by tailoring the size of cations/anions, millions of types of ILs can be synthesized with desired properties. Imidazolium cation-based ILs are commonly used due to their several advantages like low viscosity, high ionic conductivity, high diffusivity, high thermal and chemical stability compared with pyridinium ammonium and pyrrolidium cation-based ILs (Murugesan et al. 2014).

Liquid electrolytes have been used in most electrochemical devices. Using liquid electrolytes as a medium for charge transportation, DSSC achieves the highest conversion efficiency. But the problems of leakage and volatilization of solvent, photo degradation and adsorption of dye, and corrosion of counter electrodes led to developing the best alternative to liquid electrolytes. The polymer electrolyte not only eliminates the limitations of liquid electrolyte but also comes with further advantages including good electrode-electrolyte interface, lesser weight making them more portable, freedom over shape and size, etc. For the preparation of polymer electrolytes, suitable salt is dissolved in a polymer having a high molecular weight (film-forming ability). The use of salt reduces the degree of crystallinity of the polymer network and hence increases the ionic conductivity.

9.5 POLYMER ELECTROLYTE (PE) IN DSSCs

For the development of PV devices like DSSCs, polymer electrolytes are probable candidates due to their unique ionic conductivity, thermal stability, chemical stability, and mechanical flexibility. For the development of DSSCs, PEs are optimized as probable applicants to replace the liquid electrolytes. In photovoltaic cells, PEs are in the form of solid state or quasi-solid-state-like gel (Su'ait et al. 2015; Karuppannan et al. 2015; Meng et al. 200)3. The PEs have excited lower electrical conductivity at room temperature than the liquid electrolytes but one can enhance the electrical conductivity with other desired properties by adding nanofillers (ceramic fillers like Al_2O_3, SiO_2, TiO_2, carbon nano tubes (CNTs)) (Kim et al. 2009; Singh and Saroj 2021, 2015; Kumar et al. 2019), etc. Several polymer electrolytes are based on poly(ethylene oxide) (PEO), poly(methyl methacrylate) (PMMA), poly(vinyl pyrrolidone) (PVP), poly(vinyl chloride) (PVC), poly poly(vinyl alcohol) (PVA), poly(vinyl fluoride) (PVdF), poly(vinylfl uoride-hexafl uoro propylene) (PVdF- HFP), etc. In DSSCs, the PEs are used as solvent-free electrolytes and these materials have several advantages over liquid electrolytes such as being leakage-free, non-corrosive, non-volatile, thermally stable, and photo stable and having good electrochemical stability (MacCallum and Vincent 1989). In these materials charges (i.e. cations/anions) are chemically bonded with the polymeric chains whereas counter ions (solvated by a high dielectric constant solvent) are free to move in the network. In gel electrolytes, polymer and salt are mixed with an appropriate solvent having a wt% ratio greater than 50% with the additives (polymer acts as a stiffener and provides a network) and cations/anions are free to move through the free volume present in the matrix (liquid-like phase). For solar cells applications, the use of polymer electrolytes began in 1999, and the first DSSC based on polymer electrolytes was reported by Nogueira et al. in 1999 (Nogueira et al. 1999). They assembled the DSSC using poly(o-methoxyaniline) as a sensitizer and co-polymer of PEO (i.e. poly(epichlorohydrin-co-ethylene oxide)), poly(ECH-co-EO) containing NaI/I_2 as electrolyte. Its reported conversion efficiency was ~1.3%, and the I-V curve with V_{oc} = 0.71 volt, J_{sc} = 0.46 mAcm^{-2} is shown in Figure 9.4 (Nogueira et al. 1999). Later on in 2000, the same material was used for DSSC application with Ruthenium complex sensitizer and reported the overall conversion efficiency 0.22% with open-circuit voltage 0.71 Volt and short-circuits current ~0.46 mAcm^{-2} (Nogueira et al. 2000). After that, several efforts have been made to develop new polymer electrolytes that could increase the conversion efficiency and other performance of DSSCs.

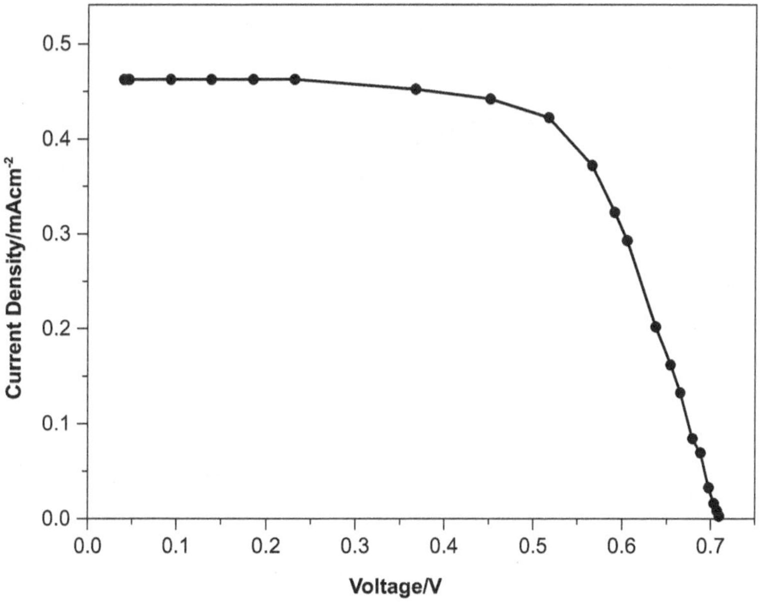

FIGURE 9.4 Current density and voltage characteristics curve of poly(ECH-co-EO)-based polymeric electrolyte photovoltaic cell under 120 mW cm^{-2}.

Source: Reproduced with permission from Nogueira et al. [2000, © Elsevier 2000].

9.6 PLASTICIZED OR GEL POLYMER ELECTROLYTES IN DSSCS

Ionic conductivity of PEs can be enhanced by increasing inorganic/organic salt concentration (lithium perchlorate, $LiClO_4$, sodium chloride, NaCl, lithium hexafluorophosphate, $LiPF_6$, lithium tetrafluoroborate, $LiBF_4$, etc.) and also by adding the low molecular weight organic liquids like low molecular weight polyethylene glycol (PEG-200/400), PC, EC, dimethyl formamide (DMF), dimethyl sulfoxide (DMSO) (Saroj et al. 2017; Teo et al. 2020). The use of plasticizers reduces the crystalline phase of the polymer matrix and hence the amorphous network results in an increase in ionic conductivity with mechanical stability. In Figure 9.5, J-V characteristics for DSSC containing polymer electrolyte of poly(ethylene-coepichlorohydrin) [P(EO-EPI)], LiI/I_2 with and without plasticizer γ-butyrolactone (GBL) is presented (Freitas et al. 2009).

9.7 NANOPARTICLES IN ELECTROLYTES FOR DSSCs

Nanoscience and nanotechnology play an important role to improve the power conversion efficiency, durability, and large-scale manufacturability of DSSCs. The materials having a size less than 100 nm are called nanomaterials. Generally, these materials can be divided into three categories: (i) metal oxide, metal carbide, sulfide, and nitride-based nanofillers; (ii) carbonaceous-based nanofillers; and (iii) clay-based nanofillers (Ali et al. 2016). Besides the use of conventional plasticizers like EC, PC, PEG, etc. and ionic liquids (ILs), other additives like metal-based nanomaterials such as SiO_2, TiO_2, aluminum oxide (Al_2O_3), ZnO, magnetite (Fe_3O_4), etc. can be explored to improve the properties of polymer electrolytes. By using these fillers,

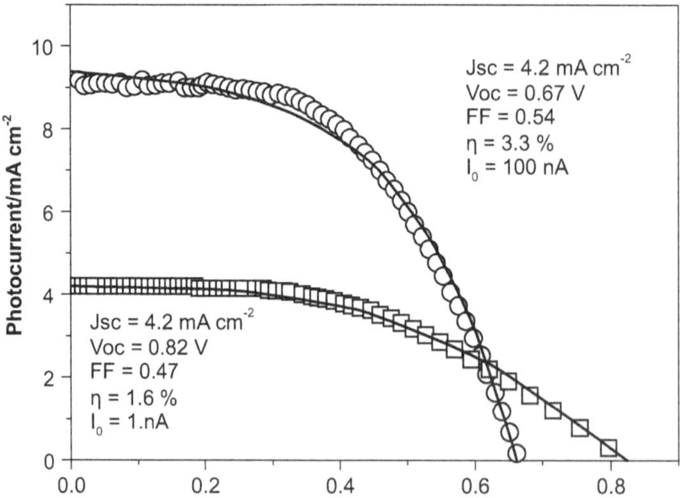

FIGURE 9.5 Photovoltaic curve (J-V) of the DSSC containing P(EO-EPI), LiI/I_2, (○) with and (□) without GBL as plasticizer at 100 mW cm^{-2}.

Source: Reproduced with permission from Freitas et al. [2009, © RSC 2009].

mechanical, thermal, electrical transport properties can be enhanced and DSSCs can be developed with high open-circuit voltage and conversion efficiency. The pioneering work was reported by Croce et al. in 1998. In this work, the addition of TiO_2 and other nanofillers was used to improve the ionic conductivity and other properties of polymer electrolytes. It is well established that the use of these nanofillers modifies the conduction mechanism and changes ionic conductivity and mechanical properties, but how these fillers act is still the subject of research. To develop DSSCs based on nanofillers, TiO_2 nanoparticles were frequently used (Venkatesan and Lee 2017; Stergiopoulos et al. 2002; Zhang et al. 2007; Kang et al. 2008). Stergiopoulos et al. in 2002 reported the DSSC based on TiO_2 nanoparticles-doped PEO-based polymer electrolyte having the V_{oc} = 0.66volt, J_{sc} = 7.2mAcm^{-2}, and η = 4.2%. In this system, PEO was used as a polymer network and TiO_2, lithium iodide (LiI), and I_2 are the additives. LiI is used as a salt and I_2 as a redox material. The nanofillers have a large surface area which prohibited the recrystallization and hence decreased the degree of crystallinity of PEO (Singh et al. 2016). TiO_2 nanotubes are also used as nanofillers [75]. Akhtar et al. investigated electrolyte based on poly ethylene glycol (PEG)—10 wt% TiO_2 (nanotubes) (TiNT)—and ionic conductivity was found to be 2.4 × 10^{-3} S/cm. DSSC fabricated using this electrolyte showed open-circuit voltage 0.73, J_{sc} = 9.4 mAcm^{-2}, and η = 4.4% and FF = 0.65 under the illumination power 100 mWcm^{-2} and DSSC performance for PEG—$x$wt%TiNT. x = 5, 10, and 20 is shown in Figure 9.6 (Akhtar et al. 2007). Other nanofillers like Al_2O_3, SiO_2, CNT, and ZnO with different compositions are also used for the developing composite polymer electrolyte and

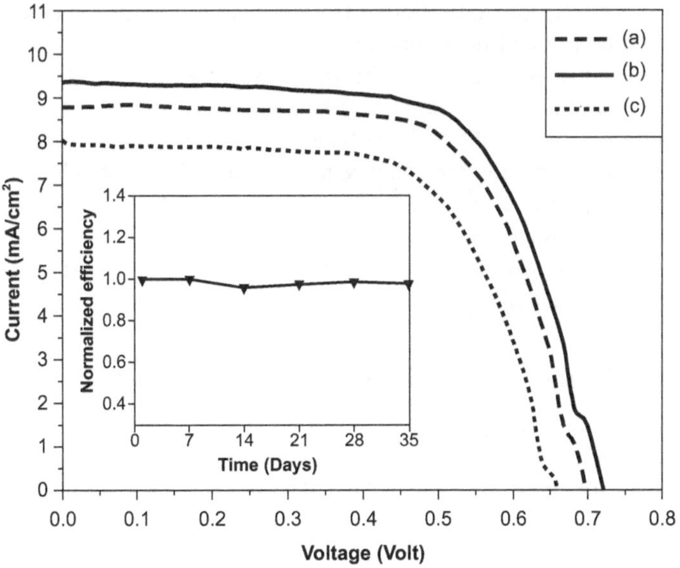

FIGURE 9.6 Current vs. voltage characteristics of DSSC assembled with composite electrolytes of (a) PEG—5 wt% TiNT, (b) PEG—10 wt% TiNT and (c) PEG—20 wt% TiNT.

Source: Reproduced with permission from Akhtar et al. [2007, © Elsevier 2007].

showed very exciting results in DSSCs applications (Venkatesan and Lee 2017). The table shows the different composite systems with other parameters of DSSCs.

9.8 POLYMER COMPOSITES (PCs)

A polymer composite (PC) is a material composed of the polymer as host and one or more additives resulting in a new material having different properties from the individual materials. To prepare PE materials one can use polymers/biopolymers as a network matrix and organic/inorganic nanofillers as an additive. These materials have several properties like light weight; high stiffness; good thermal, chemical, and mechanical properties. Polymer composites are multi-phase materials in which nanofiller integrated with the polymeric chains/functional groups of polymer network via weak interactions, resulting in modified mechanical, chemical, and thermal properties. The performance of PCs generally depends on the amount, size, and type of nanofiller. This also depends on the geometry and orientation of the constituents of polymer in the presence of additives. The use of nanofillers modifies the electrical, dielectric, and interfacial properties of ion/electron-conducting polymer electrolytes. The incorporation of inert fillers/nanofillers into the semi-crystalline/crystalline polymer matrices changes its phase (i.e. crystalline to amorphous or amorphous to crystalline). In the case of composite polymer electrolytes, the particle size and the filler concentrations play an important role because the addition of a small amount of inert filler will collapse the chain organization of the polymers which in turn facilitates higher ionic conduction. From the DSSC application point of view, the electrical conductivity of the PC electrolyte plays a significant role. In PCs electrical conductivity depends on different factors such as amorphous phase/free volume, chains' mobility and flexibility, dielectric property of the material, and also the interactions between nanofillers with the constituents of polymers. A literature survey reveals that the use of nanofillers like SiO_2, TiO_2, Al_2O_3, ZnO, Fe_3O_4, CNT, etc. reduces the crystalline phase of a polymer network, hence reducing melting and glass transition temperature with enhancement in chain flexibility (Venkatesan and Lee 2017; Stergiopoulos et al. 2002). The stability of DSSCs can be improved by using polymer composites. Devices having PCs as electrolytes have different advantages over liquid electrolytes such as minimizing leakage, long-term stability, and minimum solvent evaporation problems.

9.9 METAL OXIDE NANOFILLER-BASED PCs IN DSSCS

The performance of DSSCs depends on different factors such as particle size of nanofillers dispersed in the electrolyte, the ionic conductivity of the electrolyte, and interfacial contact between photoanode and electrolyte. A high-level interfacial contact can be obtained by taking highly mechanically flexible film (gel-like nature) and deep penetration of electrolyte material into the pores of the TiO_2 semiconductor layer (photoanode) (Lee et al. 2008; Zebardastan et al. 2016; Mohan et al. 2013). The particle size of metal oxide-based nanofillers (<100nm) should be smaller than the size of the pores of TiO_2. Metal oxide-based nanoparticles play various roles in polymer electrolytes due to the interaction between the constituent of polymer and nanoparticles. Due to these interactions the ionic conductivity, dissociation of ions, mobility of ion liquidity, and polymeric chain flexibility increase. Interfacial contact also improves because of the increase in the concentration of interfacial defects. The nanoparticle has a large surface area to volume ratio in polymer electrolytes and hence modifies the degree of crystallinity of the polymer network in the electrolyte (increase in interfacial free volume between the polymer network and another constituent of the electrolyte). Silicon dioxide (SiO_2) or silica nanoparticles can be used as nanofiller for the preparation of polymer composite electrolytes using the solution casting method. Generally, the SiO_2 nanoparticle has a three-dimensional structure, and siloxane or silanol Si-O-Si groups (hydrophilic in nature) are created on the silica surface (Yoon et al. 2014; Zhao et al. 2014b). This modified surface of silica has been used from the DSSCs application's point of view. Different methods like the solution casting method, blending method can be used for the preparation of polymer composite films. The polymer composite films prepared by the solution casting method are composed of two phases, one associated with the SiO_2 and another polymer-Si-O-Si mixed phase at the nanoscale level (Lee et al. 2008). Lee et al. [2008] fabricated quasi-solid-state DSSC based on PVdF-HFP-MPN-TBAI/I_2-TBP-SiO_2 has the J_{sc} = 14.4mAcm^{-2}, V_{oc} = 0.71volt, fill factor = 0.598 and overall conversion efficiency of 5.97% (Lee et al. 2008). A novel composite polymer gel electrolyte was prepared using PVdF-HFP-PEO-EC-PC-NaI-I_2-SiO_2 by Zebardastan et al. 2016 and it was found that the prepared sample was amorphous due to the interaction between the fumed SiO_2 and the polymers, as confirmed using XRD and IR analyses. The fabricated DSSC based on 13 wt% SiO_2 had a high efficiency of 9.44% with J_{sc} = 27.31 mAcm^{-2} and plot of J-V characteristics without SiO_2 and with different amounts of SiO_2 (shown in Figure 9.7) (Zebardastan et al. 2016). Mohan et al. 2013 reported a

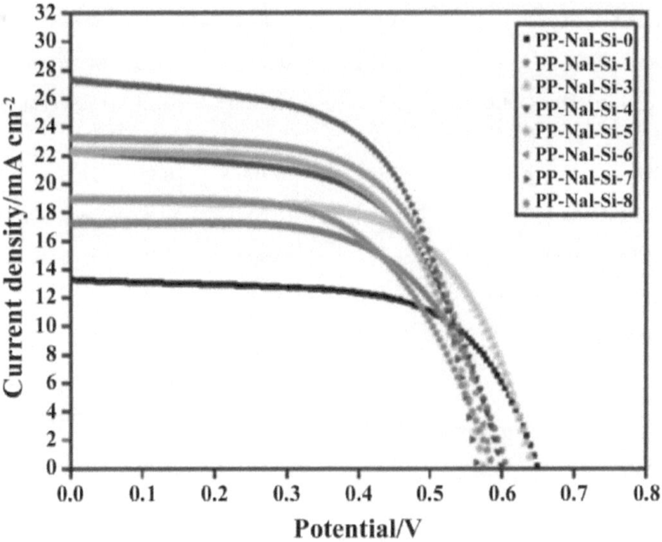

FIGURE 9.7 J-V curve for DSSC fabricated using gel polymer electrolytes without SiO_2 (PP-NaI-Si-0) and with various amounts of SiO_2.

Source: Reproduced with permission from Zebardastan et al. [2016, © Elsevier 2016].

fumed SiO_2-NFs/PAN-based polymer composite electrolyte using a hot pressing technique (Mohan et al. 2013). In this work it was reported that the different sizes and surface areas of SiO_2 nanoparticles highly influenced the properties of the polymer electrolytes. The fabricated cell by using optimized SiO_2 (9 wt%) has a higher value of J_{sc} (12mAcm^{-2}). The ionic conductivity and ion diffusivities depend upon SiO_2 concentration and the bond formation between iodide ions and surface hydroxyl groups of SiO_2 (I-Si-O-Si), as well as free volume created by the nanofillers. Yoon et al. [2014] fabricated DSSCs based on PEGDME-SiO_2 (rod-like shape) and found that the DSSC exhibited higher energy conversion efficiency with J_{sc} = 14.1mAcm^{-2} value due to enhanced diffusion of the redox couple in the presence of the SiO_2 nanoparticles (Yoon et al. 2014). Zhao et al. prepared polymer composite electrolytes based on SiO_2-PEO-PVdF-HFP and fabricated cell exhibits lowest cell impedance (20.96 Ω) and the highest solar conversion efficiency (4.85%) [Zhao et al. 2014b]. Some ionic liquid-based polymer nano-composite materials were also exhibiting very exciting performance of the cell (data are listed in Table 9.1).

TABLE 9.1
PC-Based Electrolytes in DSSCs and Their Performance.

Polymer Composites	Voc (volt)	Jsc (mAcm^{-2})	η %	References
ex-MMT- PNIPAAm	0.73	12.6	5.41	Tu et al. 2008
ex-MMT- PMMA	0.741	15.12	7.77	Lee et al. 2011
APS-ex-MMT- PVdF—HFP/PEO	0.73	7.70	3.8	Prabakaran et al. 2015
MMT-NR^{4+-} P(EO—EPI)	0.49	0.8	3.2	Ito et al. 2008
PVAc-Fe_3O_4@NCs/TiCl	0.76V	14.94	7.22	Mazloum et al. 2019
(PAN-VA)—ACN-LiI-TBP-DMPII-TiO_2-I_2	0.74	18.80	10.58	Chen et al. 2013
PEO/PVdF-TiO_2	0.77	16.10	8.91	Liu et al. 2017
P(VA-co-MMA)-CAN/or MPN	0.74	17.40	9.4	Wang et al. 2013
PEO-TiO_2—EC/PCor PEG	0.80	16.90	9.2	Seo et al. 2014
PVA-PAN- (PAN-VA)—gBL-TiO_2	0.75	15.58	7.87	Venkatesan et al. 2015a
PEO-PEGDME-SnO_2-KI-I_2	0.74	10.80	5.3	Chae et al. 2014
PEG-$CuAlO_2$	0.50	1.06	0.27	Suriwong et al. 2015
PVdF-HFP-TiC	0.77	12.45	6.29	Venkatesan et al. 2016a
PVdF-HFP-TiC -PPN	0.72	16.27	7.70	Venkatesan et al. 2016b
PVdF-HFP-AIN	0.66	12.92	5.27	Huang et al. 2011
PVdF-HFP-CoS	0.73	14.42	7.34	Vijayakumar et al. 2015
PAN-CNT-LiI-TBP-ACN-I_2 With dye $C_{26}H_{16}O_8N_6S_2Ru$	10.90	0.57	3.90	Akhtar et al. 2011

Polymer Composites	Voc (volt)	Jsc ($mAcm^{-2}$)	η %	References
PAN-SiO_2-ACN-LiI-TBP-I_2 with dye $C_{26}H_{16}O_8N_6S_2Ru$	16.12	0.74	7.85	Zhao et al. 2014a
PAN-PVdF-V_2O_5-EC-PC-HDMII-LiI-TBP-I_2 with dye $C_{26}H_{16}O_8N_6S_2Ru$	13.80	0.78	7.75	Sethupathy et al. 2014
PAN-PVdF-SiO_2-EC-PC-HDMII-LiI-TBP-I_2 with dye $C_{26}H_{16}O_8N_6S_2Ru$	11.60	0.79	5.61	Sethupathy et al. 2014
PAN-VA-LiI-ACN-DMPII-TBP-TiO_2-I_2 with dye $C_{26}H_{16}O_8N_6S_2Ru$	15.33	0.74	8.66	Venkatesan et al. 2015b
(PAN-VA)-LiI-ACN-PPN-DMPII-TBP-TiO_2-I_2 with dye $C_{26}H_{16}O_8N_6S_2Ru$	16.84	0.72	8.30	Venkatesan et al. 2015b
(PAN-VA)-LiI-ACN-MPN-DMPII-TBP-TiO_2-I_2 with dye $C_{26}H_{16}O_8N_6S_2Ru$	14.95	0.74	7.80	Venkatesan et al. 2015b
PAN-P(VP-co-Vac)-PC-EC-NaI-Co_3O_4-I_2 with dye $C_{26}H_{16}O_8N_6S_2Ru$	16.20	0.61	6.46	Saidi et al. 2019

Aluminum oxide (Al_2O_3) or alumina is one of the most complex materials and it is easily available in the form of gibbsite and bauxite. Al_2O_3 has a different crystalline structure in which α-Al_2O_3 is thermally stable and γ-Al_2O_3 is most probably used for dye synthesized solar cell applications. γ-Al_2O_3 has a honeycomb-like structure having a large surface-to-volume ratio with catalytic properties. Alumina with different shapes like spheres, rods, and surface-modified Al_2O_3 was used for the electrochemical device application point of view (Chi et al. 2013; Sacco et al. 2015).

Titanium dioxide (TiO_2) (also known as titanium or titania) is a naturally occurring oxide of titanium and a low-cost material having different crystallographic phases like brookite, rutile, and anatase. This material is used in DSSCs as a nanofiller in the form of nanosphere, nanorods, nanotubes for the preparation of polymer composite materials. Stergiopoulos et al. reported on the PEO-TiO_2-based polymer composite material for DSSC applications and it is found that titania nanofiller reduces the crystalline phase of PEO-based matrix due to interaction between the oxygen atom of PEO and the surface of the hydroxyl group of TIO_2 and hence enhances the conductivity of a polymer composite system. The fabricated DSSC based on this material has a J_{sc} =7.2mAcm^{-2}, V_{oc} = 0.66 volt with the efficiency, η = 4.2% (Stergiopoulos et al. 2002). Tiautit et al. [2014] prepared PVdF-HFP/PVA-TiO_2/SiO_2-based gel polymer composite electrolytes and used these materials for DSSCs fabrication, and it was found that the η% of the cells using liquid electrolyte and PGE were 3.49% and 3.26%, respectively (Tiautit et al. 2014). The use of ionic liquid enhances the performance of the device due to the interaction between the cations (imidazolium) and the surface (O-H group) of TiO_2 which aligns the charge carrier species by electrostatic interaction and, therefore, facilitates the charge transportation via the ion-exchange mechanism. Huo et al. reported the TiO_2 nanoparticles—PVdF-HFP-based gel electrolyte based quasi-solid-state dye-sensitized solar cells—had an efficiency of η = 7.18%, V_{oc} = 0.69 volt, and J_{sc} = 15.23 mAcm^{-2} (Huo et al. 2007). Venkatesan et al. [2018] prepared quasi-solid-state DSSC based on PEO-PMMA-TiO_2-based printable electrolyte. TiO_2 nanofiller was added into the polymer blend composite, and it was found that the efficiency of the cell was 8.48%, which was higher than the efficiency of PEO-based polymer electrolyte (7.63%) and liquid electrolyte (8.32%). The efficiency of the cell increases to 9.12% by adding TiO_2 nanofiller in PE due to an increase in electrical conductivity (Venkatesan et al. 2018).

Zinc oxide is an inorganic compound with the chemical formula ZnO (zinc and oxygen centers are tetrahedral). It is a wide energy gap semiconducting material having similar properties to TiO_2. This material is rarely used in DSSC fabrication. Zhang et al. used PEGME grafted nanoparticle-based polymer composite material for fabrication of DSSC, and the fabricated cell has the J_{sc} = 15.85 mAcm^{-2}, V_{oc} = 0.60 volt, and efficiency of η = 6.40% (Zhang et al. 2006). Xia et al. fabricated a DSSC based on PMOP-capped ZnO polymer composite electrolyte and it was found that the cell has the J_{sc} = 17.60 mAcm^{-2}, V_{oc} = 0.64volt, and efficiency of η = 6.90% (Xia et al. 2007).

Tin oxide (SnO_2), copper aluminum oxide ($CuAlO_2$), nickel oxide (NiO), and cobalt oxide (Co_3O_4) were also used in electrolytes for DSSC fabrications. The performances of DSSCs based on polymer composite materials are listed in Table 9.1. Some metal carbide, metal sulfide, and metal nitride-based nanoparticles like titanium carbide (TiC), cobalt sulfide (CoS), and aluminum nitride (AlN) were also used in the electrolyte to enhance the electrical conductivity and performance of DSSCs (Table 9.1). From the data listed in Table 9.1, it is clear that these materials play a significant role to enhance the electrical conductivity of polymer composite electrolytes and the performance of the cell.

9.10 CARBONACEOUS NANOPARTICLES-BASED PCS IN DSSCs

Recently, carbonaceous material-based polymer composite electrolytes have received much attention for their role in the development of electrochemical devices and high efficient DSSCs. The conductivity of an electrolyte plays a crucial role to enhance the efficiency and performance of the cell. Carbonaceous materials have several advantages due to their structural, electrical, electrochemical, and mechanical properties. The properties of these materials depend on particle size, shape, porosity, and functional groups attached to the surface. These materials used in DSSCs can be categorized in different types such as (i) spherical carbon nanofillers like activated carbon (AC), carbon black (CB), carbon sphere (CS); (ii) carbon nanotubes (CNTs) like single-walled CNTs, double-walled CNTs, and multi-walled CNTs; and (iii) layered carbon materials such as graphene (Gr), graphene oxide (GrO), graphite (G). Activated carbon (AC) is a crude form of graphite and highly porous material having a wide range of porosity with a large surface area to volume ratio. These materials are generally prepared by chemical or physical activation processes by using coal and wood. This is low-cost material and can be easily physically activated in the lab by using air, oxygen, carbon dioxide, or other gases as oxidizing reagent/gases. For chemical activation, some chemicals like KOH, NaOH, H_3PO_4 can be used. In these materials existence of pores makes them suitable for the preparation of polymer composite electrolytes, and porous polymer composites can be filled with an organic solvent or ILs. Due to the presence of organic solvent/IL in the pores of the polymer matrix, the electrical conductivity enhances and hence the ion-charge exchange process becomes easier. Mohan et al. [2011] reported a quasi gel-electrolyte based on PAN/PEG with activated carbon, ethylene carbonate, propylene carbonate, lithium iodide, iodine, 4-tertinary butyl pyridine, 1-N-butyl-3-hexyl imidizolium iodide along with acetonitrile and tetrahydrofuran binary solvent. The fabricated DSSC has the J_{sc} = 13.00 mAcm^{-2}, V_{oc} = 0.76 volt, and η = 6.55% (Mohan and Murakami 2011). Mohan et al. [2013] also reported composite gel polymer electrolyte based on poly(acrylonitrile) (PAN)/LiI/activated carbon and the fabricated DSSC based on this material has the efficiency of 8.42% with V_{oc} = 0.74 volt and J_{sc} = 18.70mAcm^{-2}, and DSSC performance of polymer electrolytes with different amounts of activated carbon are shown in Figure 9.8 (Mohan et al. 2013). Carbon black (CB) is a polycrystalline form of carbon with a high surface area to volume ratio and this material can be used with ionic liquid for the preparation of polymer composite electrolytes. Some other forms of carbon black—like carbon spear (CS), carbon dots (Cdts)—are also used with ionic liquid for electrolytes in DSSCs. Ikeda et al. [2006] prepared to conduct polymer-carbon imidazolium-based IL-based composite and fabricated DSSC, which have J_{sc} = 12.8 mAcm^{-2}, V_{oc} = 0.58 volt and efficiency of h = 3.48% (Ikeda et al. 2006).

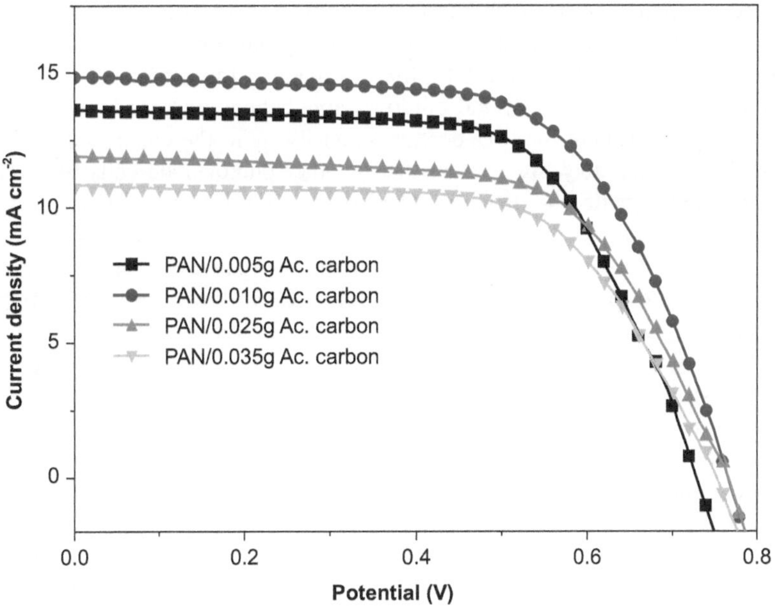

FIGURE 9.8 Current density vs. potential plot for DSSC fabricated with composite gel polymer electrolytes PAN, LiI, and different amounts of activated carbon.

Source: Reproduced with permission from Mohan et al. [2013, © RSC Publishing 2013].

9.11 CLAY NANOFILLERS-BASED PCS IN DSSCs

Clay-based nanoparticles are also used for the preparation of polymer composite electrolytes in DSSC fabrication. Some natural and synthetic montmorillonite (MMT), talc, nanomica, layered double hydroxides (LDHs), laponite (LAP), and saponite (SAP) are used to prepare PCs (Venkatesan and Lee 2017). Clay nanoparticles are naturally abundant, swelling in nature, have high ion exchangeability, high chemical stability, rheological properties, and easily synthesizing materials. The platelet-like structure, morphology, ionic nature, and particle size play various roles to enhance the desired properties of PCs as well as the performance of DSSCs (Teo et al. 2020; Venkatesan and Lee 2017). Cationic clays like MMT, mica, and talc modify the polymer network and increase the free volume/space within the matrix which provides a wider path for charge transportation, hence the movement of charged particles becomes faster (Venkatesan and Lee 2017; Gong et al. 2012). Several clays-based PCs are available in the literature which reveals the importance of these materials in DSSC applications (Table 9.1). The fabricated DSSC based on PVdF-HFP/ZnAl-CO_3-LDH nanocomposite polymer electrolyte had a high efficiency of ~8.11% (Ho et al. 2014). Table 9.1 lists the performance of DSSCs based on clay nanoparticles (Ho et al. 2014; Prabakaran et al. 2015; Saidi et al. 2019; Chen et al. 2013).

9.12 CONCLUSIONS

DSSCs are promising photovoltaic devices that have attracted many researchers for the development of next-generation renewable and sustainable energy sources because of their feasible properties such as low production cost, flexibility, eco-friendliness, ease of fabrication, long life, mechanical robustness. They also have the potential for both indoors and outdoors applications and can work under the low intensity of incident light. Under low-intensity light, the conversion efficiency of DSSCs is comparably higher than that of other photovoltaic cells like silicone solar cells and organic photovoltage. The liquid electrolyte-based DSSC has high efficiency (14.3%) but the liquidus nature of electrolytes has several problems, like leakage, chemical instability, and evaporation of the organic solvents, which reduce the durability and performance of the cell. To resolve these drawbacks, polymer electrolyte/polymer composite electrolytes can be used as electrolytes in place of liquid electrolytes due to their promising physical and electrochemical properties. Unlimited varieties of materials are possible in PCs with higher levels of conversion efficiency and durability. TiO_2 nanofiller-based polymer composite DSSC has an efficiency of about 10.58%. For a continuous power supply from the cell, the long-term performance of DSSCs plays a very important role and this can be obtained by using nanocomposite polymer/biopolymer electrolytes. In the future, biopolymer/polymer composite electrolytes with hybrid nanoparticles will play a crucial role in determining the conversion efficiency and performance of DSSCs for both indoors and outdoor applications.

List of Acronyms

CAN—acrylonitrile
AlN—aluminum nitride
APS—aminopropyl trimethoxy silane
DMPII—1,2-Dimethyl-3-propylimidazoliuum iodide
ex-MMT—exfoliated montmorillonite
GBL—gamma-butyrolactone
GuanSCN—guanidinium thiocyanate
HDMII—1-hexyl-**2**,3-dimethyl imidazolium iodide
Fe_3O_4@)NCs/TiCl—magnetic core-shell nanocellulose/titanium chloride
MPN—methoxy propionitrile
PMII—1-methyl 3-propyl imidazolium iodide
NCs/TiCl—nanocellulose/titanium chloride
Nb_2O_5—niobium pentaoxide
PMOP—1-phenyl-3-methyl-4-octadecyl-5-pyrazolone
PAN—polyacrylonitrile
(PAN-VA)—poly(acrylonitrile-co-vinyl acetate)
PEGDME—poly(ethylene glycol) dimethyl ether
P(EO-EPI)—poly(ethylene oxide-coepichlorohydrin)
PNIPAAm—poly(n-isopropyl acrylamide)
PVAc—polyvinyl acetate
P(VA-co-MMA)—polyvinyl (acetate-co-methyl methacrylate)
P(VP-co-VAc)—poly(1-vinylpyrrolidone-co-vinyl acetate)
KI—potassium iodide
PPN—propionitrile
$C_{26}H_{16}O_8N_6S_2$—rucis-diisothiocyanato-bis (2,2'-bipyridyl-4,4'-dicarboxylicacid) ruthenium
NaI—sodium iodide
TBAI—tetrabutylammonium iodide TBP-4-Tert-butylpyridine

ACKNOWLEDGMENTS

The author ALS is grateful to SERB, India for providing financial assistance through MRP-EEQ/2018/000862.

REFERENCES

Akhtar, M.S., Chun, J.M. and Yang, O.B. 2007. Advanced composite gel electrolytes prepared with titania nanotube fillers in polyethylene glycol for the solid-state dye-sensitized solar cell. *Electrochemistry Communications.* 9(12): 2833–2837.

Akhtar, M.S., Li, Z.Y., Park, D.M., Oh, D.W., Kwak D.H. and Yang O.B. 2011. A new carbon nanotube (CNTs)—poly acrylonitrile (PAN) composite electrolyte for solid state dye sensitized solar cells. *Electrochim Acta.* 56: 9973–9979.

Ali, A., Hira, Z., Zia, M., ul Haq, I., Phull, A.R., Ali, J.S. and Hussain, A. 2016. Synthesis, characterization, applications, and challenges of iron oxide nanoparticles. *Nanotechnology, Science and Applications*. 9: 49–67.

Ameri, T., Li, N. and Brabec, C.J. 2013. Highly efficient organic tandem solar cells: A follow up review. *Energy Environ. Sci*. 6: 2390–2413.

Andersen, N.A. and Lian, T. 2005. Ultrafast electron transfer at the molecule-semiconductor nanoparticle interface. *Annual Review of Physical Chemistry*. 56: 491–519.

Andrade, L., Ribeiro, H.A. and Mendes, A. 2011. Dye-sensitized solar cells: An overview. *Encyclopedia of Inorganic and Bioinorganic Chemistry*: 1–20.

Bergeron, B.V., Marton, A., Oskam, G. and Meyer, G.J. 2005. Dye-sensitized SnO_2 electrodes with iodide and pseudohalide redox mediators. *The Journal of Physical Chemistry B*. 109(2): 937–943.

Cao, Y., Liu, Y., Zakeeruddin, S.M., Hagfeldt, A. and Grätzel, M. 2018. Direct contact of selective charge extraction layers enables high-efficiency molecular photovoltaics. *Joule*. 2: 1108–1117.

Carlson, D.E. and Wronski, C.R. 1976. Amorphous silicon solar cell. *Applied Physics Lett*. 28: 671–673.

Chae, H., Song, D., Lee, Y.G., Son, T., Cho, W., Pyun, Y.B. et al. 2014. Chemical effects of tin oxide nanoparticles in polymer electrolytes-based dye-sensitized solar cells. *J. Phys. Chem. C*. 118: 16510–16517.

Chebrolu, V.T. and Kim, H. 2019. Recent progress in quantum dot sensitized solar cells: An inclusive review of photoanode, sensitizer, electrolyte, and the counter electrode. *J. Mater. Chem. C*. 7: 4911–4933.

Chen, C.L., Chang, T.W., Teng, H., Wu, C.G., Chen, C.Y., Yang, Y.M. et al. 2013. Highly efficient gel-state dye-sensitized solar cells prepared using poly(acrylonitrile-co-vinyl acetate) based polymer electrolytes. *Phys Chem*. 15: 3640–3645.

Chi, W.S., Roh, D.K., Kim, S.J., Heo, S.Y. and Kim, J.H. 2013. Hybrid electrolytes prepared from ionic liquid-grafted alumina for high-efficiency quasi-solid-state dye sensitized solar cells. *Nanoscale*. 5: 341–5348.

Choubey, P.C., Oudhia, A. and Dewangan, A. 2012. A review: Solar cell current scenario and future trends. *Recent Research in Science and Technology*. 4(8): 99–101.

Dai, Q. and Rabani, J. 2002. Unusually efficient photosensitization of nanocrystalline TiO_2 films by pomegranate pigments in aqueous medium. *New Journal of Chemistry*. 26(4): 421–426.

Dekkers, H.F.W., Carnel, L. and Beaucarne, G. 2006. Carrier trap passivation in multicrystalline Si solar cells by hydrogen from SiNx:H layers. *Appl. Phys. Lett*. 89: 2004–2007.

Dette, C., Pérez-Osorio, M.A., Kley, C.S., P. Punke, P., Patrick, C.E., Jacobson, P. et al. 2014. TiO_2 anatase with a bandgap in the visible region. *Nano Lett*. 14: 6533–6538.

Dresselhaus, M.S. and Thomas, I.L. 2001. Alternative energy technologies. *Nature*. 414: 332–337.

Eftekhari, A., Liu, Y. and Chen, P. 2016. Different roles of ionic liquids in lithium batteries. *Journal of Power Sources*. 334: 221–239.

Fakharuddin, A., Jose, R., Brown, T.M., Fabregat-Santiago, F. and Bisquert, J. 2014. A perspective on the production of dye-Sensitized solar modules, *Energy Environ. Sci*. 7: 3952–3981.

Feldt, S.M., Gibson, E.A., Gabrielsson, E., Sun, L., Boschloo, G. and Hagfeldt, A. 2010. Design of organic dyes and cobalt polypyridine redox mediators for high-efficiency dye sensitized solar cells. *J. Am. Chem. Soc*. 132: 16714–16724.

Freitas, J.N.D., Nougeria, A.F. and Paoli, M.A. 2009. New insights into dye-sensitized solar cells with polymer electrolytes. *Journal of Materials Chemistry*. 19: 5279–5294.

Fukai, Y., Kondo, Y., Mori, S. and Suzuki, E. 2007. Highly efficient dye sensitized SnO_2 solar cells having sufficient electron diffusion length. *Electrochemistry Communications*. 9(7): 1439–1443.

Gong, J., Liang, J. and Sumathy, K. 2012. Review on dye-sensitized solar cells (DSSCs): Fundamental concepts and novel materials. *Renewable and Sustainable Energy Reviews.* 16(8): 5848–5860.

Gong, J., Sumathy, K., Qiao, Q. and Zhou, Z. 2017. Review on dye-Sensitized solar cells (DSSCs): Advanced techniques and research trends. *Renew. Sustain. Energy Rev.* 68: 234–246.

Gorlov, M. and Kloo, L. 2008. Ionic liquid electrolytes for dye-sensitized solar cells. *Dalton Transactions.* 20: 2655–2666.

Grätzel, M. 2004. Conversion of sunlight to electric power by nanocrystalline dye-sensitized solar cells. *Journal of Photochemistry and Photobiology A: Chemistry.* 164(1–3): 3–14.

Han, D.W., Heo, J.H., Kwak, D.J., Han, C.H. and Sung, Y.M. 2009. Texture, morphology and photovoltaic characteristics of nanoporous F: SnO_2 films. *Journal of Electrical Engineering and Technology.* 4(1): 93–97.

Hara, K., Sato, T., Katoh, R., Furube, A., Ohga Y., Shinpo, A., Suga, S. et al. 2003. Molecular design of coumarin dyes for efficient dye-sensitized solar cells. *Journal of Physical Chemistry B.* 107(2): 597–606.

Harikisun, R. and Desilvestro, H. 2011. Long-term stability of dye solar cells. *Solar Energy.* 85(6): 1179–1188.

Ho, H.W., Cheng, W.Y., Lo, Y.C., Wei, T.C. and Lu, S.Y. 2014. Layered double hydroxides as an effective additive in polymer gelled electrolyte based dye-sensitized solar cells. *ACS Appl. Mater. Interfaces.* 6: 17518–17525.

Hosseinnezhad, M., Gharanjig, K., Moradian, S. and Saeb, M.R. 2017. In quest of power conversion efficiency in nature-inspired dye-sensitized solar cells: Individual, co-sensitized or tandem configuration. *Energy.* 134: 864–870.

Huang, K.C., Chen, P.Y., Vittal, R. and Ho, K.C. 2011. Enhanced performance of a quasi-solid-state dye-sensitized solar cell with aluminum nitride in its gel polymer electrolyte. *Sol. Energy Mater. Sol. Cells.* 95: 1990–1995.

Huo, Z., Dai, S., Wang, K., Kong, F., Zhang, C., Pan, X. et al. 2007. Nanocomposite gel electrolyte with large enhanced charge transport properties of an I^{3-}/I-redox couple for quasi-solid-state dye-sensitized solar cells. *Sol. Energy Mater. Sol. Cells* 91: 1959–1965.

Ikeda, N., Teshima, K. and Miyasaka, T. 2006. Conductive polymer—carbon—imidazolium composite: A simple means for constructing solid-state dye-sensitized solar cells. *Chem. Commun.* 1733–1735.

Ito, B.I., De Freitas, J.N., De Paoli, M.-A. and Nogueira, A.F. 2008. Application of a composite polymer electrolyte based on montmorillonite in dye-sensitized solar cells. *J. Braz. Chem. Soc.* 19: 688–696.

Ito, S., Cevey Ha, N.L., Rothenberger, G., Liska, P., Comte, P., Zakeeruddin, S.M., Péchy, P. et al. 2006. High-efficiency (7.2%) flexible dye-sensitized solar cells with Ti-metal substrate for nanocrystalline-TiO_2 photoanode. *Chemical Communications.* 38: 4004–4006.

Jun, Y., Kim, J. and Kang, M.G. 2007. A study of stainless steel-based dye-sensitized solar cells and modules. *Solar Energy Materials and Solar Cells.* 91(9): 779–784.

Kakiage, K., Aoyama, Y., Yano, T., Oya, K., Fujisawa, J. and Hanaya, M. 2015. Highly efficient dye-sensitized solar cells with collaborative sensitization by silyl-anchor and carboxy-anchor dyes. *Chem. Commun.* 51: 15894–15897.

Kang, M.S., Ahn, K.S. and Lee, J.W. 2008. Quasi-solid-state dye-sensitized solar cells employing ternary component polymer-gel electrolytes. *Journal of Power Sources.* 180(2): 896–901.

Karuppannan, R., Sambandam, A. and Kandasamy, J. 2015. Polymer electrolytes in dye sensitized solar cells. *Materials Focus.* 4: 262–271.

Kavan, L. 2017. Electrochemistry and dye-sensitized solar cells. *Curr. Opin. Electrochem.* 2: 88–96.

Kim, S.C., Song, M., Ryu, T.I., Lee, M.J., Jin, S.H., Gal, Y.S. et al. 2009. Liquid crystals embedded in polymeric electrolytes for quasi-solid state dye-sensitized solar cell applications. *Macromol. Chem. Phys.* 210: 1844–1850.

Kumar, S., Prajapati, G.K., Saroj, A.L. and Gupta, P.N. 2019. Structural, electrical and dielectric studies of nano-composite polymer blend electrolyte films based on (70—x) PVA—x PVP—NaI—SiO_2. *Physica B: Condensed Matter.* 554: 158.

Kusama, H. and Arakawa, H. 2005. Influence of pyrazole derivatives in I^-/I_3^- redox electrolyte solution on Ru(II)-dye-sensitized TiO_2 solar cell performance. *Solar Energy Materials & Solar Cells.* 85(3): 333–344.

Lee, C.H., Liu, K.Y., Chang, S.H., Lin, K.J., Lin, J.J., Ho, K.C. et al. 2011. Gelation of ionic liquid with exfoliated montmorillonite nanoplatelets and its application for quasi-solid-state dye-sensitized solar cells. *J. Colloid Interface Sci.* 363: 635–639.

Lee, K.M., Suryanarayanan, V. and Ho, K.C. 2008. A photo-physical and electrochemical impedance spectroscopy study on the quasi-solid state dye-sensitized solar cells based on poly(vinylidene fluoride-co-hexafluoropropylene). *J. Power Sources.* 185: 1605–1612.

Lewis, N.S. 2007. Toward cost-effective solar energy use. *Science* 315(80): 798–801.

Li, G., Shrotriya, V., Huang, J., Yao, Y., Moriarty, T., Emery, K. et al. 2005. High-efficiency solution processable polymer photovoltaic cells by self-organization of polymer blends. *Nat. Mater.* 4: 864–868.

Liu, I.P., Hung, W.N., Teng, H., Venkatesan, S., Lin, J.C. and Lee, Y.L. 2017. High performance printable electrolytes for dye-sensitized solar cells. *J. Mater. Chem. A.* 5: 9190–9197.

Lupan, O., Guerin, V.M., Tiginyanu, I.M., Ursaki, V.V., Chow, L., Heinrich, H. et al. 2010. Well-aligned arrays of vertically oriented ZnO nanowires electrodeposited on ITO-coated glass and their integration in dye sensitized solar cells. *Journal of Photochemistry and Photobiology A: Chemistry.* 211(1): 65–73.

MacCallum, J.R. and Vincent, C.A. 1989. *Polymer Electrolytes Reviews-II.* Elsevier, London, UK.

Mathew, S., Yella, A., Gao, P., Mathew, S., Humphry-Baker, R., Curchod, B.F.E. et al. 2014. Dye-sensitized solar cells with 13% efficiency achieved through the molecular engineering of porphyrin sensitizers. *Nature Chemistry.* 6(3): 242–247.

Mazloum, M.A., Arazi, R., Mirjalili, B.B.F. and Azad, S. 2019. Synthesis and application of Fe_3O_4@ nanocellulose/TiCl as a nanofiller for high performance of quasi solid-based dye-sensitized solar cells. *International Journal of Energy Research.* 43(9): 4483–4494.

McCandless, B.E. and Sites, J.R. 2011. Cadmium telluride solar cells, photovoltaic science and engineering. In *Chapter14 in Handbook of Photovoltaic Science and Engineering.* p. 619, Wiley, USA.

Meng, L., Zhang, Y., Wan, X., Li, C., Zhang, X. and Wang, Y. et al. 2018. Organic and solution-processed tandem solar cells with 17.3% efficiency. *Science.* 361(80): 1094–1098.

Meng, Q.B., Takahashi, K., Zhang, X.T., Sutanto, I., Rao, T.N., Sato, O. et al. 2003. Fabrication of an efficient solid-state dye-sensitized solar cell. *Langmuir.* 19(9): 3572–3574.

Miles, R.W., Zoppi, G. and Forbes, I. 2007. Inorganic photovoltaic cells the inorganic semiconductor materials used to make photovoltaic cells. *Mater. Today.* 10: 20–27.

Mohan, V.M. and Murakami, K. 2011. Dye sensitized solar cell with carbon doped (PAN/PEG) polymer quasi-solid gel electrolyte. *J. Adv. Res. Phys.* 2: 021112.

Mohan, V.M., Murakami, K., Kono, A. and Shimomura, M. 2013. Poly(acrylonitrile)/activated carbon composite polymer gel electrolyte for high efficiency dye sensitized solar cells. *J. Mater. Chem. A.* 1: 7399–7407.

Murakami, T.N., Kijitori, Y., Kawashima, N. and Miyasaka, T. 2004. Low temperature preparation of mesoporous TiO_2 films for efficient dye-sensitized photoelectrode by chemical vapor deposition combined with UV light irradiation. *Journal of Photochemistry and Photobiology A. Chemistry.* 164(1–3): 187–191.

Murugesan, S., Quintero, O.A., Chou, B.P., Xiao, P., Park, K., Hall, J.W. et al. 2014. Wide electrochemical window ionic salt for use in electropositive metal electrode position and solid state Li-ion batteries. *J. Mater. Chem. A*. 2: 2194–2201.

Nazeeruddin, M.K., Kay, A., Rodicio, I., Humphry-Baker, R., Mueller, E. and Liska, P. 1993. Conversion of light to electricity by cis-X_2 bis(2,2'-bipyridyl-4,4'-dicarboxylate) ruthenium(II) charge-transfer sensitizers (X = Cl^-, Br^-, I^-, CN^-, and SCN-) on nanocrystalline titanium dioxide electrodes. *Journal of the American Chemical Society*. 115(14): 6382–6390.

Neale, N.R., Kopidakis, N., Jao van de Lagemaat, J.V., Grätzel, M. and Frank, A.J. 2005. Effect of a coadsorbent on the performance of dye-sensitized TiO_2 solar cells: Shielding versus band-edge movement. *J. Phys.Chem. B*. 109(49): 23183–23189.

Ning, Z., Gong, X., Comin, R., Walters, G., Fan, F., Voznyy, O. et al. 2015. Quantum-dot-in-perovskite solids. *Nature*. 523: 324–328.

Nogueira, A.F., Alonso-Vante, N. and De Paoli, M.A. 1999. Solid-state photoelectrochemical device using poly (o-methoxy aniline) as sensitizer and an ionic conductive elastomer as electrolyte. *Synthetic Metals*. 105(1): 23–27.

Nogueira, A.F., Longo, C. and de Paoli, M.A. 2004. Polymers in dye-sensitized solar cells: Overview and perspectives. *Coordination Chemistry Reviews*. 248(13–14): 1455–1468.

Nogueira, A.F. and Paoli, M.A.D. 2000. A dye sensitized TiO2 photovoltaic cell constructed with an elastomeric electrolyte. *Sol. Energy Mater. Sol. Cells*. 61: 135–141.

Omara, Z.M., Kabeel, A.E. and Abdullah, A.S. 2017. A review of solar still performance with reflectors. *Renewable and Sustainable Energy Reviews*. 68: 638–649.

O'Regan, B. and Grätzel, M. 1991. A low-cost, high-efficiency solar cell based on dye-sensitized colloidal TiO_2 films. *Nature*. 353: 737–740.

Oskam, G., Bergeron, B.V., Meyer, G.J. and Searson, P.C. 2001. Pseudo halogens for dye-sensitized TiO_2 photoelectrochemical cells. *The Journal of Physical Chemistry B*. 105(29): 6867–6873.

Park, N.G., Lagemaat, J.V. and Frank, A.J. 2000. Comparison of dye-sensitized rutile- and anatase-based TiO_2 solar cells. *Journal of Physical Chemistry B*. 104(38): 8989–8994.

Pi, X., Zhang, L. and Yang, D. 2012. Enhancing the efficiency of multicrystalline silicon solar cells by the inkjet printing of silicon-quantum-dot ink. *J. Phys. Chem. C*. 116: 21240–21243.

Prabakaran, K., Mohanty, S. and Nayak, S.K. 2015. Chemically exfoliated nanosilicate platelet hybridized polymer electrolytes for solid state dye sensitized solar cells. *New J. Chem*. 39: 8602–8612.

Pradhan, S.C., Hagfeldt, A. and Soman, S. 2018. Resurgence of DSCs with copper electrolyte: A detailed investigation of interfacial charge dynamics with cobalt and iodine based electrolytes. *J. Mater. Chem. A*. 6: 22204–22214.

Sacco, A., Lamberti, A., Gerosa, M., Bisio, C., Gatti, G., Carniato, F. et al. 2015. Toward quasi-solid state dye-sensitized solar cells: effect of γ-Al_2O_3 nanoparticle dispersion into liquid electrolyte. *Sol. Energy*. 111: 125–134.

Saidi, N.M., Omar, F.S., Numan, A., Apperley D.C., Algaradah, M.M., Kasi, R. et al. 2019. Enhancing the efficiency of a dye-sensitized solar cell based on a metal oxide nanocomposite gel polymer electrolyte. *ACS Appl Mater Interfaces*. 11(33): 30185–30196.

Sapp, S.A., Elliott, C.M., Contado, C., Caramori, S. and Bignozzi, C.A. 2002. Substituted polypyridine complexes of cobalt (II/III) as efficient electron-transfer mediators in dye-sensitized solar cells. *Journal of the American Chemical Society*. 124(37): 11215–11222.

Saroj, A.L., Krishnamoorthi, S. and Singh, R.K. 2017. Structural, thermal and electrical transport behaviour of polymer electrolytes based on PVA and imidazolium based ionic liquid. *J. Non-Cryst. Solids* 473: 87.

Seo, S.J., Cha, H.J., Kang, Y.S. and Kang, M.S. 2014. Printable ternary component polymer gel electrolytes for long-term stable dye-sensitized solar cells. *Electrochim. Acta*. 145: 217–223.

Sethupathy, M., Pandey, P. and Manisankar, P. 2014. Photovoltaic performance of dye-sensitized solar cells fabricated with polyvinylidene fluoride-polyacrylonitrile-silicon dioxide hybrid composite membrane. *Mater Chem Phys*. 143: 1191–1198.

Sethupathy, M., Ravichandran, S. and Manisankar, P. 2014. Preparation of PVdF-PAN-V2O5 hybrid composite membrane by electrospinning and fabrication of dye-sensitized solar cells. *Int J Electrochem Sci*. 9: 3166–3180.

Sharma, S., Jain, K.K. and Sharma, A. 2015. Solar cells: In research and applications—a review. *Materials Sciences and Applications*. 6: 1145–1155.

Sima, C., Grigoriu, C. and Antohe, A. 2010. Comparison of the dye-sensitized solar cells performances based on transparent conductive ITO and FTO. *Thin Solid Films*. 519(2): 595–597.

Singh, P. and Saroj, A.L. 2021. Effect of SiO_2 Nano-particles on plasticized polymer blend electrolytes: Vibrational, thermal, and ionic conductivity study. *Polymer-Plastics Technology and Materials*. 60: 298.

Singh, R., Polu, A.R., Bhattacharya, B., Rhee, H.W., Varlikli, C. and Singh, P.K. 2016. Perspectives for solid biopolymer electrolytes in dye sensitized solar cell and battery application. *Renewable and Sustainable Energy Reviews*. 65: 1098–1117.

Stergiopoulos, T., Arabatzis, I.M., Katsaros, G. and Falaras, P. 2002. Binary polyethylene oxide/titania solid-state redox electrolyte for highly efficient nanocrystalline TiO_2 photoelectrochemical cells. *Nano Letters*. 2(11): 1259–1261.

Su'ait, M.S., Rahman, M.Y.A. and Ahmad, A. 2015. Review on polymer electrolyte in dye-sensitized solar cells (DSSCs). *Solar Energy*. 115: 452–470.

Suriwong, T., Thongtem, T. and Thongtem, S. 2015. $CuAlO_2$ powder dispersed in composite gel electrolyte for application in quasi-solid-state dye-sensitized solar cells. *Mater. Sci. Semicond. Process*. 39: 348–354.

Teo, L.P., Buraidah, M.H. and Arof, A.K. 2020. Polyacrylonitrile-based gel polymer electrolytes for dye-sensitized solar cells: A review. *Ionics*. 26: 4215.

Tiautit, N., Puratane, C., Panpinit, S. and Saengsuwan, S. 2014. Effect of SiO_2 and TiO_2 nanoparticles on the performance of dye-sensitized solar cells using PVdF-HFP/PVA gel electrolytes. *Energy Procedia*. 56: 378–385.

Tu, C.W., Liu, K.Y., Chien, A.T., Yen, M.H., Weng, T.H., Ho, K.C. et al. 2008. Enhancement of photocurrent of polymer-gelled dye-sensitized solar cell by incorporation of exfoliated montmorillonite nanoplatelets. *J. Polym. Sci. Part A: Polym. Chem*. 46: 47–53.

Tvrdy, K. and Kamat, P.V. 2011. Quantum dot solar cells. *Compr. Nanosci. Technol*. 1–5: 257–275.

Tzounis, L., Stergiopoulos, T., Zachariadis, A., Gravalidis, C., Laskarakis, A. and Logothetidis, S. 2017. Perovskite solar cells from small scale spin coating process towards roll-to-roll printing: Optical and morphological studies. *Mater. Today Proc*. 4: 5082–5089.

Venkatesan, S., Hidayati, N., Liu, I.P. and Lee, Y.L. 296: 2016b. Highly efficient gel-state dyesensitized solar cells prepared using propionitrile and poly (vinylidene fluoride-co-hexafluoropropylene). *J. Power Sources*. 336: 385–390a.

Venkatesan, S. and Lee, Y.L. 2017. Nanofillers in the electrolytes of dye-sensitized solar cells: A short review. *Coordination Chemistry Reviews*. 353: 58–112.

Venkatesan, S., Liu, I.P., Chen, L.T., Hou, C.W., Li, C.W. and Lee, Y.L. 297: 2016a. Effects of TiO_2 and TiC nanofillers on the performance of the dye sensitized solar cells based on the polymer electrolyte of cobalt redox system. *ACS Appl. Mater. Interfaces*. 8: 24559–24566b.

Venkatesan, S., Liu, I.P., Lin, J.C., Tsai, M.H., Teng, H. and Lee, Y.L. 2018. Highly efficient quasi-solid-state dye-sensitized solar cells using polyethylene oxide (PEO) and poly (methyl methacrylate) (PMMA)-based printable electrolytes. *Journal of Materials Chemistry A*. 6(21): 10085–10094.

Venkatesan, S., Su, S.C., Hung, W.N., Liu, I.P., Teng, H. and Lee, Y.L. 2015a. Printable electrolytes based on polyacrylonitrile and gamma-butyrolactone for dye sensitized solar cell application. *J. Power Sources*. 298: 385–390a.

Venkatesan, S., Su, S.C., Kao, S.C., Teng, H. and Lee Y.L. 2015b. Stability improvement of gel-state dye-sensitized solar cells by utilization of the co-solvent effect of propionitrile/acetonitrile and 3-methoxypropionitrile/acetonitrile with poly(acrylonitrile-co-vinyl acetate). *J Power Sources.* 274: 506–511b.

Vijayakumar, E., Subramania, A., Fei, Z. and Dyson, P.J. 2015. High-performance dye sensitized solar cell based on an electrospun poly(vinylidene fluoride-cohexafluoropropylene)/cobalt sulfide nanocomposite membrane electrolyte. *RSC Adv.* 5: 52026–52032.

Wang, C., Wang, L., Shi, Y., Zhang, H. and Ma, T. 2013. Printable electrolytes for highly efficient quasi-solid-state dye-sensitized solar cells. *Electrochim. Acta.* 91: 302–306.

Weerasinghea, H.C., Huanga, F. and Chenga, Y.B. 2013. Fabrication of flexible dye sensitized solar cells on plastic substrates. *Nano Energy.* 2(2): 174–189.

Wu, J., Lan, Z., Lin, J., Huang, M., Huang, Y., Fan, L. et al. 2015. Electrolytes in dye-sensitized solar cells. *Chemical Reviews.* 115(5): 2136–2173.

Xia, J.B., Li, F.Y., Yang, H., Li, X.H. and Huang, C.H. 2007. A novel quasi-solid-state dye sensitized solar cell based on monolayer capped nanoparticles framework materials. *J. Mater. Sci.* 42: 6412–6416.

Yoon, I.N., Song, H., Won, J. and Kang, Y.S. 2014. Shape dependence of SiO_2 nanomaterials in a quasi-solid electrolyte for application in dye-sensitized solar cells. *J. Phys. Chem. C.* 118: 3918–3924.

Zebardastan, N., Khanmirzaei, M.H., Ramesh, S. and Ramesh, K. 2016. Novel poly (vinylidene fluoride-co-hexafluoro propylene)/polyethylene oxide based gel polymer electrolyte containing fumed silica (SiO_2) nanofiller for high performance dye-sensitized solar cell. *Electrochim. Acta.* 220: 1–8.

Zhang, J., Han, H., Wu, S., Xu, S., Zhou, C., Yang, Y. et al. 2007. Ultrasonic irradiation to modify the PEO/P (VDF—HFP)/TiO_2 nanoparticle composite polymer electrolyte for dye sensitized solar cells. *Nanotechnology.* 18(29): 295606.

Zhang, W., Eperon, G.E. and Snaith, H.J. 2016. Metal halide perovskites for energy applications. *Nat. Energy.* 1: 16018.

Zhang, X., Yang, H., H.M., Xiong, F.Y., Li, Y.Y. and Xia, A. 2006. Quasi-solid-state dye sensitized solar cell based on the stable polymer-grafted nanoparticle composite electrolyte. *J. Power Sources.* 161: 1451–1455.

Zhao, J., Jo, S.G. and Kim, D.W. 2014a. Photovoltaic performance of dye-sensitized solar cells assembled with electrospun polyacrylonitrile/silica-based fibrous composite membranes. *Electrochim. Acta.* 142: 261–267a.

Zhao, X.G., Jin, E.M., Park, J.Y. and Gu, H.B. 2014b. Hybrid polymer electrolyte composite with SiO_2 nanofiber filler for solid-state dye-sensitized solar cells. *Compos. Sci. Technol.* 103: 100–105b.

Index